CONSERVATION AND CONFLICT: MAMMALS AND FARMING IN BRITAIN

LINNEAN SOCIETY
OCCASIONAL
PUBLICATIONS 4

CONSERVATION & CONFLICT

MAMMALS & FARMING IN BRITAIN

Edited by

FRAN TATTERSALL and WILL MANLEY

Fran Tattersall
Wildlife Conservation Research Unit, Department of Zoology, South Parks Road, Oxford OX1 3PS

Will Manley
The Royal Agricultural College, Stroud Road, Cirencester GL7 6JS

This book arises in a large part out of symposium on *Farming and Mammals*, jointly hosted by the Mammal Society and the Linnean Society of London, in London in November 1999.

LINNEAN SOCIETY OCCASIONAL PUBLICATIONS 4, 2003
PUBLISHED FOR THE LINNEAN SOCIETY OF LONDON
BY WESTBURY PUBLISHING

First published in 2003 by
Westbury Publishing
Ilkley Road
Otley
West Yorkshire
LS21 3JP

ISBN 1 84103 001 5

Designed, printed and bound by
Smith Settle Printing & Bookbinding Ltd
Ilkley Road, Otley, West Yorkshire LS21 3JP

CONTENTS

SECTION 3: Economic impacts

CONTRIBUTORS TO THIS VOLUME

Baker, Philip J. School of Biological Sciences,
University of Bristol, Woodland Road, Bristol BS8 1UG, UK.

Baker, Simon J. Rural Development Service, Defra
Brooklands Avenue, Cambridge CB2 2DR, UK.

Birks, Johnny D.S. The Vincent Wildlife Trust,
3 & 4 Bronsil Courtyard, Eastnor, Ledbury HR8 1EP, UK.

Bourne, F. John. Independent Scientific Group on Cattle TB,
Room 105, 1A Page Street, London SW1P 4PQ, UK.

Carbone, Chris. Institute of Zoology,
Zoological Society of London, Regent's Park, London NW1 4RY, UK.

Cheeseman, Chris L. Central Science Laboratory,
Sand Hutton, York YO41 1LZ, UK.

Cowan, David P. Central Science Laboratory,
Sand Hutton, York YO41 1LZ, UK.

Cox, Sir David R. Independent Scientific Group on Cattle TB,
Room 105, 1A Page Street, London SW1P 4PQ, UK.

Delahay, Richard J. Central Science Laboratory,
Sand Hutton, York YO4 1LZ, UK.

Dendy, Julie, A. Central Science Laboratory,
Sand Hutton, York YO41 1LZ, UK.

Donnelly, Christl A. Independent Scientific Group on Cattle TB,
Room 105, 1A Page Street, London SW1P 4PQ, UK.

Duvergé, P. Laurent. The Vincent Wildlife Trust,
3 & 4 Bronsil Courtyard, Eastnor, Ledbury, Herefordshire HR8 1EP, UK.

Fletcher, Mark R. Central Science Laboratory,
Sand Hutton, York YO41 1LZ, UK.

Fox, Sue. M. Central Science Laboratory,
Sand Hutton, York YO41 1LZ, UK.

Gettinby, George. Independent Scientific Group on Cattle TB,
Room 105, 1A Page Street, London SW1P 4PQ, UK.

Gorman, Martyn L. Department of Zoology,
The University of Aberdeen, Tillydrone Avenue, Aberdeen AB23 2TN, UK.

Harris, Stephen. School of Biological Sciences,
University of Bristol, Woodland Road, Bristol BS8 1UG, UK.

Johnson, Ian P. Rural Development Service, Defra
Brooklands Avenue, Cambridge CB2 2DR, UK.

Johnson, Paul J. The Wildlife Conservation Research Unit,
University of Oxford, Department of Zoology, South Parks Rd, Oxford OX1 3PS, UK.

Jones, Gareth. School of Biological Sciences,
University of Bristol, Woodland Road, Bristol BS8 1UG, UK.

Kjellander, Petter. Grimso Wildlife Research Station,
Swedish University of Agricultural Sciences, S-730 91 Riddarhyttan, SWEDEN.

Macdonald, David W. The Wildlife Conservation Research Unit,
University of Oxford, Department of Zoology, South Parks Rd, Oxford OX1 3PS, UK.

Manley, Will J. The Royal Agricultural College,
Stroud Road, Cirencester GL7 6JS, UK.

Mathews, Fiona. The Wildlife Conservation Research Unit,
University of Oxford, Department of Zoology, South Parks Rd, Oxford OX1 3PS, UK.

McDonald, Robbie A. The Game Conservancy Trust,
The Gillet, Forest-in-Teesdale, Barnard Castle DL12 0HA, UK.

McInerney, John P. Independent Scientific Group on Cattle TB,
Room 105, 1A Page Street, London SW1P 4PQ, UK.

Mckillop, Ian, G. Sports Turf Research Institute,
Bingley West Yorks BD16 1AU, UK.

Moorhouse, Tom. The Wildlife Conservation Research Unit,
University of Oxford, Department of Zoology, South Parks Road Oxford OX1 3PS, UK.

Morrison,Willian I. Independent Scientific Group on Cattle TB,
Room 105, 1A Page Street, London SW1P 4PQ, UK.

Oliver-Bellasis, Hugh R. The Manydown Company,
Wootton House, Wootton St Lawrence, Basingstoke, Hampshire RG23 8PE, UK.

Putman, Rory J. Behavioural and Environmental Biology Group,
The Manchester Metropolitan University, John Dalton Building, Chester Street, Manchester M1 5GD, UK.

Quy, Roger J. Central Science Laboratory,
Sand Hutton, York YO41 1LZ, UK.

Reynolds, Jonathan. C. The Game Conservancy Trust,
Fordingbridge SP6 1EF, UK.

Reynolds, Peter. Capreolus Wildlife Consultancy,
2 West Point, Garvald, East Lothian EH54 6NN, UK.

Rogers, Lucy M. Central Science Laboratory,
Sand Hutton, York YO41 1LZ, UK.

Shore, Richard F. Centre for Ecology & Hydrology,
Monks Wood, Abbots Ripton, Huntingdon, Cambridgeshire PE28 2LS, UK.

Sotherton, Nick. The Game Conservancy Trust,
Fordingbridge, Hampshire SP6 1EF, UK.

Strachan, Robert. The Wildlife Conservation Research Unit,
University of Oxford, Department of Zoology, South Parks Road, Oxford OX1 3PS, UK.

Tattersall, Fran H. The Wildlife Conservation Research Unit,
University of Oxford, Department of Zoology, South Parks Road, Oxford OX1 3PS, UK.

Trout, Roger. Forest Research Agency,
Alice Holt Lodge, Wrecclesham, Farnham, Surrey GU10 4LH, UK.

Walker, Lee A. Centre for Ecology & Hydrology,
Monks Wood, Abbots Ripton, Huntingdon, Cambridgeshire PE28 2LS, UK.

Western, Gavin J. Central Science Laboratory,
Sand Hutton, York YO41 1LZ, UK.

White, Piran C L. Environment Department,
University of York, Heslington, York YO10 5DD, UK.

Wilson, Gavin. Central Science Laboratory,
Sand Hutton, York YO41 1LZ, UK.

Woodroffe, Rosie. Independent Scientific Group on Cattle TB,
Room 105, 1A Page Street, London SW1P 4PQ, UK.

FOREWORD

One thing that stands out from this book is how little we know about the specific ecological needs of many mammals. Some of the reasons are clear; they are generally secretive and nocturnal and many occur at low density, making them almost invisible in our farmed landscape. Despite this, mammals have a high public profile and the current debate over hunting illustrates the strong emotions they can generate. Clearly if we are to secure their future we need to know more about their needs and how they interact with the farmed landscape.

With 75% of our landscape given over to farming, the activities of the agricultural sector have a significant impact on mammal populations. The current increase in agri-environment funding, which is gradually shifting the emphasis away from production, can help, but we need a much greater move towards a more diverse and less intensive agriculture if we are to deliver the biodiversity improvements we have all signed up to. It's clear that some species are hanging on by the skin of their teeth and we need urgent action at the landscape scale if we are to avoid species such as the water vole or greater horseshoe bat going the way of the beaver.

Although much of the focus of this book is about mammals and habitats, there are other problems too. TB in cattle is a significant issue in some areas and we need to seek a long term sustainable solution based on good science. It is worrying that secondary poisoning by anticoagulant rodenticides appears to be widespread in predators, and we need to learn from the mistakes of the past in tackling this insidious problem.

Finally, one message that comes through very clearly is the need for an effective integrated monitoring scheme for our mammals. At the moment, we're hardly in a position to say whether the improvements to countryside management that we are promoting will make any difference to mammal populations. We need to establish these links and begin to use mammals as indicators of the health of our countryside.

Baroness Young
Chief Executive, English Nature, Peterborough

PREFACE

It seems to us that farming is very much at the heart of both conservation and conflict with wild mammals in Britain. Agricultural land of one sort or another covers 17.9 million hectares, just under three-quarters of the UK, and is home to at least 40 of our 54 wild terrestrial mammals. The opportunities for conservation of so many species, over such a large area, through conceptually simple changes in management such as provision of field margins and buffer zones are manifold, and this is an exciting area of conservation science. But of course, where people and wildlife live together conflict is, perhaps, inevitable, and the needs of wild mammals (and the desires of those who derive pleasure from them) must be balanced against the needs and desires of farmers and others who make their living from the land. Mammals can cause a real or perceived cost to farm businesses, but may also provide economic or recreational opportunities, for example, through uptake of grant-aided management schemes or sporting interests.

The drive for self-sufficiency since the middle of the twentieth century profoundly changed agriculture and the wider countryside, and many of these changes have been generally detrimental to wildlife. They have included very significant increases in the use of agrochemicals, polarisation of agriculture into arable east and pastoral west, a doubling in the number of sheep on the land, larger field sizes and the removal of hundreds of thousands of miles of hedgerow. However, since the late 1980s, new opportunities for conservation on farmland have emerged. This period saw removal of land from production into set-aside and a significant increase in the range of attractive financial incentives for farmers wishing to benefit wildlife using agri-environment schemes, including the expansion of organic farming. The importance of farmland habitat loss or alteration to species such as water voles or greater horseshoe bats also began to be more fully appreciated. This period also saw an explosion of debate on issues surrounding conflicts between farming and mammals, such as how to manage expanding populations of badgers – linked to bovine tuberculosis – and management of foxes and deer by hunting with dogs.

One of the most obvious ways of reconciling the sometimes conflicting pressures of conservation and economically viable farming is reform of the Common Agricultural Policy. Both conservationists and the farming community are dissatisfied with all or elements of CAP and continue to press for further changes. A reformed CAP has the potential to benefit wild mammals (and biodiversity in general) more than any other single government or EU policy. As we write, farming has begun to emerge from under the shadow of foot-and-mouth disease. This latest in a series of crises to hit farming in recent years is both stimulating vigorous debate about how we manage our rural landscapes in Britain, and is likely to increase the rate of changes already occurring in the farming industry. These include restructuring and amalgamation of farm businesses, an increasing trend towards contract farming arrangements and a decrease in farm labour.

This is, therefore, a period of important change in British farming. If we wish to influence the outcome to benefit mammals, and other wildlife, then policy-makers, scientists and the farming community need to know the issues and problems involved. Understanding the effects of farming practices on wildlife is essential for policy-makers formulating strategies for restoring biodiversity. Evolving legislation and grant-aid schemes taking account of mammals also make it increasingly important for farmers and their advisors to have access to the best available information when formulating management plans. Our hope is that this book will go some way to disseminating that information.

In putting together *Conservation and Conflict: Mammals and Farming in Britain*, we have solicited a range of types of contributions, bringing together original scientific data and analyses with new syntheses of existing scientific literature, juxtaposed with discussions of policy and practice. These contributions address a range of farmland mammals and conservation issues, and in accomplishing this challenge we have drawn upon the expertise of leading specialists from the UK's top academic and research institutions.

The book is divided into three sections. The first, *'The Farmed Environment'*, is intended to provide a background to the remaining sections. It describes some of the constraints under which both farmers and conservationists operate, beginning with Macdonald & Johnson's overview of farmers' attitudes to wildlife (Chapter 1), and mammals in particular, and how this translates into actions. The agri-environmental schemes and Biodiversity Action Plans described by Johnson & Baker (Chapter 2) provide one means of channelling those actions. Some of the management prescriptions encouraged under agri-environment schemes are highly beneficial to game species, and Oliver-Bellasis & Sotherton (Chapter 3) describe the constraints on game management on a lowland farm and the implications for mammals. Finally, in this section, Shore, Fletcher & Walker (Chapter 4) illustrate the widespread and pervasive potential for agrochemicals to harm mammals, a topic notable for a lack of systematic monitoring and research.

The second section deals with *'Ecology and Conservation'*, focussing on what we know about the lives of particular species living on farmland. Otters and water voles may not immediately spring to mind as farmland species, but as Strachan, Moorhouse & Macdonald (Chapter 5) point out, nine-tenths of most river catchments drain through farmed land. Bats are also not generally considered farmland animals, but Duvergé & Jones (Chapter 6) demonstrate the vital importance of cattle-grazed pasture for one very rare species, the greater horseshoe bat. On the other hand, the considerably more abundant wood mouse is interesting because it can live in many farmland habitats, and is exposed to a variety of farming activities. Tattersall & Macdonald's (Chapter 7) review highlights the species' remarkable adaptability. The Orkney vole, while also relatively abundant, is rather less adaptable. Gorman & Reynolds' (Chapter 8) analyses suggest that habitat loss, through agricultural intensification, may have reduced the population by as much as 50% over the past 50 years. The two final chapters in this section deal with carnivores. Macdonald & Birks (Chapter 9) review the ways in which farmland management, and particularly control of mammalian pests, affects small mustelids, while Baker & Harris's (Chapter 10) synthesis of data on the diet of rural foxes is a first step towards solving the vexed question of the extent to which fox predation affects populations of their prey.

The final, largest, section of this book, 'Economic Conflict' explores some of the problems that mammals cause, or are perceived to cause, on farmland. Three chapters deal with the potential role of badgers in transmitting bovine tuberculosis. The first contribution (Woodroffe et al., Chapter 11) sets out current government research strategies to tackle the problem, including the controversial 'Krebs experiment'. Harris & White (Chapter 12) then provide a detailed analysis of past and present badger culling strategies and their efficacy (or lack of efficacy) at reducing the incidence of TB in cattle. Finally, Delahay et al. (Chapter 13) synthesise results from a long-term study of a TB-infected badger population at Woodchester Park.

The rat is perhaps the archetypal mammalian pest, and their safe and effective control presents a difficult challenge, described by Cowan & Quy (Chapter 14). While the damage that rats cause is undisputed, assessing deer damage is more difficult. Putman & Kjellander (Chapter 15) explore some of the factors associated with deer damage to agriculture in the UK and elsewhere in Europe. Rabbits are becoming an increasingly important farmland pest. Trout (Chapter 16) provides an overview of their impact and management, while Dendy et al. (Chapter 17) summarise the results of a series of experiments designed to quantify the costs of rabbit damage to silage and cereal crops. Foxes are a major predator of rabbits, raising the possibility that they might actually benefit some farmers, and Macdonald et al. (Chapter 18) end this section with an assessment of some of the economic implications of fox control.

In drawing together these varied chapters, with their different styles and approaches, we have been greatly helped by the authors' good humour and patience, and we would like to thank them all for contributing to this volume. Each chapter was independently reviewed by two referees, and we are extremely grateful for their constructive criticisms. We also benefited greatly from the encouragement and experience of Dr John Flowerdew, from The Mammal Society. Dr David Cutler from the Linnean Society graciously guided us through the editing and publishing process. We would also like to gratefully acknowledge the support provided to us by The Royal Agricultural College, Cirencester, and the Wildlife Conservation Research Unit, Oxford. Finally, our thanks to our spouses for their forbearance and enthusiasm.

Fran Tattersall & Will Manley

I
The Farmed
Environment

1

Farmers as conservation custodians: Links between perception and practice

D W Macdonald & P J Johnson

INTRODUCTION

Traditionally, farmers have been seen, by both the general public and the farming community themselves, as having a dual role in producing food and acting as stewards of the rural environment. Protecting amenity and sustaining wildlife is, then, a by-product of their husbandry of the land (Sheail, 1976). Post-war agricultural intensification introduced a novel type of conflict between farming and conservation (these conflicts were not unforeseen — Sheail, 1976, describes how the wartime committees considered the possible impacts of the coming of industrial agriculture). The Scott report of 1942 into land use in rural areas asserted that 'every agricultural acre counts', and the dominant ideas that originated in the report have shaped the post-war countryside (Blunden & Curry, 1988). The loss of habitats and the repercussions that these losses have had for biodiversity has been called a 'conservation crisis' (Pain & Pienkowski, 1997). But recent years have seen considerable changes in the pressures for intensification. By the end of the 1980s farming imperatives were under review, and the core of rural policy was aimed at reducing over-production. The principal motivation for this was economic: spending on agriculture was threatening to reach an unsustainable level. Rural conservation was emerging at about the same time as a significant force enjoying widespread political support (Blunden & Curry, 1988).

At the end of the 1990s, there was a widespread recognition that future production must be tempered with conservation (e.g. Krebs *et al.*, 1999). However, it is also undoubtedly true that these changes force on the farming community a sometimes awkward change of self-image and goals, to which its response has been ambivalent. There is a perception by farmers that many of the changes affecting their lifestyle are set in motion by 'an ignorant majority' (Clover, 1999). This is encapsulated by the debates regarding foxes *Vulpes vulpes*, badgers *Meles meles* and red deer *Cervus elaphus* (Bateson & Bradshaw, 1997; Krebs *et al.*, 1997; Burns *et al.*, 2000) where, in addition to complex biology, the increasing prominence of animal welfare issues has further complicated the traditional relationship between farming and wildlife.

In this first chapter, we examine the relationship between farmers and mammals using a number of approaches. Our intention is to show how biologists can contribute to an understanding of the debate by describing some impacts of farming practice on mammals, and how these practices are related to reported perceptions of wildlife issues. In particular, we believe that there is a middle ground between agriculture and wider environmental needs that can flourish with good will between farmers and conservationists (Macdonald & Smith, 1991). To illustrate this potential, we synthesise the results from a series of questionnaire surveys of farmers.

THE QUESTIONNAIRE SURVEYS

Two of our questionnaire surveys compared the way in which farmers reported treating non-productive habitats on their farms in 1998 compared with 1981. These were carried out at very different points in the shifting agricultural climates (dominated by very different grant schemes) of recent decades. The first followed a period when habitat destruction was continuing apace. When the second was carried out, there were increasing incentives to restore and protect wildlife habitats, from a variety of agri-environment schemes (although even by 2000 these covered only approximately 15% of agricultural land, G Taylor, personal communication; see also Baker & Johnson Chapter 2).

We have to note that the use of questionnaire data is, in general, problematical for a number of reasons. Firstly, unless the response rate approaches 100%, the respondents are likely to be a biased sample of the target population. For the 1981 survey, we partially addressed this problem by following up non-respondents (in Oxfordshire) with a reminder. We could find no evidence that farmers who responded to the reminder differed in any way from those responding unreminded. Secondly, the information provided by respondents reflects perceptions, and may also be influenced by what the respondent considers to be the 'correct' answer in terms of, for example, the purpose they assume a study is aimed at. Some types of questions are known to elicit inaccurate responses. Farmers over-estimate the number of foxes killed on their land for example (Heydon & Reynolds, 2000b).

If the behaviour of a typical farmer were to be formally modelled, we would expect them to follow something closer to unconstrained profit maximisation in the 1970s compared with the 1990s, when Common Agricultural Policy (CAP) and subsidy constraints were in place. And where this model failed to fit, one possible form of explanation is that farmers may be influenced by the amenity value of their farms for field-sports, or that they have an altruistic attitude to some aspects of the environment (many other explanations would also be possible). In this chapter, we do not aim to construct such a model, we merely seek to identify some broad patterns that would be likely to be major features of its dynamics. Specifically, we describe reported patterns of habitat loss and investigate if attitudes to wildlife, and relationship with field sports at different degree of involvement, were associated with farming practice. A total of 1,312 farmers from different parts of England provided responses. The methods are described in full in Macdonald & Johnson (2000). The response rate in 1981 was 38% (859 of 2,288 questionnaires sent out), compared with 45% (451 of 1,000) in 1998. This is a statistically significant difference (χ^2=9.5, d.f.=2, p=0.001), perhaps itself reflecting a change in attitudes between the surveys. (The questionnaires were not identical, but questions about habitat management were worded in exactly the same way). Questionnaires were circulated with the aid of Master of Fox Hounds Association fox hunts, which hold convenient lists of (all) farmers, and with a covering letter stressing our non-partisan attitude to environmental issues.

We also looked at farmers' responses to a series of questions related to their attitudes to, and treatment of, the more conspicuous mammals of farmland. For this, we used the first of the surveys described above, and drew on a survey of farmers and their vertebrate pests (Atkinson *et al.*, 1994). This was a national survey of farmers for which 460 questionnaires were circulated in 1992, including a variety of questions on damage attributable to different vertebrate pests; 157 (34%) were returned.

We pay particular attention to the relationship between farmers and foxes. Foxes are arguably one of the most 'high profile' of mammals affecting farming. A fourth questionnaire survey focused on farmers and their relationship with foxes and foxhunting in Wiltshire in 1995. The predominant land use type among responding farmers was dairy farming (64%), with 13% non-dairy stock, 17% mixed, and 6% arable. Again, we used a questionnaire survey of farmers, together with interviews with the Masters of all 13 registered mounted hunts in the county. Of 220 farmers selected at random, 46% (101) replied. A full account of these results is given in Baines *et al.* (1995) and Baker & Macdonald (2000).

ATTITUDES TO WILDLIFE, HABITAT LOSS AND FIELD-SPORTS

Firstly, we consider farmers' reported perception of wildlife in general. The first question we asked was simply: 'how interested are you in wildlife?' There was a highly significant change in the way this question was answered. Whereas, in 1981, almost 40% of farmers reported that they were 'very' interested in wildlife, the equivalent figure was over 60% in 1998 (Figure 1:1). There were significant differences among regions, but an increase was observed everywhere. We asked a similar question with respect to custody of the countryside. Farmers were asked to say who — farmers, local conservation bodies or the state — they considered principally to be responsible for conservation. We do not know precisely what each farmer understands by 'conservation' in this context. It could be wildlife protection, landscape preservation or refer to a general way of life; there is known to be a mismatch between what farmers and conservationists consider to be good environmental practice (Young *et al.*, 1995). Whatever they meant by 'conservation', an overwhelming majority of farmers said that farmers were the principal custodians, and there was remarkable consistency between the two surveys. There is no evidence that the recent expansion in government agri-environment schemes has had any impact on the number of farmers who consider outside agencies to be responsible — under 5% in both surveys.

So, while the farmers' reported interest in wildlife appeared to have changed between 1981 and 1998, their view of themselves as principal custodians remained unchanged. Can these perceptions be linked with any environmental practices? One potentially

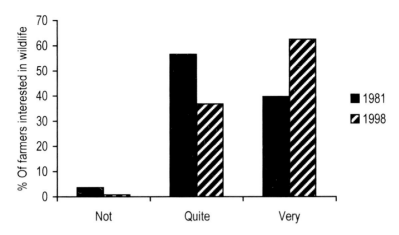

Figure 1:1 Farmers' responses to the question: "Are you 'not interested', 'quite interested' or 'very interested' in wildlife?" Data are percentages ticking each option-box.

useful index of management for wildlife is attitude towards hedgerows. Hedgerows are an important habitat for a diverse and abundant fauna, providing food, shelter and dispersal corridors for a range of small mammals, including uncommon species such as yellow-necked mice *Apodemus flavicollis*, harvest mice *Micromys minutus* and dormice *Muscardinus avellanarius* (Bright *et al.*, 1994, Tew, 1994; Tattersall *et al.*, 1999a). Farmers were asked to describe their principal strategy towards hedgerow management in the decade preceding each survey, with options including leaving alone, removal and encouragement. Their responses indicated that there had been a substantial shift from removing hedgerows to leaving them alone (Figure 1:2). This shift is highly statistically significant (χ^2=316.9, d.f.=1, p<0.001). This is likely to be at least partly due to the Hedgerow Regulations (Department of the Environment 1997), which stipulate that landowners must apply to their local authority for permission to remove a hedgerow.

However, in neither survey was the principal hedgerow management strategy associated with the farmer's reported attitude to wildlife. Farmers who said they were 'very' interested in wildlife were not significantly less likely to have removed hedge-rows (χ^2=2.5 and 2.8 for 1981 and 1998 respectively, both d.f.=2, p>0.05). For other non-productive habitats, though, there were associations. In the more recent survey, parkland trees, shelter belts and spinneys were said to have been encouraged rather than 'left alone' by significantly more farmers who said they were 'very' interested in wildlife (χ^2=9.9, d.f.=2, p=0.002; χ^2=6.2, d.f.=2, p=0.05; χ^2=18.2, d.f.=2, p=0.001 respectively). Of course, encouragement is as perceived by the farmers and may not, in fact, be related to any practical difference. The active removal of any of these habitats was very rarely reported in the 1998 questionnaire.

We also considered how habitat removal was related to field-sport participation. Figure 1:3 shows the amount of hedgerow removed by farmers with different professed involvements in field-sports, whether they hunted, shot, did neither or did both. Field-sport involvement was statistically significantly related to the mean amount of hedgerow removed per farm in the 1981 survey ($F_{3,632}$=6.2, p=0.004). In 1981, farmers whose principal field-sport involvement was in hunting reported having removed fewer hedgerows than other farmers (though the only statistically significant pairwise comparison was that between shooting and hunting farmers). The average amounts were very much lower in the 1998 survey, as we would expect from the reported change in principal management strategy. Measured in terms of amounts of hedgerow per unit area of farms, a similar pattern prevailed: the amount of hedgerow lost was lowest for hunting farmers, but this was not statistically significant.

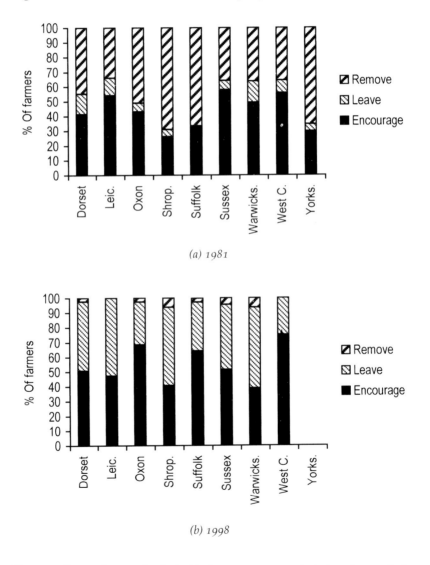

(a) 1981

(b) 1998

Figure 1:2 Farmers' reported principal hedgerow management strategy by county in
(a) 1981 and (b) 1998.

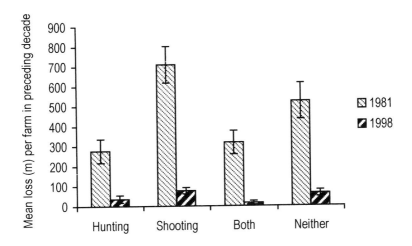

Figure 1:3 Reported hedgerow loss and field-sports interests. Data show mean (±SE) losses in metres per farm.

We also observed that farmers who said they were 'very' interested in wildlife reported having removed fewer hedgerows than other respondents, though this was not statistically significant. There is evidence that questions concerning these patterns are confounded to some extent. Farmers with an interest in hunting were more likely to describe themselves as 'very' interested in wildlife. In both surveys, there was a tendency for farmers involved in hunting or shooting or both to say they were 'very' interested in wildlife (1981: χ^2=27.8, d.f.=3, p<0.001; 1998: χ^2=18.2, d.f.=3, p=0.006).

The confounding of these predictors of hedgerow removal complicates the interpretation of statistical models. The evidence is that there is a stronger link in these data between field-sport involvement and hedgerow management, than between reported wildlife interest and hedgerow management. Interestingly, there is also evidence, albeit relatively weak, that the confounding was not as strong among the recent respondents. In statistical models predicting hedgerow removal using both of these factors as predictors, in the 1998 data only, wildlife interest was marginally significant adjusting for the effect of field-sport involvement ($F_{2,274}$=2.51, p=0.086).

Trends in habitat creation may be of more current practical relevance than past removal strategies. In our second survey, we complemented the questions concerning habitat removal by asking questions concerning the creation of hedgerow and woodland, and linked these to wildlife interest and field-sport involvement. The great majority of farmers (92%) said they had created some hedgerow in the ten years before the survey, and there was no evidence that this was affected either by field-sport involvement or by wildlife interest.

Creation of woodland was, however, associated with field-sport participation (Figure 1:4). Fewer farmers with no interest reported creating woodland than did any of the other categories ($\chi^2=18.6$, d.f.=3, p<0.001). Reported treatment of some other non-productive habitats was also linked to field-sport participation. In the recent survey, encouragement was least frequent among farmers with no field-sport involvement for several habitats (Table 1:1), including spinneys and shelterbelts, which might be expected to be desirable for those with an interest in hunting and shooting.

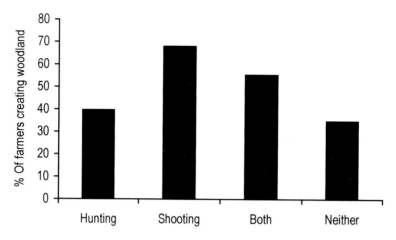

Figure 1:4 Percentages of farmers reporting creating woodland in the decade before the 1998 survey, by field-sport involvement.

In summary of this section, then, we found evidence that there were links between farmers' perceptions of wildlife and their management of habitat, that these had in some respects altered dramatically in the years between 1981 and 1998, and that there were also links between their field-sports involvement and the way in which wildlife habitats were managed. Of course, a whole range of other factors are also likely to influence farmers' attitudes to wildlife, including farm type, and we discuss this particular issue below for mammals.

Table 1:1 Relationship between the reported principal management strategy of non-productive habitats and field-sports involvement in the 1998 survey of farmers. The table shows the percentages of farmers reporting 'encouragement' as being their principal strategy. The significance of a χ^2 test of independence of strategy and involvement is indicated at end of each row (**=p<0.001; ns=not significant).

	Hunting	Shooting	Both	Neither	p
Shelter Belts	60.0	63.4	70.9	43.2	**
Spinneys	48.8	54.6	63.0	38.9	ns
Parkland trees	73.9	61.1	57.1	33.3	**
Ponds	56.4	56.7	63.5	45.8	ns

ATTITUDES TO MAMMALS

In this section, we examine attitudes to the more conspicuous mammals of farmland. For this, we use farmers' responses to the 1981 questionnaire described above, and a 1992 survey of farmers' attitudes to mammals (Atkinson et al., 1994).

How many farmers consider different species to be a pest (where this is defined as a positive response to the question 'do you believe you suffer significant damage from…'), and how many report carrying out some control? Table 1:2 presents this information. We did not attempt to establish the factors that determined whether damage was 'significant'; this is likely to be affected by a complex of factors including farm size and the economic status of the farmer.

Table 1:2 Responses of farmers in the 1981 survey of attitudes to mammalian pests (780 farmers answered these questions). Impact is the percentage of farmers reporting damage and/or control.

% Farmers reporting:	Rabbit	Rat	Mole	Fox	Deer	Weasel/stoat
Damage	58.2	56.3	41.5	30.3	12.9	5.5
Control	61.0	79.1	39.7	32.6	8.5	9.5
Impact	68.0	80.0	50.0	40.0	16.3	10.4
'Yes' to damage, but 'no' to control	10.4	0.7	22.5	22.1	55.5	11.9
'Yes' to control and 'yes' to damage	85.5	70.7	80.9	72.3	68.2	50.0

More farmers reported controlling rats than any other species. Foxes ranked behind rats *Rattus norvegicus*, rabbits *Oryctolagus cuniculus* and moles *Talpa europaea* in this 'league table'. The links between perceived damage and reported control practice varied from mammal to mammal. (We note that we have no means of assessing how accurately farmers attribute damage of any type to a particular species). The third row of Table 1:2 shows the percentage of farmers who reported either that they controlled the animal in question or (said they) sustained some damage to it (or both), and therefore can be said to be affected by the pest in question. The degree to which this percentage exceeds the higher of the figures in the first two rows describes how frequently damage occurs in the absence of control (or *vice versa*). This description does not alter the rank order of pest status.

When only those farmers who responded that each mammal was a pest (reported some damage) were considered, the proportion of these who also said they carried out no control (fourth row of Table 1:2) provides an index of the pest's relative lack of severity. While this is complicated by variation in the cost of control and by farm composition, we might still expect non-trivial pests to be controlled. Of those mammals for which we have these data, the proportion was lowest for rabbits and rats, and highest for deer. A majority of farmers reporting damage due to deer exercised no control. One explanation for this is that lethal control is not as readily accessible as for other species; rifles and the accompanying licenses are required.

The final row of Table 1:2 provides an index of the effectiveness of whatever control was in place: the proportion of farmers reporting damage by a species while at the same time reporting that some control measures were in place. We cannot judge the extent to which control reduced damage without eliminating it, but there is a high degree of variability between species in this attribute. Damage occurring in the presence of control was reported most frequently for rabbits, and was lowest for the stoat *Mustela erminea* and weasel *Mustela nivalis* category. And most of the farmers who reported controlling foxes also reported that they sustained damage they attributed to foxes.

For some mammals, the overall picture presented in Table 1:2 masks substantial regional variability. Table 1:3 presents a county-by-county breakdown of the damage and control patterns. Deer show the greatest variability, with by far the highest proportion of farmers affected in Dorset, where populations of both sika *Cervus nippon* and roe deer *Capreolus capreolus* are high (Macdonald *et al.*, 1998). Populations of several species of deer are thought to have increased their range and abundance over much of lowland Britain since this survey (Putman & Moore, 1998). In a recent ADAS survey, deer were present on the holdings of 69% of respondents and 38% believed that deer caused significant damage (Doney & Packer, 1998; see also Putman & Kjellander, Chapter 15). Perceptions of deer are also complicated by the fact that deer themselves are regarded as a resource; for example, when the price of venison is low, complaints about red deer increase (SNH, 1994).

Other species are widespread, and the frequency with which they are pests reflects land management factors. There is a significant correlation, for example, between the proportion of farmers in each county who have a gamekeeper working on their land and the proportion controlling weasels and stoats. Similarly, rabbit control is most prevalent in areas of predominantly arable cultivation.

In 1992, farmers were asked to categorise mammal species according to whether they were 'pests', 'neutral' or 'desirable' (the question was phrased so as to elicit an opinion relating to the farm in question, not a wider perception). This revealed some distinct polarities in attitudes. While rats, for example, were almost universally considered pestilential, substantial minorities of farmers perceived badgers, hares *Lepus europaeus* and bats (Chiroptera) to be desirable. The same farmers were asked to nominate their worst pest, their second worst pest and so on. While the question asked was not the same as in 1981, the rank order of pests was similar: 27% of farmers nominated the rabbit as their worst pest, and the fox was again below both the mole and rabbit. We also asked farmers who reported either that they had sustained some damage, or had expended some effort on control, how they perceived the financial impact due to different mammal pests. Their answers provided a different perspective. On those farms affected, badgers were reported to have the most severe impact (mean of £1,075), with rats and foxes at £190.70 and £199.60 respectively.

Table 1:3 *Regional variation in the reported 1981 pest status of common mammalian pests.*
West C.=West Country.

County	Dorset	Leic.	Oxon.	Salop.	Suffolk	Sussex	Warw.	West C.	Yorks.
Rabbit:									
% control	68.9	39.1	53.7	65.5	73.5	76.5	61.3	47.4	64.8
% damage	68.9	33.3	50.4	70.9	72.1	74.1	60.0	43.2	55.7
Rat:									
% control	75.6	88.4	71.5	80.0	86.8	75.3	77.3	74.7	89.8
% damage	59.7	53.6	46.3	52.7	69.1	58.8	41.3	54.7	70.5
Mole:									
% control	44.5	18.8	18.7	69.0	39.7	42.4	28.0	45.3	62.5
% damage	48.7	21.7	26.0	70.9	32.4	47.1	26.7	47.4	58.0
Fox:									
% control	38.5	27.1	31.5	22.8	26.1	39.5	40.8	37.8	21.4
% damage	31.7	25.0	27.0	29.1	19.7	40.9	26.3	41.2	27.0
Deer:									
% control	23.8	1.4	5.5	0.0	10.1	7.0	6.6	8.2	6.7
% damage	40.3	1.2	8.1	1.8	14.7	19.4	9.3	11.6	5.7
Weasel/stoat:									
% control	10.1	10.1	7.3	12.7	16.2	4.7	14.7	1.1	13.7
% damage	6.7	2.9	4.9	7.3	8.8	0	9.3	0	10.2
N	119	69	123	55	68	85	75	95	88

THE FOX

Few species better illustrate the complexities of mammal management than the fox (Macdonald, 1984; Macdonald *et al.*, Chapter 18). Here we explore the links between the farmer's control of foxes, their farming activities and their wider perceptions as to the need for fox control.

Perceptions of the fox as a pest

In Baker & Macdonald's (2000) Wiltshire study in 1995, perceived pest status of foxes varied with farm type. No arable farmers stated that the fox was a pest on *their* farm (Figure 1:5). Stock farmers who had suffered losses the previous year were more likely to control foxes than those who had not. Elsewhere, the reasons cited for fox control also follow patterns of land use: culling foxes to protect game birds is more common in the east, while protection of stock shows the opposite trend (Heydon & Reynolds, 2000a). Foxes are more frequently considered 'serious' pests on sheep farms than on dairy or mixed enterprises (Produce Studies, 1995).

The Wiltshire data suggest that the perceived threat to game birds provokes control of foxes more readily than does stock farming. There was no statistically significant link between whether a farmer kept sheep or not, and whether they controlled foxes (33% of farmers with at least one sheep reported controlling foxes compared with 32% of non-sheep rearing farmers), nor was there any evidence within individual regions. On the other hand, there was strong evidence that the presence of game shooting was linked to the control of foxes (37% of such farmers said they controlled foxes compared with 25% where game birds were not shot: χ^2=10.1, d.f.=1, p<0.001).

In the 1981 survey, farmers were asked how many foxes were killed on their farm in the twelve months before the survey. If the keeping of sheep and the presence of game shooting are used as categoric predictors in a General Linear Model predicting this number (ln+1 transformed), then both are highly significant ($F_{1,619}$=8.4, p=0.004 and $F_{1,619}$=22.7, p<0.0001 respectively). Most foxes are killed on farms with some sheep and game shooting (adjusting for region). This adds to the findings of Heydon & Reynolds (2000a), who calculated that the likelihood of fox control on a farm was an additive function of both these predictors.

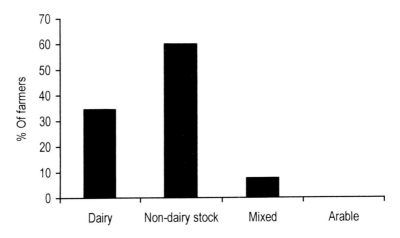

Figure 1:5 Percentages of farmers citing the fox as a pest by enterprise type in the 1995 Wiltshire survey.

Table 1:4 presents a regional analysis of the data given by Macdonald & Johnson (1996) for reasons for fox control (farmers were asked to state which reasons provoked the need for control, and the question was not phrased so as to restrict the 'need' to their own farm). Reasons varied in their frequency broadly as would be predicted from knowledge of the distribution of enterprise types. Of the two principal areas of damage associated with foxes — loss of stock and loss of game — there was much more regional variability in the 'game' response. A comparison with the 1995 data from Wiltshire (where the question was identical) did not suggest any great shift in reasons for fox control; the rank order of response frequencies was the same.

Table 1:4 Regional patterns of reported reasons for control of foxes (1981 survey, with Wiltshire 1995 in final column). West C.=West Country. Data are percentage of farmers stating reason.

County	Dorset	Leic.	Oxon.	Salop.	Suffolk	Sussex	Warw.	WestC.	Yorks.	Wilts. (1995)
Disease	53.3	49.2	50.4	50.0	53.3	60.2	43.1	29.2	38.8	29.7
Stock	68.6	76.9	61.0	72.0	71.7	74.7	66.7	71.9	71.3	53.3
'Too many'	62.9	67.7	71.3	70.0	63.3	74.7	73.6	65.2	65.2	72.8
Game	53.2	38.5	43.3	64.0	71.7	36.1	55.6	21.6	21.6	33.7
N	119	64	124	55	68	85	75	95	95	92

Farm size may also be related to fox management. Nationally, farmers were much more likely to report fox control on larger farms. In 1981, the mean (±SE) size of farms reporting control was 201.5 ha (±22.9) compared with 115.4 ha (±6.5) for farms with no fox control ($F_{1,635}=32.4$, p<0.0001, adjusting for region, in models predicting log farm area). This pattern may vary to some extent on a more local scale. In the Wiltshire data from 1995, farms where foxes were pests were not significantly different in size from those where they were not pests. The reported reasons for controlling foxes also generally varied with size. Protection of game was cited by 59% of farmers of farms larger than 200 ha, compared with 42% of small farms, while stock protection showed the opposite pattern: 64% of large farms and 81% of small ($\chi^2=12.8$, d.f.=1, p<0.001 and $\chi^2=15.8$, d.f.=1, p<0.001 respectively). This is broadly in agreement with the observations of Heydon & Reynolds (2000a), and probably a simple function of the relationship between farm size and land management. There was no evidence, however, that there was a closer link between perceived damage and fox control on small compared with large farms, as was observed by Heydon & Reynolds (2000a). Of those farmers controlling foxes on small farms, 79% said they had sustained damage in the previous year, compared with 71% on large farms ($\chi^2=1.1$, d.f.=1, p=0.30).

The susceptibility of an individual farmer to fox damage is only one element in his attitude to foxes. Whether farmers state that control of foxes is necessary in a wider context is not closely related to whether foxes are considered a pest on their own farms. In the Wiltshire study, 55% of farmers who said the fox was not a pest on their farm said they should be culled 'everywhere'. The culling of foxes, or permitting the culling of foxes, is widely cited by farmers as a 'good neighbour' policy, mainly in addition to other reasons for culling (Heydon & Reynolds, 2000a). A majority (72%) of respondents to Heydon & Reynolds' (2000a) survey said they aimed to contribute to regional control, rather than solely to their farm.

Wiltshire farmers frequently responded that there were too many foxes regardless of damage on their farm (Baker & Macdonald, 2000). Macdonald & Johnson (1996) reported that 65% of all farmers said that foxes should be culled because they were 'too numerous'. The regional reanalysis of those data shows that the majority of farmers in all regions stated that there were 'too many' foxes (Table 1:4). Farmers of arable land may report foxes as being 'serious' pests — 15% of arable farmers did so in one study, where the question was framed in terms of a wider perception of the pest status of the fox, rather than the farmers' own enterprise (Produce Studies, 1995). This may be partly attributable to secondary interests such as poultry (White *et al.*, 2000); there is a widely held perception that there are too many foxes. Table 1:3 gives a regional breakdown in control and damage patterns, based on our 1981 survey. It can be seen that patterns of damage generally mirror those of control. Respondents from Warwickshire do not conform to this pattern; almost twice as many report control as report damage. This is explained by the link with the 'too many' foxes reason. If control of foxes is treated as a binomial response, then, unsurprisingly, reported damage to foxes is a significant predictor. The only other reason that is a significant predictor of 'control' in addition to 'damage' is the 'too many' reason (logistic regression: Wald χ^2=13.2, d.f.=1, p<0.0001). This reason can be seen to be particularly prevalent in Warwickshire (Table 1:4).

There is some evidence that when farmers believe that there are too many foxes, this is related to their estimates of fox numbers on the ground. While in the Wiltshire sample, farmers estimates of earth densities and litter sizes of foxes did not differ according to whether they said the fox was a pest or not, in the much larger sample of 1981 there were significant differences. The 341 farmers who cited 'too many' as a reason for controlling foxes averaged 1.69 (±0.1) for their estimate of litters born each year on their farms, and 2.21 (±0.17) for the number of earths, compared with 1.11 (±0.10) litters and 1.41 (±0.15) earths estimated by the 184 farmers responding 'no' to the 'too many' reason ($F_{1,528}$=5.3, p=0.023 and $F_{1,579}$=11.0, p<0.001 respectively, in models adjusting for region). Of course, we have no means of knowing how farmers' perceptions relate to real numbers. The reported number of litters and earths were closely related to the size of the farms, and the 'too many' reason was not a significant predictor of the density of litters and earths in models adjusting for region ($F_{1,512}$=0.34, p=0.56 and $F_{1,562}$=2.98, p=0.08; overall the reported litter density was 1.9 per square kilometre and earth density was 2.43 per square kilometre).

Perception and the practice of control

We have established that while there are clear links between enterprise type and patterns of control, the relationship between farmers and foxes is not a simple one. We can also ask if there is a link between pest status and culling pressure, either by shooting or by mounted hunts. In the Wiltshire study, pest status had very little effect on the reported number of mounted hunt kills, but there was a statistically significant effect on shooting (Figure 1:6). This suggests that the efficiency of the mounted hunt is less responsive than shooting to pest status. These figures also indicate a tendency of farmers to overestimate the number of foxes killed by the hunt on their land. The farmers' account of the number of foxes killed by the hunt is several times that of the more reliable hunt records. This tendency for overestimation was also observed by Heydon & Reynolds (2000a).

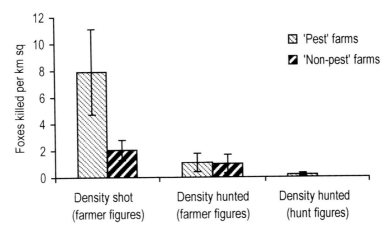

Figure 1:6 Mean (±SE) reported density of foxes killed by shooting (farmer figures) and by hunts (hunt and farmer figures) on farms in Wiltshire.

Comparing the two Wiltshire cull counts across the more specific categories of those farmers who said they had lost lambs to foxes, and those who had not, a similar pattern emerged. While there was a slight tendency for more hunt kills to be recorded where lambs were said to have been lost this was not statistically significant, and the difference was small by comparison with the effect on shooting kills.

Attitudes and actions with respect to foxes are complicated further by cultural factors, most conspicuously the field-sports interest of the farmer. In the national sample of 1981, significantly more hunting farmers said they actively approved of the conservation of foxes (27% of hunting farmers compared with 13% of non-hunting farmers: χ^2=24.8, d.f.=1, p<0.001). This is also consistent with the view that mounted hunting is largely a sport, as it is with the view that hunting farmers are more conservation minded. Both may, of course, be true. The practice of stopping fox holes on hunting days was unrelated to reported damage attributed to foxes (31% of farmers said holes were stopped on their farm on hunt days, and this figure was the same whether or not fox damage was also reported), nor was there any evidence for a link in the Wiltshire sample of 1995 (28% compared with 14%; χ^2=1.6, d.f.=1, p=0.20).

In the Wiltshire survey, we asked farmers to say which, among a list of non-exclusive options, were their principal motivations for hunting. All respondents opted for 'recreation', while 55% said 'to control foxes as a pest'. More farmers thought shooting was effective (63%) than did hunting (55%), as had been true in the 1981 survey (76% and 48%). Farmers who had participated in hunting in the earlier survey were also more likely to say that hunting was effective (77% versus 38%; χ^2=102.5, d.f.=1, p<0.001). At the same time, hunting farmers (farmers who said they had at some time participated in hunting) were also more likely to state that they approved of the active conservation of foxes (29%) than non-hunting farmers (15%; Macdonald & Johnson 1996). If we scrutinise this more closely, we find there is no evidence that, overall, farmers who approve of fox conservation are more common in any region (25% overall, χ^2=5.9, d.f.=1, p=0.65). Nor is there any regional variation if the question is restricted to hunting farmers. Similarly, there was no evidence that wildlife interest was related to approving of fox conservation (χ^2=2.1, d.f.=2, p=0.34). The Wiltshire sample also provided some evidence that tenant farmers perceived foxes to be worse pests than did owners, and that this was unrelated to reported losses (Baker & Macdonald, 2000); there was no evidence of this in the 1981 sample.

CONCLUSIONS

Our main point here has been a simple one: while the conservation of farmland mammals cannot be tackled without ecological insight, ecologists should not lose sight of the inter-disciplinarity of the issues. In particular, the inter-relations of land use, perceptions and economics are crucial ingredients — a blend that is becoming known as the 'human dimension' to conservation science. The perceptions and practices of farmers are crucial. The fox demonstrates the multiplicity of factors, and the regional complexity in their relative importance, which may contribute to the fate of any mammal inhabiting the agricultural landscape. In presenting this smorgasbord of results, we hope we have revealed the interest that lies in exploring links between perception and practice. More than that, this topic is crucial to policy formulation whereby modulation (the process whereby subsidies are transferred from production to agri-environmental and rural development goals) and cross-compliance (the process whereby receipt of subsidy for production is conditional upon adherence to environmental standards) nudge farmland towards what could be a radical future for mammal conservation in the UK (Macdonald *et al.*, 2000b). While much of the statutory effort of British conservation has hitherto been aimed at high priority sites (such as Sites of Special Scientific Interest), the majority of the countryside, and thus perhaps the greatest potential for improvement in conservation potential, lies in the agricultural mosaic. This potential will only be brought to fruition in partnership with farmers, so understanding of their perceptions, motivations and behaviour is crucial. The results presented here suggest that cross-compliance might benefit from greater focus on regional issues.

ACKNOWLEDGEMENTS

We are grateful to Jochen Langbein, Dieter Helm, Gary Taylor, Gary Kass, Ruth Feber, David Bullock and an anonymous referee for their comments.

2

The impact of agri-environment schemes on mammals

I P Johnson & S J Baker

INTRODUCTION

Farming is the dominant land use in the UK: the total area of agricultural land in the UK is 18.6 million hectares, some 76% of the total land area (MAFF *et al.*, 1999). In England alone, there are over nine million hectares of agricultural land (MAFF *et al.*, 1998). The intensification of UK agricultural practices over the last fifty years, and the effect of this on our wildlife has been the subject of considerable concern and study (e.g. Aebischer, 1991; Fuller *et al.*, 1995; Campbell *et al.*, 1997). Changes in agriculture in recent decades include increased use of pesticides (see also Shore, Fletcher & Walker, Chapter 4), intensification of grassland management (e.g. Gorman & Reynolds, Chapter 8), the decline of traditional arable/grass rotations and changes in sowing and harvesting times (Stoate, 1996; Donald, 1998). Swash *et al.* (2000) estimate that 88% of English farmland is now intensively managed.

Agricultural intensification has probably contributed to the major declines of formerly common farmland birds, and also those of some plants and invertebrates (Green, 1990; Marchant *et al.*, 1990; Wilson, 1992a). Changing agricultural practices are also implicated in the decline of some species of mammal in the UK, one of the best studied being the brown hare *Lepus europaeus* (e.g. Barnes *et al.*, 1983; Tapper & Barnes, 1986; Department of the Environment, 1995). Most agri-environment schemes have a number of objectives, such as enhancing landscapes, maintaining historical interests and promoting access to the countryside. However, one broad aim of the schemes is to seek to promote more extensive, lower-input agriculture, reversing many of the recent changes that are perceived as being damaging to wildlife.

UK agri-environment schemes were launched in 1987 with the introduction of the Environmentally Sensitive Areas (ESA) scheme. Since that date, the ESA scheme has been expanded and other schemes introduced. In England, the schemes are part-funded by the EU and part by the Department for Environment, Food and Rural Affairs (DEFRA), with scheme administration being provided by DEFRA's Rural Development Service. Prior to DEFRA'S creation in 2001, UK funding and scheme administration had been provided by the Ministry of Agriculture, Fisheries and Food (MAFF). The importance of agri-environment schemes in meeting Biodiversity Action Plan (BAP; see below) conservation targets is illustrated by the frequency of actions referring to them within individual BAPs. For example, of the 112 species BAPs produced in 1995, over one third specifically referred to agri-environment schemes as mechanisms for achieving targets (Ovenden *et al.*, 1998).

This chapter considers the impact of a number of these schemes on wild mammals in England, focusing on species identified as priorities in the UK Biodiversity Action Plan (UK Biodiversity Group, 1998a). Although here we only consider England, broadly similar schemes operate elsewhere in the UK; for example Tir Gofal in Wales, Rural Stewardship Scheme in Scotland and Countryside Management Scheme in Northern Ireland. The general conclusions of this chapter are likely to apply to agri-environment schemes in these areas.

AGRI-ENVIRONMENT SCHEMES IN ENGLAND

DEFRA has three main schemes to encourage farmers to adopt environmentally sensitive land management practices, namely the ESA, Countryside Stewardship and Organic Farming Schemes Table 2:1. In addition to these, DEFRA currently operates several other grant schemes within the England Rural Development Programe. DEFRA has also piloted other schemes that are now closed, although management agreements with individuals who entred land into these schemes will remain in force until their agreed end dates. (e.g. the Habitat Scheme). The two largest agri-environment schemes operated by DEFRA are the ESA and the Countryside Stewardship Scheme (CSS), with estimated total annual payments to landowners of £59 million during 1999/2000. Payments are calculated on the basis of income forgone for adopting more extensive, low-input practices that offer environmental benefits. As with the other agri-environment schemes, they are voluntary, that is, farmers cannot be required by DEFRA to enter into management agreements. The purpose and operation of the three main agri-environment schemes are outlined below.

Table 2:1 Agri-environment schemes operated in England by DEFREA (formerly MAFF).

Scheme	Purpose
Environmentally Sensitive Areas:	To encourage environmentally sensitive land management practices. (Limited to defined areas).
Countryside Stewardship Scheme (incorporating the Arable Stewardship Pilot Scheme):	To encourage environmentally sensitive land management practices. (Operates outside Environmentally Sensitive Areas).
Organic Farming Scheme:	To encourage conversion to organic production methods.

Environmentally Sensitive Areas

The ESA scheme aims to maintain and enhance the landscape, wildlife and historic interest of designated areas by encouraging appropriate agricultural practices. The area covered by the 22 designated English ESAs amounts to approximately 13% of the agricultural land area. They have all been identified as being discrete areas of national environmental significance. In addition, the conservation interest of these areas is dependent upon the adoption, maintenance or extension of particular farming practices which have or are likely to change, or which could, if modified, result in a significant improvement in that interest (Harrison, 1997).

Each ESA has its own specific objectives, with particular emphasis given to landscape, wildlife or historic interest as appropriate to the area. Land management agreements run for ten years and are based on a tiered structure of management prescriptions, with higher payments made to farmers for the more demanding tiers, which produce greater environmental benefits. Options within the ESAs include management of existing habitats, restoration of degraded habitats and reversion of arable land to grassland or other semi-natural habitats. The prescriptions typically cover items such as fertiliser application rates, water level management, intensity and duration of grazing and timing of hay cutting. However, as with the environmental objectives, each ESA has its own specific management requirements. In addition to the various management tiers, further funding is available for specific projects through ESA Conservation Plans. These cover items such as pond creation, hedgerow restoration and pollarding.

By 1999, the 22 English ESAs had 10,191 management agreements covering over 0.56 million hectares (MAFF, unpublished data).

Countryside Stewardship Scheme

The Countryside Stewardship Scheme (CSS) was launched by the Countryside Commission as a pilot scheme in 1991. It was transferred to MAFF (now DEFRA) in 1996 and expanded to become the main agri-environment scheme outside the ESAs. It operates throughout England. The scheme was expanded in 1999 to incorporate new and enhanced upland options, to supersede the Moorland Scheme. Following an Arable Stewardship Pilot Scheme, a range of new management options for arable land were introduced into the CSS in 2002. The CSS is a multi-objective scheme that aims to provide landscape, wildlife, historical and public access benefits. Countryside Stewardship has targets for particular landscape types and features, namely chalk and limestone grassland, lowland heath, watersides, coasts, uplands, historic landscapes, traditional orchards, old meadows and pastures, the countryside around towns, traditional field boundaries and the margins around arable fields. The scheme is targeted locally at a county level, in consultation with relevant organisations. Acceptance of applications by DEFRA is discretionary, with agreements being selected on their merit in relation to the CSS objectives, national and local targets.

CSS agreements are usually for a period of ten years. The scheme options are based on a suite of standard management prescriptions, which can be tailored to meet site-specific objectives. The standard prescriptions cover management of farmland habitats including grassland, arable land, field margins, field boundaries (hedges, walls and ditches) and semi-natural habitats including heath, fen and reedbed. In 1999 there were approximately 8,600 CSS agreements covering around 152,800 ha (MAFF *et al.*, 1999).

The Arable Stewardship Pilot Scheme offered payments to arable farmers to manage their land under a series of options designed to benefit wildlife. Following evaluation of the land management options trialled in the Pilot Scheme, a range of options including overwintered stubbles and conservation headlands were added to the CSS in 2002. Unlike the other CSS options, which relate to a number of objectives, the new arable options are aimed soelely at benefitting farmland biodiversity, in particular species identified as priorities in the UK Biodiversity Action Plan.

Organic Farming Scheme

The Organic Farming Scheme (OFS) replaced the Organic Aid Scheme in spring 1999. The principal objective of this scheme is to provide financial assistance for farmers wishing to convert to organic farming methods. The OFS also requires farmers to comply with a number of additional environmental measures, beyond those required for conversion to organic production. These measures largely relate to the retention and appropriate management of existing areas of environmental interest, including species-rich grassland, hedgerows and woodlands and any features of historical or archaeological interest. By the end of 1999, some 85,000 ha of land had been placed into organic conversion under this scheme (MAFF et al., 1999).

THE UK BIODIVERSITY ACTION PLAN

In response to the 1992 UN Convention on Biological Diversity, the UK Government published a UK Biodiversity Action Plan (BAP) in 1994 (Anon, 1994). A key feature of the UK BAP was the establishment of the UK Biodiversity Steering Group, which included representatives of governmental and non-governmental organisations, and which was charged with developing costed action plans for key habitats and species.

In their 1995 report (Department of the Environment, 1995), the Steering Group published action plans for 14 habitats and 116 species, and recommended that plans be prepared for a list of further habitats and species. On 25 October 1999, the final tranche of action plans was published (see: http://www.jncc.gov.uk/). In total, BAPs have now been published for 45 habitats (also known as Habitat Action Plans or HAPs) and 391 species (Species Action Plans or SAPs). Each BAP includes conservation targets with specified timescales. Contributing organisations and mechanisms for achieving these targets are also identified.

Species were identified as priorities on the basis of a number of criteria, including either being globally threatened or being in rapid decline in the UK (having declined by more than 25% in the last 25 years). This resulted in the identification of ten extant, non-marine mammal species as conservation priorities (Table 2:2). BAPs for these species were published in 1995 and 1998 (Department of the Environment, 1995; UK Biodiversity Group, 1998a). All action plans are now being taken forward by individual steering groups, again including government and non-government organisations, each with a specific lead partner or partners.

Table 2:2 UK BAP priority mammal species.

Species	Key farmland habitat(s)	Agri-environment schemes identified as BAP mechanism?
Greater horseshoe bat *Rhinolophus ferrumequinum*	Low-input pasture, hedges, wetlands	No
Lesser horseshoe bat *Rhinolophus hipposideros*	Hedgerows, treelines, woodland, farm buildings	Yes
Bechstein's bat *Myotis bechsteinii*	Woodland	Yes
Pipistrelle bat *Pipistrellus pipistrellus*	Old trees, hedgerows, wetlands	Yes[1]
Barbastelle bat *Barbastella barbastellus*	Woodland, old trees, low input pasture	Yes
Brown hare *Lepus europaeus*	Arable grassland, woods, hedges	Yes
Red squirrel *Sciurus vulgaris*	Woodland	No
Water vole *Arvicola terrestris*	Riparian	No
Dormouse *Muscardinus avellanarius*	Woodland, hedges	No
Otter *Lutra lutra*	Riparian	Yes

CAN AGRI-ENVIRONMENT SCHEMES CONTRIBUTE TO MEETING MAMMAL BAP TARGETS?

Although ESAs and CSS were developed prior to the Biodiversity Convention, the scheme objectives relating to wildlife species and habitats fit well with BAP objectives. The BAP process has identified UK conservation priorities and this has enabled more focused targeting of agri-environmental wildlife objectives. Indeed, the potential contribution to biodiversity objectives is specifically taken into account in the assessment of applications to the CSS. Of the ten terrestrial BAP priority mammals, agri-environment schemes are specifically referred to in six of the plans (Table 2:2). Even where agri-environment schemes have not been identified as a mechanism, there may still be some benefits for priority species.

Although agri-environment schemes are predominantly aimed at habitat management, there is scope within them for adjusting agreements to benefit species of particular

---

[1] BAP included reference to incentive schemes run by various bodies including MAFF

conservation importance. Thus, priority mammal species may benefit in two ways from the implementation of agri-environment agreements: either from the broad habitat management measures of the schemes, or from site-specific measures aimed at individual species known to occur on or near the area under agreement. Table 2:3 summarises how priority mammals are most likely to benefit from agri-environment schemes.

ESA and CSS agreements are promoting lower-input farming practices on over 7% (0.71 million hectares) of the agricultural land in England. Mammals such as brown hares, water voles and pipistrelle bats, are particularly likely to benefit from the broad habitat management measures in place over this area. Other species which are likely to benefit from specifically targeted management at selected sites include the greater horseshoe bat. Table 2:3 indicates that there is potential for at least some benefits for all ten species.

Table 2:3 Potential benefits of agri-environment schemes to BAP priority mammal species.

Agri-environmental measures likely to provide benefits:	Fertiliser & pesticide restrictions	Restricted times for hay/silage cuts	Tree & woodland management	Field boundary management (hedges, walls etc)	Restoration of semi-natural habitats (e.g. wet grassland, heathland)	Building and historic parkland restoration
Greater horseshoe bat	X		X	X	X	X
Lesser horseshoe bat			X	X		X
Bechstein's bat			X			X
Pipistrelle bat	X		X	X	X	X
Barbastelle	X		X		X	X
Brown hare	X	X	X	X	X	
Red squirrel			X			
Water vole	X			X	X	
Dormouse			X	X		
Otter				X		

Examples of the range of ways in which mammals can benefit from agri-environment schemes are illustrated by the three species considered below.

The brown hare

The brown hare is a widespread species of farmland landscapes (Tapper, 1991; Harris *et al.*, 1995). In the UK, brown hares are most numerous on arable land where cereals predominate, but with grass fields available for summer feeding. Copses and hedgerows are also frequently used as daytime resting areas (Tapper & Barnes, 1986). Although still common, this species has undergone a substantial decline since the 1960s (Tapper & Parsons, 1984; Tapper. 1992; Department of the Environment, 1995). Their decline has been particularly marked in pastoral areas in the south-west of Britain. Numerous explanations have been proposed, but no single one is in itself satisfactory. The most widely accepted cause of the decline is change in the agricultural ecosystem (Harris *et al.*, 1995). Agricultural changes implicated include conversion of grassland to arable; loss of habitat richness in the agricultural landscape; intensive management of grassland for silage production; high densities of livestock; and the shift to autumn sowing of cereals (Department of the Environment, 1995; Hutchings & Harris, 1996). Tapper & Barnes (1986) concluded that the simplification of arable systems was a major factor.

ESA management prescriptions have the effect of stemming or reversing many of these changes, shifting farming towards more extensive, lower-input practices. Some ESAs, such as Breckland, hold significant populations of brown hare (Rothera, 1998). However, given their widespread distribution, the CSS has, by virtue of its wider geographical availability, even more scope for benefiting brown hares. The CSS offers payments to land managers to revert arable land to grassland. This includes options for whole fields, and for grass margins (2 m or 6 m wide) around arable fields and for 2 m wide 'beetle banks' within fields. In areas of intensive arable production these measures could locally increase the habitat richness of the farmland landscape. In the period 1991 to 1998, over 11,700 ha of arable land was reverted to grassland or other semi-natural habitats, such as heathland. In addition, during the same period, 3,100 km of 6 m wide margin and 6,400 km of 2 m wide margin or beetle bank were created through CSS agreements (MAFF, unpublished data). The arable reversion area will normally be sown with a mixture of at least four native grass species appropriate to the locality, although in some circumstances swards are established through natural regeneration. Hutchings & Harris (1996) refer to a reduction in the diversity of food sources available to brown hares as perhaps the most important single factor in the species' decline. Arable reversion to grassland through the CSS increases the diversity of sites available for summer feeding in areas of intensive cereal production.

After the sward has been established under the CSS, it must be managed extensively, in a manner similar to existing grassland on agreement land. Typically, grassland management prescriptions prohibit applications of inorganic and organic fertilisers. Pesticide application is normally restricted to spot treatment of injurious weeds (as defined in the Weeds Act 1959) and nettles. Restrictions on fertiliser and herbicide usage will, in the medium to long term, tend to increase the diversity of the sward and hence the diversity of foods available to hares. Work in Sweden has demonstrated that, where they are available, wild grasses and herbs are preferred to cultivated varieties (Frylestam, 1986). Thus it seems likely that grassland managed under a CSS agreement will be more attractive to brown hares than typical, intensively-managed agricultural grassland.

Grassland management prescriptions require fields to be managed by grazing or cutting. If cut, the aftermath is normally grazed. Suitable grazing species are specified, together with either maximum stocking densities or target sward heights. The time of year when grazing is permitted would also be specified. Stocking rates on grassland under agreement would be lower than those typically found on non-agreement grasslands. However, it is not known whether stocking at the rates typically specified in CSS agreements will significantly reduce the disturbance apparently caused by livestock to hares (Barnes *et al.,* 1983).

Cutting of grass and other crops may result in significant mortality of brown hares using those fields (Hutchings & Harris, 1996). Within the CSS, where fields are cut for hay or silage, earliest permissible cutting dates are specified. Restrictions on dates of cutting, and restrictions on rolling and harrowing, seem likely to reduce hare (particularly leveret) mortality during the first half of the year. An additional frequently used prescription is to cut meadows from the centre outwards, specifically to help dispersal of young birds and mammals to the field edges (Andrews & Rebane, 1994). Tyler *et al.* (1998) demonstrated the value of this technique for corncrake *Crex crex* chicks. The brown hare is a species that may also benefit from this practice, but this has not yet been demonstrated in the UK.

Until the introduction of new arable options in 2002, management of arable land under the CSS was largely restricted to the creation of grass (or uncropped) margins, beetle banks and conservation headlands in cereal crops. A wider range of options is now available, including retention of cereal stubbles over winter, conservation headlands (now available as a stand-alone option) and the sowing of 'wildlife seed mixture' (a mixture of seed-bearing species to provide food for wild birds). The over wintered stubble option may be of particular benefit to brown hares.

Farm woodlands and hedgerows are utilised by hares as diurnal resting areas, particularly during the winter (Tapper & Barnes, 1986). CSS prescriptions relating to these habitats are also likely to provide benefits to brown hare populations. There is a requirement that CSS agreement holders protect and maintain important features on all their holding land (i.e. not just the land for which CSS payment is being received). These features include woods, hedgerows and hedgerow trees. As well as protecting and managing existing hedgerows, restoration is an important feature of many CSS agreements. By 1998, 5,530 km of hedgerow were being actively restored, by laying, coppicing or planting, through CSS agreements (MAFF, unpublished data).

Agri-environment schemes such as Countryside Stewardship are reversing many of the changes that have been implicated in the decline of the brown hare. Direct evidence of benefits to hares has yet to be demonstrated. However, in general, they are likely to benefit from the grassland management regimes required under the ESA and CSS through improved foraging opportunities, provision of cover, and through reduced mortality and disturbance from agricultural operations.

The dormouse

The dormouse is largely restricted to the southern half of England where it has a widespread but localised distribution. It is chiefly found in deciduous woodland, but occurs in other habitats including species-rich hedgerows (Morris, 1991). Changes in woodland management, including problems from stock incursion, together with fragmentation, are identified as causes of the dormouse's decline (Department of the Environment,1995; Bright & Morris, 1996).

Because it is primarily a woodland species, the dormouse has probably been less affected by agricultural change than other species more closely associated with farmland habitats. However, where dormice occur in farm woodlands or in hedgerows, there is scope for agri-environment schemes to provide benefits. At present agri-environment schemes are not identified as a mechanism for this BAP.

The majority of ESAs now have a Woodland Tier, in which the agreement holder adopts a woodland management plan. Although annual payments for woodland management are relatively modest in the lowland ESAs, Conservation Plan funding can assist with capital items. These could include, for example, fencing to prevent stock incursion into woodlands or even the provision of dormouse nesting boxes. Because ESA woodland management is based on site-specific management plans, it can be tailored to benefit dormice or other BAP priority species on the sites where they are known to occur.

Hedgerows are important to dormice, not just as a habitat, but also as arboreal corridors linking woodlands (Bright & Morris, 1989). In this latter respect, they are important in minimising the effects of woodland fragmentation, linking what might otherwise be isolated, non-viable populations. Hedgerow restoration and management requirements under the CSS have the potential to provide benefits, although there is also potential for CSS-funded hedgerow coppicing to be detrimental in the short-term. A number of ESAs also offer payments for hedgerow management.

To maximise benefits of agri-environment agreements, and to avoid the risk of inadvertently causing harm to rare and localised species such as the dormouse, it is important that the farmer and the Rural Development Service (RDS) Project Officer are aware of their presence when an agri-environment agreement is being negotiated. Project Officers do not have the time to undertake detailed wildlife surveys of land entering agri-environment schemes; neither are there the resources to proactively target land holdings of particular biodiversity interest. English Nature, the Wildlife Trusts and other organisations have an important role to play in ensuring that information on the location of priority species is made available to Project Officers. Ideally, data on the location of species of particular conservation importance would be provided in a form that can readily be transferred to RDS Geographical Information System. Partner organisations can also play an important part by actively encouraging high quality applications from owners of sites of particular interest.

The greater horseshoe bat

This species is associated with areas of deciduous woodland adjacent to cattle-grazed pasture, where insects are particularly abundant (Duvergé & Jones, Chapter 6). Caves, disused mines or other sites suitable for hibernation are also needed (Ransome, 1991).

The greater horseshoe bat has declined throughout its range within northern Europe, and in England is now restricted to the south-west. In England and Wales, there are only 35 recognised maternity and all-year roosts. Control of insects in agriculture and forestry, together with the loss of undisturbed roosting and hibernation sites, are considered likely to have contributed significantly to the decline of this species (Ransome, 1991; Department of the Environment, 1995).

The UK BAP for the greater horseshoe bat (Department of the Environment, 1995) does not specifically identify agri-environment schemes as a conservation mechanism. Nonetheless, one of the actions is 'to encourage favourable habitat management (aiming for up to 4 km around each roost), seeking to implement these through voluntary or informal agreements'. Agri-environment schemes can provide a mechanism for achieving favourable management.

This is illustrated by a CSS agreement in North Somerset, which came into effect in October 1999. This ten-year agreement concerns farmland that is adjacent to a known maternity roost of greater horseshoe bats. The agreement, which also meets a number of landscape and historical/archaeological objectives, has a specific wildlife objective of providing a favourable foraging habitat for this species. This objective is to be achieved principally by a combination of hedgerow restoration and grassland management, but also by some wetland creation and woodland planting.

All existing hedges on the holding are maintained at a minimum height of 2 m where the hedge is free standing, or 1 m of shrubby growth where the hedge is on a hedge bank. Hedge trimming is carried out during the winter months, with no individual hedge being cut more than twice in every five years. Cutting is rotated around the holding, to avoid all hedges being cut in any one year. Existing hedgerow trees are retained and protected. Hedgerows on the agreement land have become gappy and in need of restoration, and there will be over 400 m of planting to infill gaps in existing hedges. The hedge plants used will be of native stock and the same broadleaved species as those which occur naturally in the area. To ensure diversity, no one species may comprise more than 75% of the total number of hedgerow plants. In addition, approximately 800 m of hedgerow will be restored by laying or coppicing over the period of the agreement. This programme of work will produce a network of well-managed hedgerows, which will provide sheltered flightpaths for foraging bats, including the greater horseshoes.

The permanent pasture will be managed to maximise the number of invertebrates available to foraging bats. The pasture will receive no inorganic or organic fertilisers. Herbicide applications are restricted to weed wiper or spot treatment of injurious weeds. No other pesticides are permitted. The grass fields will be grazed by a combination of cattle and horses, the dung of these animals being favoured by beetles of the genera *Geotrupes* and *Aphodius* (Chinery, 1973; Harde & Hammond, 1984). Dietary studies (Ransome, 1991) have shown these groups of insects to be important prey items for greater horseshoe bats. Fields will be grazed for at least the period from March to September, thus encouraging a supply of dung community invertebrates throughout the bats' principal flight period. Grazing levels are specified, with maximum grazing from July to September

In addition to the hedgerow and grassland management, a pond will be dug, again to increase the supply of invertebrates, and an even-aged woodland stand will be diversified by new planting.

Although this agreement has been tailored specifically to benefit a BAP priority species, all of the management prescriptions are common features of CSS agreements. Although agri-environment scheme prescriptions are very largely aimed at management of habitats rather than species, the presence of UK BAP priority species will be taken into account when preparing management agreements. There is even a certain amount of flexibility within schemes to adapt standard prescriptions to the needs of particular species, where this is considered necessary.

BENEFITS OF AGRI-ENVIRONMENT SCHEMES TO OTHER FARMLAND MAMMALS

Many of the species of mammal found in the UK make use of farmland to a greater or lesser extent. For example, Gurney *et al.* (1998) list 30 species of British mammal that they consider to have a significant association with farmland. These include a variety of small mammals (mice, voles and shrews), which are all likely to benefit from the lower input, extensive farming practices promoted by agri-environment schemes (see also Strachan, Moorhouse & Macdonald, Chapter 5, for a discussion of potential benefits to riparian mammals).

The creation of uncut 2 m grass margins to arable fields and beetle banks within fields, through the CSS, are particularly likely to benefit field voles *Microtus agrestis*, which favour rough, ungrazed grassland (Gipps & Alibhai, 1991). The value of grassy margins to small mammals is recognised obliquely in agri-environment schemes, with margins often established for the benefit of avian predators such as barn owls *Tyto alba*, which are known to favour rough grassland boundary habitat for hunting (e.g. Brazil & Shawyer, 1989). However, it seems likely that mammalian predators, such as weasels *Mustela nivalis* and foxes *Vulpes vulpes*, will also benefit from an improved food supply.

CSS beetle banks and 2 m margins are typically sown with a grass mixture containing a high proportion of cocksfoot *Dactylis glomerata* which, due to its tussocky nature, is a valuable habitat for overwintering invertebrates. Studies by the Game Conservancy Trust (Boatman & Stoate, 1999) have shown that the dense tussocks also provide a valuable habitat for the harvest mouse *Micromys minutus*, with harvest mouse nests recorded at an average density of 47 per kilometre of beetle bank.

Harvest mice and field voles are also likely to benefit from the introduction into the CSS during 1999 of 'Wildlife Strips' – zero input grass margins around intensively farmed grass fields. These margins, averaging 6 m wide, are not usually grazed or cut, although occasional cutting or light grazing may be permitted under an agreed management plan. The aim is to create a heterogeneity of structure within areas of intensive grassland, producing strips of tall, tussocky grassland which will benefit birds, mammals and invertebrates. The Wildlife Strips will also act as buffers to protect adjoining habitats or features of environmental interest from the effects of potentially damaging agricultural inputs or mechanical operations.

Wood mice *Apodemus sylvaticus* in arable fields have been shown to favour foraging in conservation headlands, strips of crop around the field margins where pesticide inputs are restricted (Tew *et al.* 1992, Tattersall & Macdonald, Chapter 7). Conservation headlands in cereal crops are another frequently applied management prescription in the CSS. Similar conditions may be expected on arable land in the Organic Farming Scheme, during the period of conversion to organic. However, thorough mechanical weed control could mean that, in practice, the benefits from organic cropping are minimal in this respect.

Extensive management of grassland under agri-environment schemes is likely to produce more structurally diverse swards, with higher botanical diversity than intensive grasslands. This again might be expected to benefit mice, voles and shrews, as would the retention, planting and sympathetic management of hedges.

For other mammals utilising agricultural land, such as badgers *Meles meles*, rabbits *Oryctolagus cuniculus* and various species of deer, agri-environment schemes may be only marginally beneficial and of little significance to their conservation status. However, it seems unlikely that any mammal species will be detrimentally affected by the uptake of agri-environment schemes.

CONCLUSIONS

The agri-environment schemes operating in England have varying but broadly related objectives. These objectives include, but are not restricted to, conserving and enhancing farmland biodiversity. Management agreements under the schemes are not be based solely on the needs of individual species; indeed, agri-environment wildlife objectives are largely based on habitats rather than species. Nonetheless, for recognised conservation priorities such as UK BAPs, management may be adjusted to take account of individual species' needs.

Agri-environment schemes were introduced before the setting of national biodiversity targets. However, the schemes have always had habitat and wildlife enhancement objectives, and there is enormous potential for them to contribute; this is recognised within the individual species and habitat BAPS (Department of the Environment, 1995; UK Biodiversity Group, 1998ab). The schemes can contribute to the conservation of priority mammals, with species closely linked to farmland habitats, such as the brown hare, likely to benefit the most. For other species, such as the dormouse or the greater horseshoe bat, the benefits will be less significant, because of their lower dependence on agricultural practice. Even here, though, land management funded by agri-environment schemes can sometimes be of benefit, often through the tailoring of individual management agreements to meet species' particular requirements. There are potential benefits to most priority species from some of the broad measures of agri-environment schemes, such as reduced pesticide and fertiliser inputs (Table 2:3).

In addition to the priority species, agri-environment measures will benefit a wide range of more common mammals associated with farmland, through the creation and sensitive management of habitats and features including grassland, beetle banks, hedges and a variety of semi-natural habitats.

Agri-environment schemes have been identified as a major implementation mechanism for many of the UK species and habitat BAPs. Already influencing management on over 7% (MAFF, unpublished) of agricultural land in England, agri-environment schemes are likely to contribute significantly to restoring our farmland biodiversity.

ACKNOWLEDGEMENTS

We thank colleagues in DEFRA for assistance with drafting this paper and colleagues in the RDS GIS Unit for providing uptake data for agri-environment schemes.

$$3$$

Mammals and game management: A farmer's view

H R Oliver-Bellasis & N W Sotherton

INTRODUCTION

The UK's game animals, including several species of mammal such as deer and hares, have been hunted by people since pre-historic times. Their often substantial size means that they need large areas to accommodate them, so, with the exception, perhaps, of wintering grounds for waterfowl, they cannot be accommodated on nature reserves or protected through land designation measures such as Sites of Special Scientific Interest or Special Areas of Conservation. They are animals of the wider countryside, and as such, live, breed and forage over land used to grow crops, graze livestock or produce trees. Indeed, they do this in the crops, fields and woods of the Manydown Company Estate. In this chapter, we seek to explain how farmland mammals fit into the philosophy of game conservation management on the Manydown Company Estate, a commercial farm in southern England. The philosophy includes the need to control predators, and we illustrate this with a case study of the brown hare *Lepus europaeus*.

The material and perspectives presented in this chapter reflect the views and personal experiences of one of the authors (HR O-B), a working farmer and owner of the Manydown Company Estate. These may not necessarily be representative of the British farming community as a whole, if indeed such a representation can exist at all. However, these views and experiences certainly reflect the philosophies that underpin working practices at Manydown, and we have some confidence that such philosophies operate over a very large acreage of farmland in Britain.

BACKGROUND

Manydown is a family farm and a land-owning enterprise, which includes a mixed livestock/arable farm and let farms and houses. There is no dairy and no surface water features such as ponds or rivers. Just over 2,000 ha are farmed in hand. The soils are thin chalk and flint with outcrops of clay cap. Fields are bounded by 140 km of hedges. There are 156 ha of ancient semi-natural woodland, composed of oak *Quercus robur* standard and hazel *Corylus avellana* coppice although there is currently little economic use for the coppice. The farm always tries to make use of a diverse cropping rotation producing seed cereals, grass seed, malting barley, oilseed rape, potatoes, linseed and beans. Recently the rotation has included soya and linola.

Manydown has been a scientific study area for many years, hosting a range of large-scale, long-term, scientific studies pioneering management techniques to ameliorate the deleterious impact of modern farming practices of farmland game and wildlife. Examples include the development of 'conservation headlands' (Sotherton, 1991) and 'beetle banks' (Thomas *et al.*, 1991), the conservation of rare arable annual wildflowers (Wilson, 1993), the management of field boundaries (Boatman & Wilson, 1988), the conservation of farmland butterflies (Dover *et al.*, 1990), the identification of the broad-spectrum activity of insecticides and fungicides against beneficial, non-target, arthropods (Sotherton *et al.*, 1987) and the first experimental evidence in the field for the indirect effects of pesticides on farmland birds (Rands, 1985). Manydown provided one of six study sites involved in the MAFF LINK project (Ogilvy, 1996) to quantify the impact of Integrated Farm Management systems.

At Manydown, all the land is keepered, and supports a variety of species including about 60 pairs of wild grey partridges *Perdix perdix*, about 400 brown hares and perhaps 60 pairs of peewits *Vanellus vanellus*. Most predators, both common and protected species, occur on the estate.

Unlike some countries with substantial areas of state-owned land where game management is controlled by government, most of the UK is privately owned. Owner/occupiers of private land, such as the Manydown Estate, bear the responsibility of land management and its impact on mammals, be they game, pests or of conservation concern. Although owners and other land-occupiers are often keen to conserve game and other wildlife (e.g. see Macdonald & Johnson, Chapter 1), the economics of land use enterprises frequently force decisions to be made that are detrimental to these interests. Farmland, particularly outside protected areas, is under pressure from the demands we make on our environment to achieve higher living standards and produce inexpensive food. These pressures include agricultural intensification, the impact of pesticides, the removal or degradation of semi-natural cover and human development (Hindmarch & Piekowski, 2000). In many cases, this is a consequence of the Common Agricultural Policy, which can prejudice the economics of crop growing and livestock rearing to such an extent that wildlife conservation interests can be seriously jeopardised. Landowners could choose to disregard such policies and farm outside the subsidy payments they deliver, but such an approach would be likely to be economically unviable.

MANYDOWN'S GAME MANAGEMENT POLICY

The general public may sometimes find it difficult to understand how harvesting game can be beneficial to wildlife conservation. This is not surprising given the many examples of species' extinction caused by hunting around the world. However, harvesting game provides the Manydown Company with the incentive to help counter the economic or managerial pressures imposed by agricultural activities. This approach is encapsulated in the phrase 'conservation through wise use' and this underpins the management philosophy at Manydown. In managing game, the Manydown Company seeks to produce a game harvest that is sufficiently high to produce a benefit, either for their own sport or an economic return (for example hares can fetch £4.00 each), yet not so high as to reduce future harvests. This is the 'wise use' (or, more formally, a maximum sustainable yield; Caughley & Sinclair, 1994). The yield of the species is directly related to its survival and reproductive rate; both can be increased by management. At Manydown, better habitats are produced for game, more abundant wild foods for game are provided, and less predation and less poaching on the farm are ensured.

Predator control does not mean that game shooting need dominate countryside management to such an extent that their continuance is put in jeopardy. However, predators, if very abundant, can depress game stocks sufficiently for there to be no possibility of shooting (Tapper et al., 1996). Therefore, our objective in game conservation must be to reduce predation pressure on game populations while at the same time enhancing the community of predators as a whole.

Practical examples of habitat management policies include: targeted reduction of pesticides on cereal crop edges; the use of otherwise productive land to plant food and cover crops for game; the active management (and associated costs) of maintaining and managing woodland and hedgerows; and the creation of wildlife habitats on our set-aside rather than opting for the simpler, cheaper management options that destroy wildlife (Sotherton, 1998). Such management is primarily undertaken for the conservation of gamebirds, although other species of game and wildlife, particularly the brown hare, have also benefited. This is achieved at a considerable price, and usually at the expense of agricultural production.

THE COSTS OF GAME MANAGEMENT

The costs of such sympathetic management for game are very variable, and the methods of accounting differ considerably between estates. Many costs associated with game management are absorbed in the budgets of other estate enterprises such as farming or forestry. However, since 1992, The Game Conservancy Trust has been helping to run an estate at Loddington, where the game management prescriptions carried out at Manydown are implemented, monitored and fully costed (Boatman, 2000). The farm, run by the Allerton Research and Educational Trust (ARET), comprises 333 ha of heavy clay land near Loddington in Leicestershire, where a range of arable crops and sheep grazing on permanent pasture contribute to the farm income.

At Loddington, conservation headlands (the practice of using less pesticides and more selective compounds at the crop edge to encourage wildlife) suffer an average yield loss of 0.65 tons/hectare of cereal crop compared to headlands receiving the full complement of pesticides, an average loss of nearly 7%. The monetary value of this yield difference in any one year depends on grain prices and the relative proportion and market value of barley and wheat grown. The mean five-year cost per hectare of conservation headlands was £81.

The major costs of beetle banks (strips of sown tussocky grass placed across large arable fields) are incurred at establishment and from the value of the cropped ground lost to this non-cropped cover. At present, we estimate the establishment of a 2m by 500m beetle bank to be £25, with the average loss of the grass margin of crop foregone of £53. Beetle banks can be grown on set-aside, so area payments can be claimed to offset costs. They are also eligible for grant aid under the Countryside Stewardship Scheme (see Johnson & Baker, Chapter 2).

Much of the sympathetic management for game on the farm can be centred on the set-aside. However, costs are involved if the farm chooses to actively manage such land (e.g. plant cover crops, delay cutting and cultivation) rather than carry out a minimum maintenance programme. Costs per hectare of the use of beneficial cover, planted under the Wild Bird Cover Option (MAFF, 2000b) at Loddington, have varied over the period 1995 to 1999 from £406 to £2,501, with an annual average cost (±SE) of £1482 (±352) per hectare. The significant variation in costs reflected whether the farm's entire allocation of wild bird cover can be accommodated on the total eligible areas of set-aside, which themselves have varied over this period.

Many other costs are also involved in game management. These include avoiding blocked cropping and thereby creating a mosaic of crop types across the farm, managing hedgerows and grass verges and the hidden costs of extra management. These have all been reviewed by Boatman (2000). At Loddington, total profit foregone on the farm has varied between £1,782 and £3,825 between 1995 and 1999 with an average annual cost of £2,674 (±372). This represents between £6 and £12 per hectare. In terms of conservation costs as a percentage of funds available for reinvestments on the farm, these range from 3% to 48%. In an economic climate of reducing farm prices and declining incomes in all sectors, such costs are becoming more difficult to justify if a farm enterprise is to remain economically viable.

MANAGEMENT OF THE BROWN HARE AT MANYDOWN

Although annual harvesting of game removes a sizeable portion of a population (typically 10-30% of gamebirds in Britain) this can be expected to be offset by benefits due to a lower natural mortality and/or better reproduction rate, arising both from density-dependant effects and from management to improve habitats. The populations of game and other species on managed land are therefore generally much greater throughout the year than those on unmanaged land (e.g. Reynolds & Tapper, 1995a). As well as indirectly benefiting non-game species, management clearly aims to increase numbers of some game species. For example, the Manydown Company has been managing the brown hare population on its land since records began in the last century.

At Manydown, hares are shot, but only when population densities allow. Brown hares are managed by controlling foxes *Vulpes vulpes* and other ground predators, and by providing a mosaic of crop patches on the farm, avoiding large blocks of the same crop, maintaining an understorey of cover in woods, hedge-bottoms and beetle banks, incorporating grassland in the rotation, maintaining a mixed arable/livestock enterprise, and providing weeds for food in cereal crop edges using conservation headlands. As a result, numbers of hares shot annually at Manydown have not declined

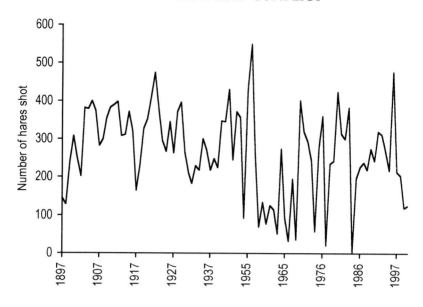

Figure 3:1 Numbers of hares shot (the bag) per 100 ha on The Manydown Estate, where game management practices are in operation; 1897 to 2000.

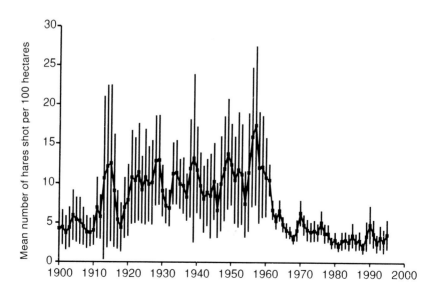

Figure 3:2 Mean numbers of hares shot (the bag) per 100 ha on estates in the southern counties of England where game management practices are less rigorously applied, 1900 to 1995. National Game Bag Census Copyright.

substantially this century (Figure 3:1) even though, perhaps, the annual numbers killed are more variable than they were before the Second World War. By contrast, over the same period the numbers shot in the south of England has dropped by more than half (Figure 3:2). We believe this is because the farm is positively managed — albeit primarily for gamebirds – as outlined above. However, it must always be remembered that shooting effort is rarely quantified alongside bag record data, and therefore some care in interpretation of such data may be necessary. Paradoxically, although they are a game species, hares also damage arable crops, and shoots are designed to reduce hare numbers significantly in late winter, in addition to providing sport and game meat. The Game Conservancy Trust conducted research on hares at Manydown in the late 1980s. In the seasons 1988 and 1989, before shooting, hare densities on the farm were 23 and 29 hares per square kilometre and this was reduced to 7 and 14 per square kilometre after shooting (Stoate & Tapper, 1993). It is interesting to note that the number of hares left on Manydown after shooting was actually about the same as for most arable districts in mid winter where shooting does not take place (Hutchings & Harris, 1996).

At Loddington, where game conservation management is also practised, hare numbers have also increased. Numbers have risen from about seven per square kilometre, when the farm was taken over by ARET and an active programme of game management implemented, to nearly 84 per square kilometre in 1999. Numbers peaked at 93 per square kilometre in 1995, after which a programme of shooting began to prevent crop damage. Since then numbers again increased to 99 per sqaure kilometre in 2001. Throughout this period, hare counts on adjacent farms, where game conservation is not practised, did not reach 10 hares per square kilometre.

BENEFITS OF GAME MANAGEMENT TO OTHER WILDLIFE

Other non-quarry mammal species also benefit from game management measures undertaken at Manydown. For example, it has been shown that conservation headlands managed at Manydown are favoured foraging areas of wood mice *Apodemus sylvaticus* (Tew *et al.*, 1992). The distribution of harvest mice *Micromys minutus* have also been mapped at the ARET's demonstration farm at Loddington, and were found on areas of the farm managed for game, either as beetle banks or field margin strips (Bence *et al.*, 1999).

There are wider benefits to wildlife too. At Manydown, one may find some of the rarest annual arable wildflowers in areas of the farm managed for game. Examples include red hemp-nettle *Galeopsis angustifolium*, corn buttercup *Ranunculus arvensis* and shepherd's needle *Scandix pecten-veneris*. Under intensive arable production such land would support very few plant species other than the crop itself.

Even landowners who have no direct interest in the game on their land will find that its management may provide potential and additional income from letting the shooting or stalking rights. At the same time, the value of their property increases if it has an improved sporting potential as a result of improved habitat. Whatever the economic driver, other wildlife species will indirectly benefit.

CONCLUSIONS

Many farmland mammals need sympathetic habitat management to survive, be they quarry, predator or protected species. However, sometimes their numbers need to be regulated either upward (if they are a game species) or downward (if they are a predator) to allow the game-shooting that provides the motivation for much beneficial management.

The philosophy of game management at Manydown is, we believe, typical of a much larger acreage where outright economic criteria for production are moderated to favour game species. In return, a harvest of game is expected for personal or economic returns. The sympathetic management of crops, the creation, management and maintenance of non-cropped habitats and the amelioration of the impacts of modern farming combined with a level of predator control, creates an environment for a sustainable level of game production. Such management approaches also often produce a favourable environment for other farmland wildlife (Sotherton, 1991).

This philosophy of conservation through managed exploitation is not new, but now also comes under the banner of the ethic described as 'conservation through wise use'. This is not solely wise use, nor is it just sustainable use. The keyword, often omitted by others, is 'conservation', that is, the active intervention by man to improve the habitats, food resources and environmental conditions of the species in question (Tapper, 1999). Such action often carries with it benefits for many other non-game species from wild flowers and butterflies (Sotherton, 1991) to songbirds (Stoate, 1999) and harvest mice (Bence *et al.*, 1999).

However, the biggest obstacle to the wise use of game on farmland in Britain is intensive agriculture. The wise use of, for example, arable fields or upland rough pasture, could be encouraged by a more discriminate use of agricultural subsidy payments. Currently what is considered sustainable and non-sustainable agricultural practices are similarly rewarded, and the costs of the detrimental effects of these practices are not borne by the farmers themselves. These subsidy payments need to be re-balanced and further reformed to improve grants for wildlife habitat conservation and creation. This would better serve the interest of the public, who are increasingly concerned about the status of wildlife on farmland. Some progress is continuing in attempts to reform the functions of the Common Agricultural Policy for example, and there is a growing consensus that compromise between farm production and conservation is essential.

As a farm that wishes to remain economically viable, the Manydown Company is not opposed to modern agricultural systems. By integrating its agricultural production with conservation and game management it is confident in what can be achieved on 2,000 ha of land in Hampshire. We would like to consider the potential and total impact on mammal species by game conservationists across, for example, the 3.4 million hectares of cereals in the UK, and conversely, the potential losses should such motivation be removed!

Agricultural pesticides and mammals in Britain

R F Shore, M R Fletcher & L A Walker

INTRODUCTION

The widespread agricultural use of organochlorine (OC) pesticides during the 1950s to 1970s demonstrated that pesticides could have an impact on wild vertebrates at the individual and population level. Insecticides, such as dieldrin and DDT, caused mortality and impaired reproduction in birds of prey. The sparrowhawk *Accipiter nisus* and peregrine falcon *Falco peregrinus* were so severely affected that they became virtually extinct in certain regions of Britain and populations decreased nationally (Ratcliffe, 1980; Newton & Wyllie, 1992). OC pesticides caused fox *Vulpes vulpes* and badger *Meles meles* deaths (Blackmore, 1963; Jefferies, 1969), and contamination of bats (Jefferies, 1972) and small mammals was common (Jefferies *et al.*, 1973; Jefferies & French, 1976). DDE and dieldrin were detected in an individual wildcat *Felis silvestris* and water vole *Arvicola terrestris* respectively (Jefferies, 1991; Jefferies *et al.*, 1989) and many other mammal species were probably also exposed. The atmospheric transport of OCs and their subsequent persistence in the environment and in animal tissues could result in contamination occurring many miles from the source, even in different countries and continents.

A lack of quantitative data on population numbers from the 1950s to 1970s means that we do not know whether OC pesticides significantly affected the numbers of foxes, badgers and other terrestrial mammals. However, it is known that otter *Lutra lutra* numbers, which were already reduced by persecution and hunting in the eighteenth and nineteenth centuries, underwent a catastrophic decline that started simultaneously in England, southern Scotland and Wales in 1957 and 1958 (Jefferies, 1989; Jefferies, 1996; see also Strachan, Moorhouse & Macdonald, Chapter 5). This mirrored the OC pesticide-related declines in birds of prey that occurred in the same areas. The loss of otters was thought to have been largely caused by dieldrin and aldrin contamination of river systems which started with the use of these compounds as cereal seed dressings and sheep dips in 1956 (Jefferies, 1989, 1996; Strachan & Jefferies, 1996). However, it has also been argued that polychlorinated biphenyls (PCBs) may have been the major cause for the loss of otters (Mason, 1989) and it is possible that OCs and PCBs both had an impact on otter populations. Following voluntary and then legal bans on the agricultural use of dieldrin, aldrin, DDT and other OC pesticides, residues in avian predators declined and populations subsequently recovered (Ratcliffe, 1980; Newton & Wyllie, 1992). It is likely that residues similarly declined in terrestrial mammals and any effect that dieldrin may have had on populations disappeared. Although dieldrin levels are still somewhat elevated on certain rivers that drain industrialised catchments (Meharg *et al.*, 1998), the threat to otters from dieldrin contamination has reduced greatly; PCB residues in otters also declined between 1983 and 1992 (Mason, 1998). Otter numbers started to recover from the early 1980s onwards (Strachan & Jefferies, 1996).

Regulatory testing of novel pesticides has been influenced by the lessons learned from OC pesticides. As a consequence, hazard and risk assessment procedures have been improved. However, such assessments often have poorly quantified exposure terms and, by necessity, involve extrapolation of data between species and from the laboratory to the field; such extrapolations can lead to errors. Risk assessments therefore reduce but do not eliminate the risk of pesticides to wildlife. Indeed, pesticides are still often cited as potential threats to mammals (Morris, 1993; Harris *et al.*, 1995); although supportive evidence is often lacking, circumstantial or takes years to gather.

The aim of this chapter is to review the current impacts that agrochemicals have on mammals in Britain. Specifically, we (i) describe the current scale of agrochemical use in Britain; (ii) indicate which terrestrial mammal species may be exposed to and directly affected by pesticides; (iii) examine the evidence that such exposure and effects actually occur and identify which agrochemicals are involved; and (iv) highlight specific groups of pesticides about which there are current uncertainties and concern over their impacts. The indirect impacts that agrochemicals may have, such as insecticide use reducing invertebrate food supply or herbicide use reducing vegetation cover, are not considered.

CURRENT USE OF AGROCHEMICALS IN BRITAIN AND PERCEIVED THREATS TO MAMMALS

Agrochemicals are used in a wide range of habitat types, including woodland, pasture and urban areas. Arguably, it is intensively farmed arable land that receives most agrochemical input, and we have examined data for applications on arable farms to provide an indication of the current scale of agrochemical use in Britain. However, some compounds, such as rodenticides, that may have important impacts on mammals are widely used in pastoral areas (Garthwaite *et al.*, 1999) and the importance of these inputs should not be underestimated (see, for example, McDonald & Birks, Chapter 9 and Cowan & Quy, Chapter 14).

Approximately 20% of the land area in the whole of Britain is under arable crops but the proportion differs significantly between England, Scotland and Wales (χ^2=21,240, d.f.=2, p<0.0001), being four and eleven times greater in England than in Scotland and Wales respectively (Figure 4:1a). Thus, if various mammal species are significantly affected by agrochemicals, any impacts might be expected to be most widespread in England. Over 200 different agrochemical compounds or mixtures are applied to arable land in Britain. Most are fungicides and herbicides (Figure 4:1b), but

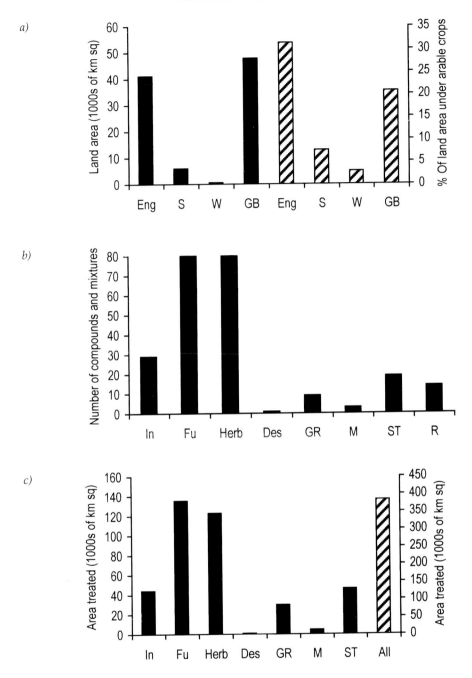

Figure 4:1 (a) Total arable land area (solid bars, left hand axis) and as a percentage of all land area (hatched bars, right hand axis) in England (Eng), Scotland (S), Wales (W) and Britain (GB). (b) Number of compounds and mixtures of each agrochemical type. (c) Area treated with different types of agrochemicals (solid bars, left hand axis) and with all agrochemicals combined (hatched bar, right hand axis). Data are from Barr et al. (1993), Thomas et al. (1997) and De'Ath et al. (1999). In=insecticides, Fu=Fungicides, Herb=herbicides, Des=desiccants, GR=growth regulators, M=molluscicides, ST=seed treatments, R=rodenticides.

Table 4:1 Rodenticide usage on farms growing arable crops in 1994 and 1996
(data from De'Ath et al., 1999).

	1994	1996
Total number of holdings in Great Britain	75,135	73,625
Estimated number of holdings using rodenticides	59,048	57,159
Percentage of all holdings using rodenticides	78.6%	77.6%
Average number of rodenticides used on holdings where rodenticides are used	1.6	1.4

there are almost 30 different formulations that contain one or more insecticides; these include organochlorine, organophosphate, carbamate and pyrethroid compounds (Thomas *et al.*, 1997). Other agrochemicals, for which there are fewer different individual active ingredients, include molluscicides and rodenticides amongst others (Figure 4:1b).

Many of these different agrochemicals are applied to the same area of land, and specific active ingredients may be repeatedly applied in the same year. If each application is treated as a separate treatment, the equivalent area to which agrochemicals are applied varies between 958 km² (desiccants) and 135,000 km² (fungicides) (Figure 4:1c). The sum area on arable farms treated by all agrochemicals (excluding rodenticides) in 1996 was estimated to be approximately 383,000 km² (Thomas *et al.*, 1997), equivalent to more than one and a half times the total area of Britain. Rodenticides are not evenly distributed over arable land and so their usage is not quantified by the amount of active ingredient per square kilometre but rather by the number of holdings in and around which they are used. In 1996, almost 80% of farms growing arable crops used rodenticides, some using more than one active ingredient (Table 4:1), and over one million kilogrammes of bait were applied (De'Ath *et al.*, 1999). Thus, it is evident that there is a high and diverse degree of agrochemical use in Britain on arable land, and so a potential risk of exposure to mammals in such habitat. This is likely to be more pronounced in England because of its high proportion of arable land relative to Scotland and Wales.

In their review of the status of British mammals, Harris *et al.* (1995) identified possible or potential threats to different mammal species from various agrochemicals (Table 4:2). Relatively few groups of compounds currently in use were considered a potential problem. Those that were identified were the OC pesticides, insecticides generally, molluscicides, rodenticides and, in the case of the brown hare, unspecified agrochemicals (Table 4:2). However, these pesticides were perceived as a potential threat to 14 non-target terrestrial mammal species and three target species, brown rat, ship rat and house mouse. The grey squirrel could be added to the list of target species as it too is subject to control by poisoning in some areas. The non-target species included insectivores, lagomorphs, rodents and carnivores (Table 4:2) and comprised 53% of the species in those orders that are currently extant in mainland Britain. With the exception of the wildcat, all are relatively widely distributed and occur in arable areas, or systems draining arable areas (Arnold, 1993; Strachan & Jefferies, 1996; Birks & Kitchener, 1999); arguably, it is these areas where exposure to agrochemicals is most likely to occur. On this basis, the badger, rabbit *Oryctolagus cuniculus*, rodents such as the bank vole *Clethrionomys glareolus* and field vole *Microtus agrestis*, and deer could be added to Table 4:2 because of their frequent occurrence in arable areas and their margins. The only other British mammals that were perceived by Harris *et al.* (1995) to be at risk from agrochemicals were various bat species.

Table 4:2 Mammals that have been suggested as under threat from agrochemicals (Harris et al., 1995).

Species	Agrochemicals that may pose a threat
Insectivora:	
Hedgehog *Erinaceus europaeus*	Agrochemicals, pesticides, molluscicides
Mole *Talpa europaea*	Insecticides
Common shrew *Sorex araneus*	Insecticides
Pygmy shrew *Sorex minutus*	Insecticides
Chiroptera:	
Bats; various species	Organochlorine pesticides (various sources)
Lagomorpha:	
Brown hare *Lepus europaeus*	Agrochemicals
Rodentia:	
Water vole *Arvicola terrestris*	Organochlorine pesticides
Wood mouse *Apodemus sylvaticus*	Molluscicides, seed treatments, herbicide sprays
Harvest mouse *Micromys minutus*	Insecticides
House mouse *Mus domesticus*	Rodenticides and other agrochemicals
Brown rat *Rattus norvegicus*	Rodenticides
Ship rat *Rattus rattus*	Rodenticides
Carnivora:	
Stoat *Mustela erminea*	2° poisoning by rodenticides and molluscicides
Weasel *Mustela nivalis*	2° poisoning by rodenticides and molluscicides
Polecat *Mustela putorius*	2° poisoning by rodenticides
Otter *Lutra lutra*	Organochlorine pesticides
Fox *Vulpes vulpes*	Deliberate poisoning
Wildcat *Felis sylvestris*	Organochlorine pesticides (dieldrin)

EVIDENCE OF EXPOSURE AND EFFECTS IN MAMMALS

Although there is a perception that a wide range of mammalian species in Britain may be at risk from agricultural pesticides, there is often little or no evidence that this is the case. Theoretically, quantification of exposure and effects can be carried out using post-registration schemes that monitor exposure and/or pesticide-induced mortality in wild vertebrates, through specific investigations, and by monitoring effects on population numbers and distribution. There are currently no national post-registration schemes that monitor the exposure of wild mammals to pesticides in Britain, and determining cause and effect relationships for the impacts of pesticides on population numbers is difficult. There is, however, the Wildlife Incident Investigation Scheme (WIIS) that monitors pesticide-related mortality incidents. There have also been specific investigations into the exposure of mammals to pesticides and the resultant effects. We examined data from the WIIS and from specific investigations to determine which mammals have been lethally exposed to what pesticides, to quantify whether mortality incidents have varied in frequency over time, and to identify chemicals for which there is evidence of widespread sub-lethal exposure.

The WIIS is a national post-registration scheme (Hardy *et al.*, 1986) run by the Department for Environment, Food and Rural Affairs. It investigates wildlife mortality incidents where there is evidence to suggest that pesticides may have been the cause of death. Each incident can involve one or more individual animals. The scheme largely relies on corpses that have been found in suspicious circumstances being handed in by the public. Investigation of each incident involves post-mortem examination of carcasses and, when cause of death is consistent with poisoning, analysis of pesticide residues in appropriate tissues. Over the ten-year period 1988 to 1997, 1,156 incidents involving more than 16 mammal species from six different classes were investigated. Pesticides were the cause of mortality in 187 (16%) incidents. Carnivore incidents, mainly involving foxes and badgers, dominated the WIIS during this time and comprised 68% of all investigations (Table 4:3). The proportion of incidents in which pesticides were the cause of death varied significantly between foxes (26%), badgers (11%), other carnivores (5%), insectivores (16%), bats (8%), rodents (17%) and lagomorphs (11%) (χ^2=47.9, d.f.=6, p<0.0001; deer data are excluded because of small sample size). This was due to the relatively high proportion of fox mortalities that were caused by pesticides (Table 4:3); when foxes were excluded from the analysis, there was no significant difference between mammal groups (χ^2=8.75, d.f.=5, p>0.05).

The WIIS data were broken down by pesticide class and, where there were sufficient data, by mammal class/species, namely hedgehog, bats, fox, badger and other mustelids (Figure 4:2). The data indicated that pesticide-related mortality amongst mammals predominantly involved five pesticide groups: organophosphates (OPs) and carbamates (classed together because of their common anticholinesterase activity), rodenticides, OC insecticides, pyrethroids and others (a miscellaneous set of compounds). OPs and carbamates were implicated in some incidents in all the different mammal classes/species. Rodenticides were similarly found in a broad range of species although not in bats. OCs were implicated in the deaths of foxes and mustelids other than badgers, although the number of these incidents was few (less than 1% of the total fox incidents investigated and just one 'other mustelid' incident) and the fox incidents were due to abuse (Fletcher *et al.*, 1991). More than half of the OC mortality incidents involved bats and were probably due to the use of OCs in remedial timber treatment (Mitchell-Jones *et al.*, 1989), although an agricultural source cannot be completely ruled out. It is probable that pyrethroid-induced mortalities of bats, the only mammals in which these compounds caused death, were likewise due to remedial timber treatments. However, permethrin-treated timber has low toxicity to bats (Shore *et al.*, 1991) and mortalities were presumably the result of either direct spraying of bats in the roof space or animals coming into close contact with recently sprayed timber. Of the miscellaneous compounds, paraquat, strychnine and the molluscicide metaldehyde occurred relatively frequently. Paraquat and strychnine are both used in illegal poisoning of foxes and badgers. Metaldehyde, when it occurred, was often found in hedgehogs and may have been due to accidental ingestion of bait laid to control slug damage.

Table 4:3 Mammal incidents investigated by the Wildlife Incident Investigation Scheme from 1988 to 1997. Data from Northern Ireland from 1992 onwards are included.

Species	Incidents:		
	Total number	Where pesticides caused death	% Involved pesticides
Insectivora:			
Hedgehog *Erinaceus europaeus*	42	8	19%
Mole *Talpa europaea*	10	0	0%
Various species of shrews	3	1	33%
Chiroptera (bats):	135	11	8%
Rodentia:			
Grey squirrel *Sciurus carolinensis*	54	5	9%
other rodents	23	8	35%
Lagomorpha (rabbist and hares):	88	10	11%
Carnivora:			
Badger *Meles meles*	264	30	11%
Otter *Lutra lutra*	57	1	2%
Pine marten *Martes martes*	3	1	33%
Polecat *Mustela putorius*	7	2	29%
Stoat *Mustela erminea*	6	0	0%
Weasel *Mustela nivalis*	4	1	17%
Mink *Mustela vison*	8	0	0%
Fox *Vulpes vulpes*	422	108	26%
Wildcat *Felis silvestris*	10	0	0%
Artiodactyla (deer):	20	1	5%

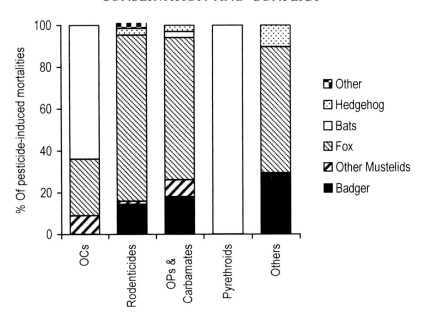

a. Percent of pesticide induced mortalities for different species or groups.

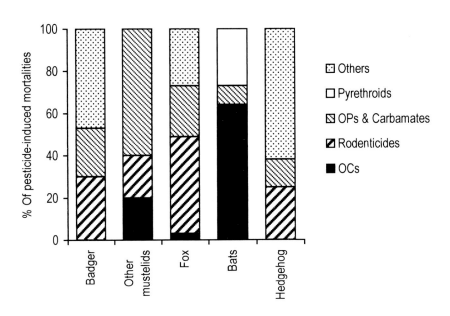

b. Pesticides involved in pesticide-induced mortalities.

Figure 4:2 Wildlife Incident Investigation Scheme incidents between 1988 and 1997. The upper graph (a) shows for each pesticide class the percentage of pesticide-induced mortalities that involve different mammal species or groups. The lower graph (b) shows for each mammal species or group the percentage of pesticide-induced mortalities that are attributable to each class of pesticide.

Of the species examined by the WIIS, only fox and badger incidents were investigated in sufficient numbers to permit analysis of changes in the frequency of pesticide poisoning over the 1988 to 1997 period. There was no year on year trend for foxes or badgers in either the number of incidents investigated or the proportion of incidents that were pesticide related (Figure 4:3). There was some evidence that the pattern of pesticide incidents varied between the two species. The numbers of pesticide-related and total incidents were weakly correlated for badgers (Pearson's Product Moment Correlation Coefficient: r=0.616, d.f.=8, p=0.058), indicating that approximately the same proportion of badger incidents were due to pesticides each year, irrespective of the number investigated. There was no evidence of any such correlation for foxes (r=0.336, d.f.=8, p=0.34) and it was the actual number, rather than the proportion, of pesticide incidents that remained constant each year. In years when there were large numbers of fox incidents, deaths due to trauma, disease and starvation tended to be greater in number (Figure 4:3). The cause(s) of this difference between foxes and badgers is uncertain but there was no evidence of any progressive decrease or increase in poisoning incidents between 1988 and 1997 in either species.

Although it is evident that the WIIS provides valuable data with which to assess some of the impacts that pesticides have on wild mammals, conclusions drawn from the data have to be treated with some caution. This is partly because it is unknown what proportion of all lethally exposed animals are discovered and reported, and whether reporting effort varies between years. Superficially, it could be concluded that the lethal effects of pesticides are most marked in foxes and badgers because they dominate the WIIS data. However, this is probably a sampling artefact and reflects the high degree of public interest in these two species and their large size, which makes carcasses easy to spot. Insectivores, rodents and small mustelids that have been poisoned may die out of sight or be overlooked and so under-reported, although there are no quantitative data to prove that this is the case. Arguably, the most robust information provided by the WIIS is the identification of those pesticides that, even when used correctly, cause some mortality in wild species and may present a wider environmental problem. Such information can be used to focus specific studies. Two such groups of pesticides that can be identified from the WIIS data (Table 4:3, Figure 4:2), and are of current environmental concern, are rodenticides and molluscicides.

RODENTICIDES

It is beyond the scope of this chapter to describe in detail all the issues relating to the potential impacts of rodenticides and molluscicides on non-target mammals in Britain. Here, we briefly highlight only the major issues involved with both pesticide groups, and use these to illustrate some generic areas of uncertainty that exist when evaluating the risk to mammals from pesticides.

Rodenticide levels have been monitored in barn owls *Tyto alba* collected from across Britain for over 15 years. The proportion of birds examined that contained levels of second-generation rodenticides has increased from 5% in 1983 to approximately 35%

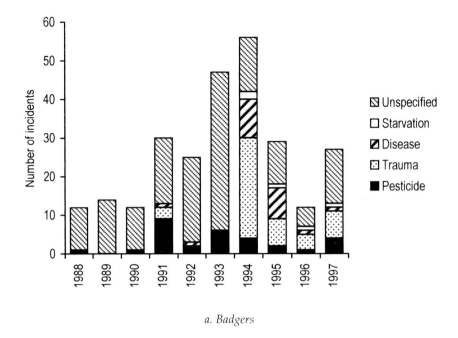

a. Badgers

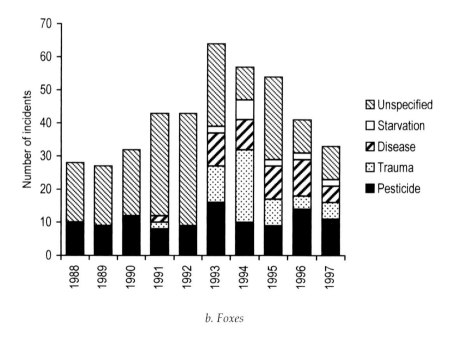

b. Foxes

Figure 4:3 *Annual number of mortality incidents and causes of death for (a) badgers and (b) foxes submitted to the Wildlife Incident Investigation Scheme between 1988 and 1997.*

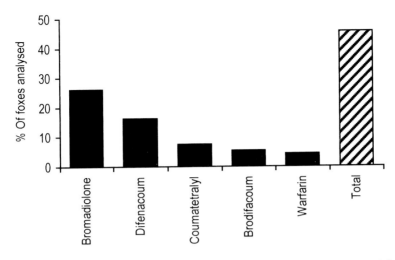

Figure 4:4 Percentage of foxes examined (n=92) that contained detectable levels of different rodenticides. Some foxes contained more than one rodenticide.

in 1996, concurrent with increased usage of these compounds (Newton *et al.,* 1999). This evidence of substantial exposure among barn owls prompted studies on non-target mammalian carnivores in Britain that may feed on contaminated rodents, and so be at risk. Subsequently, it was found that approximately 25-30% of the 40 stoats, 10 weasels and 50 polecats that were examined contained rodenticides (Shore *et al.,* 1996; McDonald *et al.,* 1998; Shore *et al.,* 1999). A sample of 92 foxes submitted to the WIIS, but not thought to have died from rodenticide poisoning, were also analysed and 46% contained detectable levels of rodenticide (Figure 4:4). These independent studies have provided growing evidence that a large percentage of various species of mammalian carnivores are exposed to rodenticides in Britain. It might be expected that carnivores that concentrate on rodent prey, and in particular prey that are subject to human control by poisoning, would be especially vulnerable to rodenticides because both predators and pest controllers target the same seasonal and geographical concentrations of rodents. The similarity between different mammal species in the proportion of the population exposed to rodenticides is surprising, given the differences in their diet and levels of association with agricultural premises. The data suggest that the level of contamination of prey species may be much more wide-ranging than previously suspected.

The effects of exposure to rodenticides in terms of mortality and potential population threats are uncertain. Experimental feeding trials (Townsend *et al.,* 1984; Grolleau *et al.,* 1989) and field studies (Berny *et al.,* 1997) have demonstrated that secondary poisoning of wild mammals by rodenticides can occur. Such effects have been particularly evident in New Zealand, where wide-scale outdoor baiting with the second-generation rodenticide, brodifacoum, is carried out (Alterio *et al.,* 1997; Brown *et al.,* 1998; Murphy *et al.,* 1998ab). The WIIS data indicate that, in Britain, rodenticides have caused badger, rabbit, polecat and hedgehog deaths and are one of the most frequent causes of pesticide-related mortality incidents in foxes (Shore *et al.,* 1999). However, there is no quantitative evidence from the WIIS of large numbers of mammal mortalities due to rodenticides (the maximum number of fox incidents attributed to rodenticides in any one year between 1988 and 1997 was nine) although many individuals, particularly small species, may die undetected. In terms of effects on populations, there is no evidence of widespread declines of mammalian carnivores in Britain. The population status of some, such as the weasel and stoat, is uncertain (McDonald & Harris, 1999) but the polecat is expanding in numbers and distribution in Britain (Birks & Kitchener, 1999). If rodenticides are affecting polecats, they are not currently doing so in a way that prevents some increase in the population.

Although there is no evidence that rodenticides cause any significant mortality or impact upon population numbers, there is little knowledge about how many poisoned animals are discovered. Thus, the real scale of rodenticide-induced mortality amongst non-target carnivores may be under-estimated. This is a generic problem when attempting to determine the extent of mortality caused by any pesticide. Information on population status, if available, cannot be relied upon to provide evidence of significant mortality because a change in population number may be a relatively insensitive measure, only revealing effects that are catastrophic. For example, population data may fail to reveal substantial mortality or impairment of recruitment if density-dependent processes govern population numbers. In such cases, pesticide-induced mortality may normally have no effect on population density but could become significant if, in some years, it acts in tandem with mortality caused by an intermittently occurring environmental stressor. Furthermore, in populations that are expanding, such as the polecat, there is no way of knowing whether pesticide-induced mortality may limit, although not prevent, the rate of expansion. Again, these issues are pertinent for most pesticides, not just rodenticides.

MOLLUSCICIDES

Molluscicides (specifically slug pellets) are applied widely, being both broadcast and drilled on fields, and used in domestic gardens. They are non-specific and exposure of mammals may be either primary, through direct ingestion of bait, or secondary, through consumption of poisoned prey. The active ingredients in molluscicides that are widely used in Britain are usually metaldehyde, often used in gardens, or methiocarb, which is largely used in agriculture. Molluscicide-related mortalities of various mammal species have been detected by the WIIS (Figure 4:2) and there is concern over the impacts of these compounds. Hedgehogs have been a particular focus of concern (Morris, 1993) because they may take pellets directly and eat poisoned slugs. Studies on metaldehyde concentrations in slugs suggest that it would be unlikely that hedgehogs could eat sufficient slugs to ingest a lethal load (P.A. Morris, personal communication) but there do not appear to be any published investigations into the primary or secondary toxicity of these molluscicides on hedgehogs.

Most investigations on the impacts of methiocarb on mammals appear to have focussed on the wood mouse (see also Tattersall & Macdonald, Chapter 7). This is because it is the commonest species in arable fields (Green, 1979) and will eat methiocarb pellets (Tarrant & Westlake, 1988). There have been three field studies to determine the effects of broadcast applications of methiocarb on wood mice. The first (Tarrant et al., 1990) found that application had no effect on capture rates whereas the other two reported a decline in numbers, although the extent to which this occurred varied between trials (Johnson et al., 1991; Shore et al., 1997). Where recovery of the population after methiocarb application was monitored, it was found to be rapid and sustained (Johnson et al., 1991). The impact that methiocarb had on mouse numbers may have differed between trials for various reasons. These include the numbers of mice on and around fields and the availability of food other than slug pellets at the time of application, how long pellets remained available to mice and the proximity of source areas from which immigration could occur. In general, timing of application in relation to population dynamics, landscape ecology and factors (such as climatic conditions) that alter exposure, may all influence the effects any pesticide can have on populations of both target and non-target species.

Shore et al. (1997) assessed the hazard to predators from eating wood mice poisoned by methiocarb. They concluded that the predicted LD_{50} doses for methiocarb in tawny owls Strix aluco, kestrels Falco tinnunculus and weasels were within the range of the amount estimated to be taken in by a single mouse that itself had ingested the equivalent of an LD_{50} dose. This suggests that there may be potential for secondary poisoning of predators from methiocarb. However, the risk of secondary poisoning by methiocarb or any other pesticide will partly depend upon whether poisoned mice are available for capture. Wood mice given a sub-lethal dose of dimethoate, another anticholinestase pesticide, reduced their locomotor activity immediately after exposure (Dell'Omo & Shore, 1996ab). Reduced activity may decrease the probability of mice being taken by 'sit-and-wait' predators, such as tawny owls, but may increase the likelihood of capture by predators that actively search for their prey, such as weasels. Furthermore, if mice die on the surface rather than in burrows, they are likely to be available to scavengers. Thus, the toxic action of a pesticide on the prey species and the hunting strategy of the predator are both likely to have an important influence on the secondary exposure of predators to pesticides and need to be considered in pesticide risk assessments.

CONCLUSIONS

It is clear that there is potential for a wide variety of terrestrial mammals in Britain to be primarily and/or secondarily exposed to a range of agrochemicals. Because of the pattern of land-use, the risk of this exposure is probably greater overall in England than in Scotland or Wales. Examination of data from the WIIS indicates that fungicides and herbicides, which are the pesticides used most extensively in Britain, do not generally appear to be a major direct hazard, although they may have indirect effects by altering habitat structure and quality. Of the pesticides currently in use, molluscicides and rodenticides arguably pose the greatest risk to mammals. Primary exposure to molluscicides may be a risk to some species, such as rodents and hedgehogs. Both molluscicides and rodenticides may be a secondary poisoning hazard to mammalian predators and indicate that pesticides do not necessarily have to biomagnify through the food chain to present such a threat. Further studies are needed to quantify the risk to mammals and other predators of secondary poisoning by these compounds.

There is no evidence from the WIIS data of widespread mammal mortality in Britain that is attributable to pesticides. Furthermore, the reported number of pesticide-related mortality incidents has remained fairly constant over approximately the last decade, for foxes and badgers at least. However, the WIIS is not a monitoring scheme and provides few data on many of the mammal species that occur in Britain. Therefore, the data from the scheme cannot be used to quantify the extent or magnitude of either pesticide exposure or pesticide-induced mortality in mammal populations. This requires specific investigations or a national monitoring scheme, although it is debatable whether any such national scheme would be the most effective means of preventing unintentional impacts on mammals. Pre-registration risk assessments of pesticides require accurate estimates of exposure and toxicity. Some generic areas of uncertainty, related to predicting the occurrence and severity of exposure, have been described in this paper. Others include the influence of exposure pattern on toxicity, intra- and inter-species variability in toxicity and interactions between the effects of pesticides and those of other environmental stressors. Reduction of these uncertainties will enhance risk assessments and may be the most effective means of preventing pesticides having significant effects on mammals.

II
Ecology and
Conservation

5

Enhancing habitat for riparian mammals on agricultural land

R Strachan, T Moorhouse & D W Macdonald

INTRODUCTION

Crop production and livestock rearing are the dominant land uses bordering watercourses in Britain. In fact, over 90% of most river catchments drain and flow through farmed land (Raven *et al.*, 1998). Changes in farming practice this century have had a major impact on the biodiversity of riparian habitat, and have also increased nutrient, agrochemical and sediment levels in watercourses. Intensive grassland or arable management close to the water's edge, as well as stock access to the watercourse, may reduce the amount of cover available and the structural and botanical diversity of the vegetation. Riparian habitat enhancement or creation, together with the establishment of buffer zones along the water fringe by withdrawal of agricultural production, can not only improve water quality, but can also have a substantial impact on the availability and suitability of waterside habitat for a wide range of wildlife, such as water voles *Arvicola terrestris* (Strachan, 1998), wildfowl and waders (Treweek *et al.*, 1997) and dragonflies (Andrews & Rebane, 1994).

To assess either the conservation requirements of an ecosystem, or the success of restoration work, it is important to be able to identify appropriate indicators of the health of that ecosystem. One such measure is often the presence and persistence of rare species within the biotic community: not only do rare species often comprise the majority of species present, but the causes of the rarity, once identified, may be used to formulate appropriate strategies for ecosystem management (Temple, 1997). With reference to riparian habitats, the otter *Lutra lutra* has been suggested as an indicator of ecological health as it is highly vulnerable to both pollution and habitat destruction (MacDonald & Mason, 1983). The synergistic effects of habitat fragmentation and American mink *Mustela vison* predation have been implicated in the decline of water voles (Barreto *et al.*, 1998), and in British riparian habitats, the water vole may therefore also be taken as an appropriate indicator of ecosystem health in that it too has proved sensitive to habitat loss. Although the otter is now showing signs of a recovery, the water vole continues to decline to critical levels (Strachan *et al.*, 2000).

This chapter explores various habitat management options that might be used to restore these mammals to the wider countryside. At a time of acute turbulence in the agricultural economy, when change seems inevitable and with increasing focus on modulation (redirection of resources from production subsidies to agri-environment payments), cross-compliance (linking environmental conditions to agricultural support payments) and agri-environmental mechanisms, perhaps the greatest opportunity for radical reform in biodiversity conservation in Britain lies in this sector (e.g. Macdonald *et al.*, 2000b; see also Johnson & Baker, Chapter 2).

DECLINES OF RIPARIAN MAMMALS ON FARMLAND

The otter

Otters have declined sharply in many European countries, including Britain, since the 1950s, and habitat destruction and pollution have been widely implicated in this decline (Foster-Turly *et al.*, 1990; Kruuk, 1995; Strachan & Jefferies, 1996). By 1979, the otter survived at only 6% of its former range in England (Lenton *et al.*, 1980), 20.5% in Wales (Crawford *et al.*, 1979) and 57% in Scotland (Green & Green, 1980).

The British otter population 'crash' has been well reported and examined in detail (see e.g. Chanin & Jefferies, 1978; Jefferies, 1989; Jefferies, 1996; Strachan & Jefferies, 1996). Its timing and geographic variation in severity depended on the usage of organochlorine insecticides as seed-dressings (to control wheat bulb fly) or sheep dip (Strachan & Jefferies, 1996) during the 1950s and 1960s (this also coincided with the decline in raptors such as the sparrowhawk *Accipiter nisus* and peregrine falcon *Falco peregrinus*). A ban on the use of the persistent organochlorine insecticides (e.g. aldrin and dieldrin), and legal protection since 1978, has helped reverse the decline (Jefferies, 1996). The otter is now showing signs of a recovery and may even be near carrying capacity on some river systems in south-west England (Strachan & Jefferies, 1996), Wales (Andrews *et al.*, 1993) and Scotland (Green & Green, 1996). However, because of the low reproductive potential of the otter (small litters of one to five young, with an average of two or three produced only once a year), it is still absent as a breeding species from over half of the waterways it used to occupy prior to 1950 (Stephens, 1957).

National surveys for otters in Britain and elsewhere in Europe have demonstrated that there is direct correlation between the amount of bankside cover and the numbers of otter spraints encountered (Delibes *et al.*, 1991; Strachan & Jefferies, 1996). However, food availability also plays an important role: fish are the major dietary component for otters and otters respond more to prey availability than habitat features (Prend & Granada-Lorencio, 1996; Carss *et al.*, 1998). Acidification, increased Biological Oxygen Demand and other pollutants has led to the reduction or extinction of fish populations, and the absence of otters from these sites (Mason & MacDonald, 1986; Strachan & Jefferies, 1996).

Notwithstanding other impediments to their recovery (e.g. road traffic: Philcox *et al.*, 1999), Adrian *et al.* (1985) state that European otter declines have been particularly due to agricultural processes. They suggest three conservation measures: the protection of riparian vegetation, control of pesticides and a reduction in water extraction. However, although these concerns are important, the protection of fish resources will probably be the most beneficial.

An analysis of causes of fish kill for salmonids, cyprinids and eels on an Irish river showed that 68% were caused by agricultural discharge (55% slurry/silage liquor run-off and 13% by herbicide/pesticide spillage), 21% completely or in part by industrial discharge with 11% from domestic sources (O'Sullivan, 1996). It therefore follows that any management that would improve water quality should benefit the otter.

The water vole

Water voles have been declining in Britain since at least 1900 (Jefferies et al., 1989; Strachan & Jefferies, 1993). Two national surveys carried out by The Vincent Wildlife Trust from 1989 to 1990 (Strachan & Jefferies, 1993) and 1996 to 1998 (Strachan et al., 2000) have shown that the decline has now developed into a serious population crash with a reduction in population by 88% in only seven years. The water vole is Britain's fastest declining mammal, and since 1998 has received some legal protection under the Wildlife & Countryside Act 1981. The decline has been attributed to a combination of factors, including:

- Direct loss of habitat.

- Sensitivity to habitat fragmentation.

- Sensitivity to population fragmentation.

- Fluctuations in water level — flooding and drought conditions.

- Predation, especially by the introduced American mink.

Unlike the otter, the water vole has a much greater potential for a rapid recovery of numbers given the right conditions. It can produce up to five litters of six young in a season and sexual maturity of the first litters can occur in the first year of birth (Boyce, 1991; Strachan, 1997).

The implication of habitat loss, degradation or fragmentation in the decline of both the otter and the water vole is, perhaps, encouraging in that the possibility remains that habitat restoration could mitigate for and perhaps reverse these trends (Strachan, 1998; Macdonald & Strachan, 1999).

INTENSIFICATION OF AGRICULTURE AND CHANGES TO RIPARIAN HABITAT

The process of post-war agricultural intensification has led to large scale changes in the countryside, and it has been argued that the most important single factor determining the requirements of conservation in Britain today is change in the agricultural landscape (Macdonald & Smith, 1991).

An indication of the extent of land use change within floodplain habitat can be gained through the examination of dated aerial photographs. Barreto et al. (1998), examined ten 2 km by 2 km quadrats across the Thames floodplain near Oxford, for which there had been photographic reconnaissance in 1947, 1961 and 1991. Careful measurements were made from these photographs and each parcel of land was assigned to a category of land use. Their analysis clearly demonstrated that there has been a change in coverage of semi-natural floodplain grassland from 49% of the whole area in 1947 to 12% in 1991 and a concomitant rise in tilled acreage from 11% to 53%.

The replacement of semi-natural grassland by tilled fields not only reduces bio-diversity but also removes from the floodplain its water-holding capacity (its capacity to act as a natural sponge, that would slowly release water into the river channel). Tilled land is efficiently drained and, as a result, adjacent watercourses have been engineered to cope with changed patterns of water flow and flooding. There are few farms which do not contain at least one wetland habitat, and in low lying areas, almost all rivers and streams have been modified to aid land drainage to the extent where '...Of the 18 million hectares of agricultural land in Britain today, about half is dependant upon artificial drainage, and about two million hectares have been won by draining marshes and enclosing saltmarsh' (Andrews & Rebane, 1994).

A typical engineering solution to flow conveyance on agricultural waterways is the trapezoidal channel, where banks are re-profiled so that they each have an angle of between 45-60°. This approach became standard riverside management in the 1940s to 1980s, and has often been described in characteristic British understatement as 'limiting to wildlife', eroding and often degrading available habitat for species such as water voles and otters (Macdonald & Strachan, 1999).

The scale and extent of river engineering, the over-widening and over-deepening of waterways in the lowlands, has recently been assessed by the Environment Agency in its River Habitat Survey. This involved visits to some 5,600 riverine reference sites, each 500 m in length, throughout Britain in the period 1994 to 1997 (Raven *et al.*, 1998). This showed that 80% of lowland sites exhibited evidence of past river re-profiling, through widening, deepening or straightening (often all three). Many of these also showed evidence of bank reinforcement for bank protection or erosion control. One fifth of the sites showed significant bank damage by the trampling of cattle, sheep or horses. A mere 4% of the sites in the database that lay below 50 m altitude were found to support areas of adjacent wetlands, open water or reedbeds.

INFLUENCE OF HABITAT QUALITY ON RIPARIAN SPECIES ON FARMLAND: THE WATER VOLE AS A CASE STUDY.

A detailed examination of the effects of various riparian habitat features upon water vole populations is found in Macdonald & Strachan (1999). Although limited to only one species of the many that live in the riparian corridor, the detrimental factors highlighted represent many of the general consequences of the recent alterations to riparian habitat.

Table 5:1 shows the results of Mantel regressions testing for statistically significant associations between water vole distribution and bankside habitat in seven river catchments surveyed between 1996 and 1998. Although the significance of many of the dependant variables varies between watercourses, some trends among the data

Table 5:1 Significant associative variables that explain the distribution of water voles in relation to available habitat in various rivers. Dashes represent variables which were not significant at the p=0.05 level. A + or a – indicates that the variable was significant and the direction of the relationship. Significant t values are given. Data are from Macdonald & Strachan (1999).

Dependant Variable	Teifi	Isle of Wight	Itchen	Sheppey	Windrush	Bure	Oxford City
Watercourse type:							
Running water	–	+6.87	+8.25	+1.02	+1.01	–	+1,83
Ditch	–	+1.20	+1.92	+10.73	-	+2.84	+1.14
Marsh/Pond	+1.58	+1.67	–	–	–	+1.69	
Other habitat (canal, esturine)	–	+3.33	–	–	–	–	+2.25
Land use (50 m):							
Rough pasture	+1.25	+1.32	+2.11	+3.74	+1.83	–	+1.16
Improved/ tilled	-3.14	-1.08	–	+1.48	-2.29	-1.98	–
Wetland	+1.68	+3.84	+4.43	–	–	+3.64	–
Woodland	-3.08	-0.88	-1.58	–	3.36	–	–
Scrub	–	–	+0.96	–	–	+1.02	-1.67
Grazing	-2.09	-2.31	-1.49	+1.07	-1.12	–	–
Vegetation:							
PC1 (herbs, tall grass, reeds)	+5.26	+2.12	+4.95	+2.78	+3.97	+3.33	+3.37
PC2 (trees and bushes)	-4.54	-1.18	-0.98	-0.25	+2.27	-1.08	-3.28
Aquatics	+4.55	–	+2.72	+1.58	+2.13	–	+4.51
Bank features:							
Rock	-2.87	–	–	–	–	–	–
Earth cliffs	-2.07	-2.76	–	-0.59	–	–	-1.47
Steep profile	+2.24	+1.73	–	+3.35	–	+1.10	+1.09
Cattle poached	-4.93	-3.14	-1.31	-1.19	-1.76	-0.69	-2.72
Bank fenced	+4.74	+2.60	+1.29	+1.34	+0.56	–	–
Channel features:							
Rapid current	–	–	+0.38	–	–	–	–
Slow current	+5.06	+6.23	+2.27	2.34	+1.77	+2.12	+0.13
Static current	–	+1.11	–	+1.98	–	+2.27	+2.21
Canalized	–	+0.78	–	+4.38	–	+1.98	+1.86

are apparent. Rough pasture and wetland as predominant land uses within 50 m of the watercourse have mainly positive influences upon water vole distribution, whereas the influences of improved/tilled land and grazing are generally negative. Herbs, tall grasses, reeds and aquatics have positive loadings, whereas trees and bushes have a negative effect. In keeping with the effects of grazing as a land use, cattle-poached banks exhibit an invariably negative influence, whilst fenced banks have a positive influence. Finally, the presence of a slow, as opposed to rapid, current has a significant positive effect upon water vole presence.

The analysis of these variables was extended by creating two General Linear Models (GLM) for the catchments of the Teifi, Itchen, Windrush and the Isle of Wight (Macdonald & Strachan, 1999). Model A took into account the habitat variables and the distribution of other riparian mammals, especially mink. Model B incorporated the habitat variables with spatial information, taking into account the number of nearby river sections occupied by water voles. When applied to each of five rivers separately, on average 33% (range: 23-39%) of the deviance in presence and absence of water voles was explained by model A. Incorporating the spatial information (model B), however, increased this figure to 63% (range: 47-75%). That is, the performance of the model, as measured by deviance explained, was doubled if data describing the number of accessible stretches of river occupied by voles was added to the habitat predictors.

For each river in the GLM models, the vegetation of the bank and/or river always emerged as important in determining water vole distribution. However, the particular habitat variables that emerged as important varied widely between rivers (as evidenced by Table 5:1). For example, water voles were positively associated with steep banks on the Teifi and the Isle of Wight, but these were unimportant on other rivers. In both rivers in which mink were present (the Teifi and Windrush), the presence of water voles was negatively correlated with that of mink.

The relationship between habitat parameters and water vole distribution is not simple. It is important to have 'suitable' habitat in order for the area to support water voles, but the definition of suitable habitat may vary between catchments. Moreover, the occupation of 'suitable' habitat will also depend upon considerations of spatial separation and dispersal ability of the voles. Lastly, occupation of habitat will also be determined by the presence or absence of mink.

Habitat management is inevitably a compromise. Creating detailed models in order to predict the exact needs of a single target species in a given location may not be feasible. Indeed, to incorporate the precise needs of all species that may use the riparian corridor into a management regime is manifestly impossible. Conversely, as demonstrated by Table 5:1, it is relatively easy to pick out factors which are likely to be of importance in the majority of cases, such as cattle grazing and presence of tall herb/aquatic vegetation, and it is upon these that restoration efforts should focus.

Restoring large stretches of high quality riparian habitat, concentrating upon those features which are likely to be of universal importance for the target species may, therefore, be the best way to conserve otters, water voles and a wealth of associated flora and fauna. As such, a large part of riparian management on farmland is conceptually relatively simple and may take the form of buffering the waterways from livestock or farming operations as outlined below. Optimal restoration for water voles should be based upon the most commonly significant habitat features (preventing cattle poaching, replacing tilled/improved fields with rough pasture and wetland, decreasing water velocity and restoring vegetation), combined with attempts to restore habitat connectedness and facilitate colonisation. In the specific case of the water vole, it may also be necessary to eradicate mink from the riverside, since in a linear ecosystem even one breeding female mink can decimate a water vole population (e.g. Barreto & Macdonald, 2000).

BUFFERING WATERWAYS

Protection of waterways from livestock damage

Poaching of river banks by livestock due to heavy grazing pressure not only reduces bank-side and emergent vegetation, thus representing a substantial loss of riparian habitat for all species, but specifically may physically destroy burrow entrances to water vole tunnel systems (Baretto et al., 1998; Macdonald & Strachan, 1999) and cause soil erosion (Summers, 1994; Wood-Gee, 1994). Here, habitat restoration may often be managed by fencing the banks of the watercourse. One such restoration was successfully executed at the River Coln in the Upper Thames Catchment using electric fencing, and with the additional construction of boulder weirs to improve the fishery interests of the river. The site was quickly recolonised by vegetation and water voles, and the existence of the fishery was effectively safeguarded (Strachan, 1998).

Protection of waterways from diffuse pollution and sedimentation

The above is an example of 'buffering' a waterway, creating a vegetated strip between the water and the surrounding land-use, to aid wildlife conservation. Buffering has potential for general application to many arable waterways. Rehabilitation of river margins using buffer zones (or 'riparian zones') has two important roles in river management. These are to enhance the value of the river for nature conservation and to improve the water quality of the river, by reducing diffuse pollution and stabilising river banks (Large & Petts, 1997). The latter is of pivotal relevance to the conservation of the otter — improvement of water quality and a concurrent increase in fish stocks are essential steps in ensuring that an area is made habitable for otters (see above).

Buffer zones can reduce the amount of nitrogen and phosphorus passing into a stream. Estimates range from 68% to 100% reduction of nutrient concentration across buffer strips, depending on the input concentration and other factors such as width, drainage or soil type (Petersen et al., 1992). There is clear evidence that buffer zones reduce nutrient leakage from agricultural fields; the recommended width of the strip is, however less clear (Driver, 1997; Large & Petts, 1997). One recommendation for both nutrient reduction and habitat improvement is for a buffer strip at least 10 m wide on either side of the stream (Petersen et al., 1992). However, a more demanding view is that at least a 50 m grassland zone is required to have a significant impact on the concentration of nitrates and phosphates reaching artificially deepened water courses in agricultural land (Driver, 1997). In practice, it seems likely that the actual width required for effective nutrient removal will depend upon the nutrient inputs and factors such as the soils and vegetation used. For wildlife conservation purposes, buffer zones should be made as wide as possible: Large & Petts (1997) note that the majority of mammal, reptile and amphibian species concentrate within 60 m of the water, so relatively wide buffers are required to provide enough habitat for wildlife and to act as wildlife corridors. An additional argument for this is seen in Table 5:1. Since the land use within 50 m of the watercourse has a significant effect upon the distribution of water voles, a buffer strip of this width should automatically mitigate for the effects of agricultural intensification

Of course, further refinements are possible. One suggestion is that a 'horseshoe' semi-circular ditch dug into the buffer strip allows water from drained fields to flow through wetland before joining the main stream. A 10 m-wide horseshoe, of about 8 m diameter, is estimated to give a nitrogen reduction of 4 kg per year, and create 80 m^2 of stream-bank wetland (Peterson et al., 1992).

The above argues for the reinstatement of well-vegetated buffer zones for both water quality and wildlife conservation, the habitat created benefiting both otters and water voles. However, the following problems with aquatic vegetation may arise:

- Reduced water velocities and raised water levels in the channel and in the water table of the surrounding land, causing it to become waterlogged.

- Increased incidence of flooding.

- Deposition of suspended sediment.

- Changes in the magnitude and direction of currents, thus causing local erosion or reducing bank erosion.

Thus, especially in nutrient-rich riparian systems, and watercourses required for drainage functions, it may prove necessary to manage the vegetation to maintain the functioning of the channel in concert with improving the wildlife value of the habitat (Wade, 1997). We return to the subject of best practice for riparian habitat management below.

Protection of waterways from pesticides

Some pesticides are extremely toxic to fish and other aquatic organisms, and can be lethal at concentrations as low as one part per billion (MAFF, 1999). Pesticides may particularly affect otters, through either loss of fish or bioaccumulation in fish tissue. Concern over the water column concentration for some pesticides by the then National Rivers Authority (now the Environment Agency) led to the labelling of pesticides and a buffering of waterways with a 'no-spray zone' of 6 m. Such buffer zones were required to ensure that spray drift fall-out into watercourses did not reach toxic levels. Since March 1999, new arrangements have been brought in by MAFF (now DEFRA) placing a legal obligation on farmers to carry out and record the results of a Local Environmental Risk Assessment for Pesticides (a LERAP).

The LERAP scheme allows for a reduced buffer zone (5 m width), depending on the category of product (especially organophosphate or synthetic pyrethroid insecticide), the size of the water course, dose rate applied and the use of low-drift spray equipment. The zone is measured from the top of the bank (rather than from the water's edge) and establishes a grassy strip of at least 1 m width (MAFF, 1999). The permanent establishment of grassy bank-top strips not only protects the watercourse from farming operations and stabilises the bank, but also provides extra riparian habitat for water voles.

AGRI-ENVIRONMENT SCHEMES AND RIPARIAN HABITAT ENHANCEMENT

A review of biodiversity conservation in Britain by the House of Commons Select Committee (HMSO, 2000ab) judged that agri-environmental schemes were crucially important (outlined by Johnson & Baker in Chapter 2). They provide many opportunities and financial incentives for managing, protecting, enhancing and creating wetlands, riparian corridors, arable margins, buffer strips and fenced-off riverbanks. No one scheme has been specifically designed to incorporate the needs of water voles or otters, but all can be used to expand, link or buffer existing habitats or create new ones suitable for these emblematic riparian species.

The Habitat Scheme

In England, management agreements set up under the (now closed) Habitat Scheme Water Fringe Option included the establishment of riparian buffer strips alongside grassland or crop lands. The Habitat Scheme had six designated areas for water fringe conservation and, of these, two areas have been specifically examined to determine their potential for restoring habitat for water voles (Critchley et al., 1998). The areas assessed were the Upper Avon, Wylye and Nadder in Wiltshire, covering 163 km of watercourses, and 28 km of the River Beult in Kent.

In the Water Fringe Option (Table 5:2), the creation of a buffer strip by the withdrawal of agricultural production (Option 1A & 1B) was likely to be most beneficial to water vole populations. Unfortunately, uptake of these various options by the farmers in these two areas was very poor with less than 5% of the eligible land put into the scheme (Joyce et al., 1998).

Assessment of the establishment and development of riparian vegetation and its potential suitability for water voles is shown in Table 5:3. Wetland, grassland and ruderal plants dominated the riparian buffer zone vegetation in the Upper Avon, Wylye and Nadder, and comprised more than half the species recorded in the River Beult. The riparian zone could therefore support an improved floristic and structural composition of a type shown be favoured by water voles elsewhere. Large populations of water voles are known to occur where there is a well structured bank vegetation composed of grasses and taller wetland species (Zejda & Zapletal, 1969; Lawton &

Table 5:2 *Summary of management options in the Habitat Scheme Water Fringe Option (modified from MAFF, 1996).*

Option	Management
1A	Withdrawal of a 10-30 m wide strip of permanent grassland from production. Limited cutting only. No fertilisers or pesticides, selective herbicides only.
1B	Withdrawal of a 10-30 m wide strip of arable from production. Low productivity grass sward to be established and managed as for 1A
2A	Extensive grassland management on existing permanent grass. Moderate grazing intensity required. Cutting before 1 July prohibited and other restrictions as for 1A
2B	Extensive grassland management on existing arable land. Low productivity grass sward to be established and managed as for 2A
3	Maintain raised water levels on any of the above

Woodroffe, 1991; Macdonald & Strachan, 1999). As yet, there has been no specific monitoring of how water voles actually respond to these improved riparian zones but the potential benefits are high.

Countryside Stewardship and schemes to create wetlands

Agri-environment schemes, such as the Countryside Stewardship Scheme and Environmentally Sensitive Areas, include provisions to enhance wildlife habitat in line with the UK's Biodiversity Action Plan strategy (Department of the Environment, 1995; Johnson & Baker, Chapter 2). Site-specific recommendations aimed at individual priority species such as the otter and water vole can become the focus of an environmental farm plan. For instance, where there are areas that are seasonally difficult for the farmer to plough due to excess water, these could be designated as suitable for wetland habitat enhancement or creation. Such swamp areas are usually former wetlands that have great potential for enhancing wildlife conservation. For example, the establishment of reedbeds on such areas can be achieved relatively simply in many cases; the main requirements are that there must be sufficient water supply to maintain summer water levels, and basic water control structures to drain the bed for winter cutting. Reedbed vegetation can be easily established by transplanting clumps of rhizome in the winter, spacing these about 1 m apart and just above the water table (Andrews & Rebane, 1994; Harding, 1994).

Such reedbeds can benefit both otters and water voles and create habitat for many other species of conservation interest, especially birds (such as bitterns *Botaurus stellaris*, bearded tits *Panurus biarmicus*, savi's warbler *Locustella luscinioides*, cetti's warbler *Cettia cetti* and reed buntings *Emberiza schoeniclus*) and invertebrates (five Red Data Book species are closely associated with reedbeds). In general, almost all wetland areas found on farms can be extremely valuable, and without exception there is potential to improve

Table 5:3 *Riparian buffer strip vegetation cover in two Water Fringe Areas (estimated lengths under Option 1 and percentage cover of vegetation type; modified from Critchley et al., 1999).*

Category	Water Fringe Area:	
	Upper Avon, Wylye & Nadder	River Beult
Option 1A (length in scheme)	4,785m	3,520m
Option 1B (length in scheme)	1,485m	1,345m
Species composition:		
(i) wetland plants	40.0%	21.0%
(ii) tall grasses	27.0%	20.0%
(iii) ruderals	06.0%	15.0%

or create them. In particular, if networks of floodplain ditches, connected backwaters, ponds and oxbows can be created or re-instated – all with a dense fringe of aquatic vegetation – as a compliment to or surpassing the provision of buffer zones, this is likely to be of great benefit to both otters and water voles (Scholey, 1995; Strachan, 1998). In recognition of the importance of flood meadows as havens for biodiversity, additional funds to the value of £40 million were made available by MAFF to implement agri-environment schemes in the UK between 1999 and 2002 (Dalyell, 1999). However, this represents only 2.5% of the UK Common Agricultural Policy income, whereas member states are permitted to channel 20% of this budget into such schemes – there is thus enormous potential to go further (HOC, 2000).

RIPARIAN HABITAT MANAGEMENT: BEST PRACTICE FOR RIPARIAN MAMMALS

Unsympathetic management techniques such as mechanical control and herbicide usage can have deleterious impacts upon the wildlife present, especially herbivores such as water voles (Beardall, 1996; Strachan, 1998). Best practice methods for weed cutting and waterway de-silting are described in *The Water Vole Conservation Handbook* (Strachan, 1998) and include rotational cuts of vegetation along alternative banks, and patchwork clearances of channel weed and silt. In each case, refuge areas for water voles can be created. A preferred management option may be the use of shading with deciduous vegetation to limit aquatic/riparian macrophyte growth. Shading with trees and shrubs may control macrophyte growth effectively (MacDonald & Mason, 1983; Wade, 1997) and may cost less than half the amount of weed-cutting and dredging. The aim of macrophyte control by shading is to reduce the amount of incidental light reaching the centre of channel by two-thirds. By leaving gaps in the tree canopy at regular intervals (for example 50 m), thus allowing stands of aquatic vegetation to grow, a good balance of habitats may be achieved (Wade, 1997). This arrangement has the advantage of incorporating habitat for both otters and water voles, in that the wooded sections may provide breeding holts for otters (Jenkins, 1981), whilst the stands of aquatic and emergent vegetation would be of benefit to both species.

In addition to vegetation management, other features of the habitat may be restored. For example, it is conceivable that den sites may limit the populations of some species (indeed, there is some evidence that mink density is affected by the availability of pollarded willows in which to den, or rabbit warrens in which to breed – see Macdonald & Strachan, 1999). For example, an otter may use up to 30 holts within its territory (Mason & Macdonald, 1986) and artificial holts can be constructed as log piles or more elaborate pipe and chamber systems (Ward *et al.,* 1994). The extent to which holts are indeed a limiting factor for otter populations is largely untested, but we have collated information on the use of a sample of 60 artificial holts in southern England and Wales. Of these, 72% had otter signs within the entrance tunnels after five to seven years, and 37% had been used within the first year of construction. At least 28% had been used for breeding.

CONCLUSIONS

Both water voles and otters are emblematic species for which public enthusiasm has the potential to foster widespread riparian conservation. Both, also, have been shown to be sensitive to the effects of habitat loss and alteration in agricultural areas. However, they have different habitat requirements in that while water voles require steep-sided banks with a good, well layered cover of emergent macrophytes (Strachan, 1998), many of the preferred natal holts of otters are wooded stretches of river, especially in the root systems of trees such as ash and sycamore, or amongst dense cover of shrubs and brambles (Anon, 1999).

Insofar as they prefer different vegetation, yet both benefit from a wide zone of riparian vegetation, the desire to accommodate others and water voles illustrates the need for a mosaic strategy in riparian habitat enhancement. Both can be accommodated on farmland, and in their wake an entire ecosystem will benefit. While much remains to be learnt, enough is already known to implement these changes. Further, while future agri-environment schemes may radically extend the possibilities, current schemes are sufficient to enable most farmers to benefit from getting on with the job.

Creating appropriate riparian habitat by way of buffer zones and improving water quality (of indirect benefit to otters by improving the fish populations and potentially reducing the likelihood of fish kills) is essential for the restoration of both otters and water voles to their former distributions.

ACKNOWLEDGEMENTS

We are grateful to members of the WildCRU for stimulating discussion, and especially to those who have participated in our riparian research programme. Laura Bonesi and Guillermo Barreto spearheaded the analyses of water vole habitat requirements summarised here, in which context the statistical guidance of Paul Johnson and Steve Rushton has been crucial. Work referred to here has been supported largely by the Environment Agency, the Peoples' Trust for Endangered Species and British Waterways, to all of whom we are grateful.

6

Use of farmland habitats by greater horseshoe bats

P L Duvergé & G Jones

INTRODUCTION

The greater horseshoe bat *Rhinolophus ferrumequinum* is one of the rarest mammal species in the United Kingdom (Stebbings, 1988; Whitten, 1990; Hutson, 1993; Harris *et al.*, 1995), and is listed in the UK Biodiversity Action Plan. The population may have fallen as low as 2,200 individuals by 1983 (Stebbings & Griffith, 1986), but has since been recovering, albeit very slowly (Harris *et al.* 1995). Approximately 5,000 bats can now be accounted for at breeding colonies during the summer (P L Duvergé, unpublished data), and the total British greater horseshoe bat population may number up to 6,600 individuals (Harris *et al.*, 1995; Mitchell-Jones, 1995). The stronghold of this species remains south-west England and Wales, a reduction of about 50% of its original range (Stebbings, 1988).

Although greater horseshoe bats have been extensively studied in the laboratory and at breeding and hibernation sites, studies of free-flying bats in the wild had proved more difficult. Advances in electronics and the rapid development of radio transmitters weighing less than a gram remedied this situation, allowing the first field study of greater horseshoe bats to be carried out during the early 1980s (Stebbings, 1982). Several radio-tracking studies have since been undertaken on this species, both in the UK and abroad. These have shown that greater horseshoe bats may use farmland habitats for three main purposes: as travelling routes (between roosts, to feeding areas, and between winter and summer quarters); as roosting sites (day-roosts, night-roosts and breeding roosts); and for foraging.

Greater horseshoe bats travel long distances through the wider landscape, by taking advantage of connected habitats. These may include strips of woodland, the wooded edges of rivers and estuaries, and smaller watercourses that retain some degree of vegetation cover (Duvergé, 1997; Jones & Billington, 1999). In some cases, these travel routes may extend to more than 10 km from the roost (Duvergé, 1997). Greater horseshoe bats often use buildings associated with estates, farms and farming. Country houses, coach houses and barns may be used as breeding quarters. Those, and stables, storage barns, lean-tos, stock shelters, old milking sheds and disused workshops may also be used as day or night roosts by one or more bats at a time. Farmland is also important for foraging. Work carried out in south-west England in the late 1980s and early 1990s showed that both adult and juvenile greater horseshoe bats relied heavily upon insects emanating from farmland habitats as main prey items (Jones, 1990; Duvergé & Jones, 1994).

In this chapter, our aim is to highlight the importance of farmland habitats to greater horseshoe bats by presenting the results of work centred on the foraging activity of greater horseshoe bats in south-west England, between 1991 and 1993. We show that greater horseshoe bats selected specific habitat types over which to forage, and identify cattle-grazed pasture and ancient semi-natural woodland as the two most important ones. We also describe the foraging behaviour of free-flying, radio-tracked, greater horseshoe bats, and describe their favoured feeding perches. Finally, we discuss the importance of the various types of farmland habitats, their structural features and land management, to the future maintenance of a healthy greater horseshoe bat population in the UK.

METHODS USED

Radio-tracking

Greater horseshoe bats were radio-tracked at three breeding sites within 50 km of Bristol, in North Somerset and Gloucestershire: King's Wood, Woodchester Park and Iford Manor. All bats were captured and handled, under licence, according to the guidelines set out in 'The Bat Worker's Manual' (Mitchell-Jones, 1987). Once captured, ring details as well as morphological and reproductive details were recorded for each bat, which was then tagged with a transmitter weighing 0.6-1.3 g. R D Ransome provided the precise age of most ringed individuals. Further details can be found in Duvergé & Jones (1994) and Duvergé (1996). Over the three years of our study, we tagged 64 individuals from a range of reproductive groups including immature males and females, sexually mature males and females, females in early pregnancy and those in lactation and post-lactation, as well as juveniles (mainly males) aged between 21 and 80 days old.

Individual bats were followed continuously from dusk to dawn, during several periods between April and October, in order to obtain as much detailed information as possible; we also recorded additional information on other bats as we came across them (this technique was used successfully by Jones & Morton, 1992). Bats were followed from the moment they left the roost, using a vehicle to move large distances. 'Peak signal' bearings (White & Garrott, 1985) were taken, by sighting down the antenna, to keep track of their general direction, and close-range radio tracking was carried out on foot.

Although tagged bats were tracked from the moment they were released, the first night of data was not used in data analysis, except where bats were caught in the morning at Woodchester Park, and thus had time to adjust to the transmitter package.

Defining the habitats available in the study areas

All habitats within a 4 km radius of King's Wood and Woodchester Park, and within 2.3 km of Iford Manor, were mapped at a scale of 1:10,000, using Phase 1 Habitat

Surveys carried out by the respective county Wildlife Trusts. The status of fields and woods, at the time of our study, was recorded and verified by ground-truthing, in accordance with the Phase 1 Habitat Survey methodology set out by the then Nature Conservancy Council (Anon, 1990). The types of field boundaries and grazing animals were recorded and added to the maps. Hedgerows 2-5 m high and wide were considered to be thick and well developed, and were differentiated from intensively trimmed hedges (1.5-2 m high by 1-1.5 m wide, with a rectangular cross-section). They were considered to be discontinuous if gaps of over 10 m were present in the hedge. Tree lines were classified according to Cresswell *et al.* (1990). Additional information about the grazing regime used on pastures (and meadows from late summer onwards) was gathered from local farmers and landowners, and changing patterns of land use were monitored yearly by re-visiting each area. These changes were taken into account when analysing habitat use and availability.

Overall, 14 major habitat categories were recorded, but this increased to 23 categories when all permutations of grazing regimens and field boundaries were considered separately. Twenty-two of these appeared in feeding areas used by bats radio-tracked successfully for full nights. The areas of all available habitats were digitised on two Summagraphics bit pads (models '*Bit pad*' and '*Summa Sketch III*') at a scale of 1:10,000, and converted into hectares. Before analyses were carried out, the data set was further concentrated into ten main habitat groups. These were:

- Ancient semi-natural woodland (ASNW), henceforth referred to as 'ancient woodland', which has been in existence since the seventeenth century (Peterken, 1977).

- Coniferous woodland (CW).

- Other woodland (OW) including broad-leaved, mixed and other woodland, primarily copses.

- Parkland (PK), which was open planted woodland.

- Urban (UR) including scrub and amenity areas.

- Arable (AR).

- Meadows (M), defined as grass used for hay or silage during the year, and divided into those grazed primarily by sheep and/or horses in late summer and early autumn (henceforth referred to as 'sheep/horse-grazed meadow', MOG), and those grazed by cattle in late summer and early autumn ('cattle-grazed meadow', MCG).

- Pastures (P), defined as grass kept solely for grazing, and divided into those grazed by cattle ('cattle-grazed pasture', PCG) and those grazed primarily by sheep and/or horses ('sheep/horse-grazed pasture', POG).

Given the categories above, we were left with nine habitat groups in spring (as meadows would not have been cut by then, and therefore no grazing would have taken place), and ten groups in summer and autumn. Some of these habitat groups had to be merged further (outlined below), where too few animals were present in the data set.

Greater horseshoe bats shift their foraging sites over a number of nights, and the area over which they search for food changes gradually over time (Jones & Morton, 1992; Beck *et al.*, 1994). Therefore, we calculated habitat availability as a function of the flight capability of each individual bat, based on the distance each flew when radio-tracked. A piece of habitat was considered 'available' to a bat if it fell within a circle whose radius was equal to the 'foraging radius' of the bat (Jones *et al.*, 1995). Where bats used widely spaced day-roosts, availability was determined independently for each roost, as described above, and an average used in calculations.

Measuring habitat use by foraging bats

Using radio tracking, as described above, we recorded the activity of, and total amount of time spent by bats at 'feeding areas' (travelling time was not allocated to a habitat). Feeding areas were defined as any areas used by one or more bats, and over which bat activity lasted for at least 30 minutes (Fleming & Heithaus, 1986). Areas used by a single bat, where activity totalled less than 30 minutes, were also considered feeding areas if we could determine, either visually or through the radio signals of 2-position tags, that bats were foraging there. Although the vast majority of such foraging events lasted between 10 and 30 minutes, some perch-feeding episodes lasting as little as two minutes were recorded. For practical purposes, existing physical boundaries, such as forest tracks and trails, field boundaries, woodland edges, roads and fence lines were used to delineate feeding area boundaries, and the habitat type and land use within each were recorded.

The precise location of specific areas used by hunting bats was possible by homing in on the signal (White & Garrott, 1985; Wai-Ping & Fenton, 1989). These were designated as 'focal areas of activity'. Within these, further searches were carried out on foot to try and observe tagged, and occasionally untagged, greater horseshoe bats. This technique is particularly useful for detailed studies of habitat selection, especially when habitat patches are small (Harris *et al.*, 1990). Both feeding areas and focal areas of activity were mapped at a scale varying between 1:921 and 1:10,000 and then digitised as described above.

Statistical analyses

Compositional analysis (Aebischer & Robertson, 1992; Aebischer *et al.*, 1993ab) was used to compare the proportional availability of each habitat with its proportional use by each bat. The data collected for each bat, summed over all the full nights of information available, were used as one sampling unit (or case) in all analyses, and missing values of habitat availability and use were treated as suggested by Aebischer *et al.* (1993b). The effects of reproductive status, sex, year, site and season were investigated by entering them as covariate factors in the MANOVA (Aebischer *et al.*, 1993ab; McNeilage, 1995), and all analyses were performed on program suites

Table 6:1 Wilks's Lambda values, F-statistics and signifcance levels for overall habitat selection. The latter was computed by comparing the overall time spent by individual bats on the various types of habitats available, against the availability of those habitats within a circle of radius equal to the foraging radius of each bat. Juveniles and adults were considered separately. All significant covariate effects are illustrated. The analyses were then repeated with selected subgroups, only including animals which had all available habitats within their foraging radius, and the results shown accordingly. Probability levels were not Bonferroni-corrected. Habitat ranks show the most selected on the left and least on the right. Triple arrows indicate a significant difference between the habitat categories. **ASNW**: ancient semi-natural woodland; **CW**: coniferous woodland; **OW**: other woodland types (including broad-leaved and mixed woodlands); **PK**: parkland; **UR**: urban; **AR**: arable; **M**: meadow; **MCG**: meadow grazed by cattle in late summer/autumn; **MOG**: meadow grazed by sheep and/or horses in late summer/autumn; **PCG**: cattle-grazed pasture; **POG**: pasture grazed by sheep and/or horses primarily.

Bat group	Statistics			Habitat ranking order
All adults	$\Lambda= 0.073$	$F_{8,27}= 42.97$	$p<0.001$	PCG > ASNW >>> POG > M > PK >>> OW > CW > AR > UR
Selected adults	$\Lambda= 0.057$	$F_{8,24}= 49.31$	$p<0.001$	PCG > ASNW >>> POG > M > PK >>> OW > CW > AR > UR
All males	$\Lambda= 0.032$	$F_{8,3}= 11.32$	$p=0.036$	PCG > ASNW >>> POG > PK > M > CW > OW > AR > UR
Selected males	$\Lambda= 0.009$	$F_{7,2}= 31.64$	$p=0.031$	PCG > ASNW >>> POG > OW/PK > M > CW > AR > UR
All females	$\Lambda= 0.062$	$F_{8,16}= 30.16$	$p<0.001$	PCG > ASNW >>> POG > M > PK > OW > CW >>> AR > UR
Selected females	$\Lambda= 0.056$	$F_{8,15}= 31.529$	$p<0.001$	PCG > ASNW >>> POG > M > PK > OW > CW >>> AR > UR
All juveniles	$\Lambda= 0.007$	$F_{9,5}= 73.630$	$p<0.001$	PCG >>> ASNW > MCG > OW >>> PK > POG > M/MOG > AR > UR
Selected juveniles	$\Lambda= 0.009$	$F_{9,2}= 25.031$	$p=0.039$	PCG >>> ASNW > MCG > OW > PK > POG > M/MOG > AR > UR

expanded from those developed, and generously supplied, by P A Robertson of The Game Conservancy Trust.

Having carried out compositional analyses on the ten condensed groups of habitats, different procedures were used to try and develop a broader picture of habitat use, by assessing which specific habitats within those groups were used extensively. Ivlev's Electivity index (E) was used to measure the 'selection' of each available habitat, defined by Johnson (1980) as *'use either more or less than availability'*. Ivlev's E varies between -1 and +1, such that values between -1 and 0 reflect avoidance, values equal to zero indicate use in proportion to availability, and values between 0 and +1 indicate habitat preference (Johnson, 1980). Manly's *a* (Chesson, 1983) was used to measure 'preference' for a habitat type, defined by Johnson (1980) as *'the degree to which each habitat is selected relative to the selection of all other available habitats'*. Manly's *a* varies primarily as a result of behavioural changes in the subject under study (Chesson, 1983), and is considered a more reasonable measure of 'preference' than Ivlev's E (Krebs, 1989). As its value ranges from 0 to +1, summing to unity for all habitats under consideration, direct comparisons can be made between individuals, reproductive classes, sexes and sites. Averaged indices were obtained from the indices calculated for each bat within groups (Chesson, 1983; Storch, 1993) to investigate group-specific selection and preference.

Seasonal analyses were carried out at King's Wood, by isolating the total amount of time spent foraging on the various habitat types in spring, summer and autumn, and then comparing those with habitat availability in each season by means of χ^2 tests.

HABITAT USE BY GREATER HORSESHOE BATS IN SOUTH-WEST ENGLAND

Cattle-grazed pasture and ancient woodland were the two most important habitats for greater horseshoe bats, and arable and urban habitats were the least important (Table 6:1). Compositional analyses showed that habitat use was highly significantly different to availability, both when considering all adults, and only those with all habitat categories present within their foraging radius ('selected' bats, Table 6:1). The habitat rankings produced by these analyses were also identical, maintaining the statistically significant breaks (identified as >>> in Table 6:1) between the same elements of the sequences. According to Ivlev's Electivity index (Table 6:2), cattle-grazed pasture was the only 'selected' habitat when considering the ten main categories. However, ancient woodland, cattle-grazed meadow and cattle-grazed pasture were all 'preferred' habitats, according to Manly's *a* index. Expanding to 22 habitats revealed that ancient woodland and several types of pastures and cattle-grazed meadows next to woodland and/or surrounded by thick hedgerows were 'preferred' habitats for this species. Season and site had a significant effect on habitat use (season: $F_{16,50} = 1.97$, p=0.035; site: $F_{16,50}=1.98$, p=0.034), but there was no effect of reproductive status, sex or year.

During our study, 226 feeding areas, averaging 6 ha in area, were recorded (Table 6:3) and these differed significantly in size between sites ($F_{2,22\ 3}= 15.07$, p<0.001). A total of 297 focal areas of activity were recorded (Table 6:3), with an average of 1.3

Table 6:2 Ivlev's Electivity (E) and Manly's a (A) indices for all male, female, adult and juvenile bats for the ten main habitat categories (for abbreviations of habitat types see Table 6:1). At the bottom of each column are the values at which habitats were used according to their availability. **Bold** *figures represent positively selected and/or preferred habitats, and plain figures represent habitats that were selected against and/or avoided.*

Habitats	All adults (n=35)		All males (n=11)		All females (n=24)		All juveniles (n=14)	
	E	A	E	A	E	A	E	A
ASNW	-0.03	**0.20**	**0.05**	**0.20**	-0.06	**0.20**	-0.41	**0.10**
CW	-0.80	0.03	-0.83	0.04	-0.79	0.03	-0.50	0.09
OW	-0.70	0.05	-0.67	0.07	-0.71	0.04	-0.61	0.07
PK	-0.71	0.05	-0.55	0.08	-0.78	0.04	-0.75	0.09
URBAN	-0.90	0.02	-0.87	0.02	-0.92	0.02	-0.92	0.01
ARABLE	-0.88	0.02	-0.90	0.01	-0.87	0.02	-0.95	0.00
MMOG	-0.69	0.05	-0.70	0.06	-0.69	0.05	-0.91	0.01
MCG	-0.44	**0.12**	-0.51	**0.11**	-0.37	**0.12**	-0.50	**0.11**
PCG	**0.29**	**0.37**	**0.33**	**0.33**	**0.27**	**0.39**	**0.45**	**0.51**
POG	-0.49	0.09	-0.57	0.08	-0.46	0.10	-0.85	0.01
Level for Use=Avail	0.00	0.10	0.00	0.10	0.00	0.10	0.00	0.10

per feeding area. The overall mean area covered by a focal area was 0.35 ha (range: 0.02-2.7 ha), and the mean area of focal areas was significantly different across sites ($F_{2,294}$ = 25.07, p<0.001).

Seasonal and site-specific habitat use by adults

Compositional analyses were carried out on data from all adults and from females radio-tracked during spring and summer/autumn periods; data for males were insufficient. The use of habitats during spring and summer/autumn was highly significantly different to availability in all cases (Table 6:4), and the same trends were observed for both the overall and female data sets. In spring, bats tended to use ancient woodland more than cattle-grazed pasture, although there were no statistically significant differences between them. Sheep/horse-grazed pasture and meadow were the next most used habitats, with coniferous woodland and urban habitats used least of all. In summer/autumn, bats used cattle-grazed pasture significantly more than any other habitat, followed by ancient woodland, cattle-grazed meadow and sheep/horse-grazed pasture. Arable and urban areas were used least of all.

The results of the compositional analyses accorded well with measures of habitat preference and selection for the ten major habitats (Table 6:5). Electivity indices

Table 6:3 Estimated area of all the recorded (a) feeding areas and (b) focal areas of activity, surrounding the main nursery roost at the three sites investigated during the course of the study. The mean ratio (±SD) of focal areas of activity (FAA) to focal areas (FA) is also shown for each site.

Site	N	Min	Max	Mean (±SD)
King's Wood	112	0.37	35.71	7.67 (±6.42)
Woodchester Park	76	0.22	24.00	5.21 (±4.51)
Iford Manor	38	0.15	10.94	2.43 (±2.26)
Total	226	0.15	35.71	5.97 (±5.64)

a. Feeding areas

Site	N	Min	Max	Mean (±SD)	FAA/FA (±SD)
King's Wood	203	0.02	2.04	0.26 (0.26)	1.83 (±1.96)
Woodchester Park	67	0.10	2.70	0.51 (0.43)	0.88 (±1.03)
Iford Manor	27	0.09	2.37	0.65 (0.54)	0.71 (±0.65)
Total	297	0.02	2.70	0.35 (0.36)	1.32 (±1.61)

b. Focal areas

produced for summer and autumn separately revealed that although selected habitats were identical in both seasons, preferred ones were more varied in summer than autumn. Expanding to all 22 habitats, Manley's *a* revealed that pastures and meadows sharing an edge with woodlands or having well developed field boundaries, and scrub habitats, were preferred to other types. In summer and autumn, habitats close to woodland edges or having one or more thick and/or well-developed boundaries were once again preferred. Mixed woodland (in the 'other woodland' category) and parkland were also preferred.

Habitat use differed significantly between sites in summer/autumn ($F_{18,2}=4.84$, $p=0.006$), but not at other times. Within sites, data were sufficient for analysis only at King's Wood. Here, cattle-grazed pastures were used more than ancient woodland, and both were used significantly more than all other habitats ($F_{8,19}=43.38$, $p<0.001$; Figure 6:1a). Habitat use at King's Wood was significantly biased towards ancient woodland and cattle-grazed pasture in spring (Figure 6:1b), cattle-grazed pasture, ancient woodland and cattle-grazed meadow in summer (Figure 6:1c), and cattle-grazed pasture in autumn (Figure 6:1d). None of the covariates tested had any significant effect on habitat selection at King's Wood.

Table 6:4 Wilks's Lambda values, F-statistics and significance levels for seasonal and overall site-specific habitat selection. Habitat selection was computed by comparing the overall time spent by individual bats on the various types of habitats available against the availability of the said habitats within a circle of radius equal to the foraging radius of each bat. All adults, and remotes only, were considered for these analyses. All significant covariate effects are illustrated. Probability levels were not Bonferroni-corrected. Habitat ranks show the most selected on the left and least on the right. Triple arrows indicate a significant difference between the habitat categories. For abbreviations of habitat types, see Table 6:1.

Bat group	Statistics			Habitat ranking order
Spring:				
Adults	$\Lambda = 0.055$	$F_{8,10} = 21.524$	$p < 0.001$	ASNW > PCG >>> POG > M > PK >>> OW > AR > CW > UR
Females	$\Lambda = 0.044$	$F_{8,5} = 13.705$	$p = 0.005$	ASNW > PCG > POG >>> M > PK > AR > OW > CW > UR
Summer/autumn:				
Adults	$\Lambda = 0.012$	$F_{9,8} = 71.377$	$p < 0.001$	PCG >>> ASNW >>> MCG > POG > PK > CW > OW > M/MOG >>> AR > UR
Females	$\Lambda = 0.009$	$F_{9,2} = 24.563$	$p = 0.040$	PCG >>> ASNW > POG > MCG >>> PK > OW > CW > M/MOG >>> AR > UR

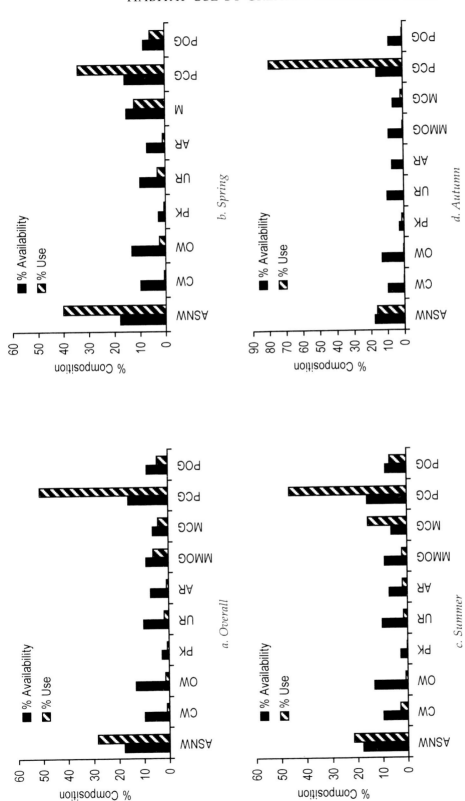

*Figure 6:1 Comparison of habitat availability (% Availability) and use (% Use) at King's Wood (a) overall ($\chi^2=4533.87$, d.f.=9, p<0.001), (b) spring ($\chi^2=4361.57$, d.f.=8, p<0.001), (c) summer ($\chi^2=4954.42$, d.f.=9, p<0.001), and (d) autumn ($\chi^2=4984.21$, d.f.=9, p<0.001). **ASNW:** ancient semi-natural woodland; **CW:** coniferous woodland; **OW:** other woodland types (including broad-leaved and mixed woodlands); **PK:** parkland; **UR:** urban; **AR:** arable; **M:** meadow; **MOG:** meadow grazed by sheep and/or horses in late summer/autumn; **MCG:** meadow grazed by cattle in late summer/autumn; **PCG:** cattle-grazed pasture; **POG:** pasture grazed by sheep and/or horses primarily.*

Habitat use by juveniles

Overall, juveniles aged less than 80 days old spent most of their time on cattle-grazed pastures, which they used significantly more than expected from availability (Table 6:1). Cattle-grazed pastures were the most favoured habitat type, and arable and urban areas the least. When the analysis was repeated only for juveniles containing all ten major habitat groups within their foraging radii, the results remained significant, with the same habitat ranking (Table 6:1).

Ivlev's Electivity indices showed that, of the ten main habitat groups, only cattle-grazed pasture was selected, but juveniles also preferred two other major habitat groups, ancient woodland and cattle-grazed meadow (Table 6:2). When expanded to include all 22 categories, ancient woodland, coniferous woodland, other woodland, parkland, cattle-grazed meadows adjacent to woodland, and cattle-grazed pastures adjacent to woodland and/or surrounded by thick hedgerows and/or tree lines were all preferred habitats.

Complete nights of activity were obtained for nine adults during the periods when the 16 juveniles were radio-tagged (of which 14 were tracked outside the nursery roost). The mean proportion in which each habitat type was used by the two groups can be seen in Figure 6:2. Juveniles spent significantly less time than adults foraging on sheep/horse-grazed pasture ($F_{1,21}=5.73$, p=0.026 for arcsin-transformed data), but no other differences were found.

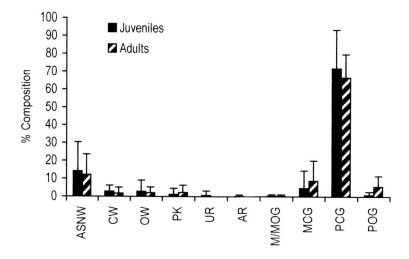

Figure 6:2 Comparison of habitat use by juvenile and adult greater horseshoe bats radio-tracked simultaneously. The data are expressed as a proportion of the respective total amounts of time spent over the various types of habitats. Data were pooled across years, from bats tracked at King's Wood and Woodchester Park. Proportions represent means (±SD, n=9 adults and 14 juveniles). Habitat categories are defined as for Table 6:1.

*Table 6:5 Ivlev's Electivity (E) and Manly's a (A) indices for all adult bats followed during the main radio-tracking seasons for the ten main habitat categories (for abbreviations of habitat types see Table 6:1). At the bottom of each column are the values at which habitats were used according to their availability. **Bold** figures represent positively selected and/or preferred habitats, and plain figures represent habitats that were selected against and/or avoided.*

Habitats	Spring (n=18)		Summer (n=12)		Autumn (n=5)		Summer & autumn (n=17)	
	E	A	E	A	E	A	E	A
ASNW	**0.13**	**0.27**	-0.18	**0.17**	-0.21	**0.12**	-0.19	**0.15**
CW	-0.82	0.03	-0.67	0.06	-1.00	0.00	-0.78	0.04
OW	-0.77	0.03	-0.49	0.10	-0.93	0.01	-0.62	0.07
PK	-0.78	0.03	-0.61	0.10	-0.73	0.06	-0.64	0.08
URBAN	-0.84	0.03	-0.96	0.00	-1.00	0.00	-0.97	0.00
ARABLE	-0.82	0.03	-1.00	0.00	-0.84	0.01	-0.95	0.00
MMOG	-0.58	0.09	-0.75	0.02	-1.00	0.00	-0.83	0.02
MCG			-0.33	**0.15**	-0.71	0.07	-0.44	**0.12**
PCG	**0.13**	**0.35**	**0.36**	**0.34**	**0.67**	**0.72**	**0.45**	**0.45**
POG	-0.42	**0.14**	-0.50	0.07	-0.77	0.02	-0.58	0.06
Level for Use=Avail	0.00	0.11	0.00	0.10	0.00	0.10	0.00	0.10

Behaviour of foraging bats

Within woodland, bats flew significantly lower when flying within rides (averaging 1.46 m above ground; n=9), compared to within the vegetation structure of the wood, where they averaged 3 m above ground (n=3) (t=4.12, d.f.=10, p=0.002).

Outside woodland, 47 observations of flying and foraging bats were grouped into one of three distance categories: less than 5 m from a field boundary; 5-10 m from a boundary; and more than 10 m from a boundary. Greater horseshoe bats preferred flying within 5 m of field boundaries (χ^2=48.03, d.f.=2, p<0.001), being found on average at 2 m from the boundary. Bats flying less than 5 m from a field boundary showed no significant difference in flight height when the data were analysed according to whether the boundaries were woodland edges or other boundaries (t=1.29, d.f.=36, p=0.206). However, bats flying against a woodland edge or copse flew significantly further out into the fields than when flying along other types of boundaries (t=2.63, d.f.=24, p=0.015)

Although greater horseshoe bats tended to fly closer to the ground the further they went from field boundaries, these differences in flight height were not statistically significant ($F_{2,44}$=2.07, p=0.139). On average (±SD), bats flew 1.8 m (±1.45) above ground when outside woodland, and the distribution of recorded flying events showed that bats preferentially flew along the side of hedgerows, rather than above them (χ^2=34.04, d.f.=1, p<0.001).

Perching and roosting sites

Thirty-two perch-feeding sites were identified for greater horseshoe bats. Of the 26 occasions when the perch position, relative to prevailing winds, was recorded, 13 were found on the leeward side of a tree or bush, three were on the windward side and ten were in windless conditions. In so doing, greater horseshoe bats show a preference for perch sites in the lee of vegetation (χ^2=6.25, d.f.=1, p=0.012).

Twenty-four perches were identified at close range. All were on pasture; 13 were found in trees emerging from hedgerows or along woodland edges and eight were under the canopy of large bushes, which formed part of thick, well-developed hedge-rows. A further three were found on isolated bushes, in the middle of a large expanse of unimproved pasture. Overall, 18 perch sites were found under the canopy of grazed trees and/or thick bushes, two were found in the archway at the entrance of forest rides, three were at the periphery of isolated bushes and one was a broken branch against the main stem of a tree along a woodland edge. They were positioned some 2.1 m (±0.54) above ground (range: 0.9-3.35 m), and 1.4 m (±1.56) away from the base of the tree, bush or hedge where they were found.

Perch sites were classified as 'bare' or 'leafy', and assigned to one of three size classes (0-5 mm, 6-10 mm and over 10 mm), according to their diameter at the point where the bat was hanging. Greater horseshoe bats significantly preferred bare branches (χ^2=13.50, d.f.=1, p=0.0002), in the 5-10 mm diameter category (χ^2=18.25, d.f.=2, p<0.001) as perch sites.

Of the 44 roost sites recorded during our study, 14 (32%) were directly related to farm buildings in some way, and this is typical for other areas as well (Table 6:6). Another 19 (eight in trees and 11 in caves, totalling 43%) were also found in farming areas or land parcels surrounded by farming activity.

GREATER HORSESHOE BATS ON FARMLAND: A DISCUSSION
Foraging habitats

In south-west England, greater horseshoe bats preferred to use ancient semi-natural woodland and cattle-grazed pasture sites in spring, above all others. However, as the seasons progressed through the summer and autumn, a strong selection and preference was shown for cattle-grazed pastures, followed by ancient woodland, cattle-grazed meadow and sheep/horse-grazed pasture. Our observations broadly confirm those from a number of other studies in the UK and abroad.

For example, in Switzerland, Beck et al. (1994) found that greater horseshoe bats flew and foraged mainly in areas where there was a high proportion of woodland edge, between two and three different types of habitat elements (e.g. woodland edge,

Table 6:6 Description and distribution of the greater horseshoe roosts recorded during a number of recent projects. The data for south-west England are from our study, those from Cheddar were re-worked from Jones & Billington (1999), with the kind permission of the authors and English Nature, and the data for Wales are from Slebech (Duvergé, 1997).

Type of roosts	SW England	Cheddar	Wales
Farms, barns, outbuildings, stock shelters & coach houses	32%	56%	60%
Other buildings	25%	13%	40%
Caves	25%	26%	0
Trees	18%	5%	0
Total number of roosts recorded	44	39	10

hedgerows, river banks and individual trees) and three or more different types of habitats per hectare. Such areas were selected at all times of the year. Bontadina *et al.* (1995, 1997) reported that greater horseshoe bats in Switzerland hunted mainly in forested areas in spring, and they considered deciduous woodland and permanent meadows to be critical habitats for these bats.

Similar areas to those used in south-west England were used by greater horseshoe bats in Wales (Duvergé, 1997). Some were also found along the edges of large estuaries and lakes at the two sites studied there (P L Duvergé, personal observation). More recently, Jones & Billington (1999) showed that of 20 large feeding areas located around a roost near Cheddar, all had areas of high, overgrown hedgerows, 12 had ancient woodland habitats, 15 had cattle-grazed pasture, 15 contained meadows, 13 had sheep-grazed pastures and 11 contained extensive tree-lines.

Although ancient woodland and cattle-grazed pasture were preferred at all our study sites, a range of habitat types were all selected or preferred at one or more of the sites during the summer, suggesting a greater flexibility in the foraging behaviour of greater horseshoe bats than had been reported previously.

Our observations also suggest that, overall, there is no significant difference in habitat use by the two sexes, nor between juveniles and adults. However, whilst adults and juveniles use the same type of habitats when foraging in some parts of the year, our observations have shown that they do not forage together. Mothers usually leave the breeding roost before their offspring, and travel much further than juveniles before settling at foraging areas (Jones *et al.*, 1995), thus avoiding competition for the same territories. Juveniles, therefore, develop their foraging abilities virtually on their own, and have to learn to forage independently. It is perhaps for this reason, as well as juveniles not having a fully developed echolocation system (Jones *et al.*, 1995), that some degree of dietary segregation has also been observed between mothers and young (Ransome, 1996).

Greater horseshoe bats forage on the wing (hawking) and in a fly-catching style (from perches) between April and October (Duvergé, 1996). Vertical structures, preferably with access to free air space for take-offs and landings, are required to forage from perches. All the habitats selected or preferred by greater horseshoe bats, including woodland edges, thick hedges and tree lines, shared this feature.

Most of the observations of foraging greater horseshoe bats indicate that they remain close to woodland edges and field boundaries, and this tallies well with what can be inferred from their echolocation and flight morphology (Aldridge, 1985, 1987; Norberg & Rayner, 1987; Jones & Rayner, 1989; Neuweiler, 1989). The strength of echolocation calls attenuates rapidly in the atmosphere, and the rate of attenuation increases with increasing call frequency (Griffin, 1971). Therefore, one might expect that greater horseshoe bats would remain relatively close to field boundaries, if they were using these as linear landscape elements along which to travel and forage (Limpens & Kapteyn, 1991). Using the call intensity model developed by Waters & Jones (1995), we were able to calculate that greater horseshoe bats should be able to detect a large moth (e.g. the large yellow underwing *Noctua pronuba*) at 4-10 m. These values closely match the observations of distances at which greater horseshoe bats fly from vegetation structures.

Why do greater horseshoe bats concentrate so much of their activity on such specific habitats? Our strongest clue may lie in the diet of these bats. Previous studies have shown that there is a definite succession in the prey eaten by greater horseshoe bats between April and October (Jones, 1990; Duvergé & Jones, 1994; Ransome, 1997), and that the main prey items are encountered yearly at all ten roosts studied across the UK (Ransome, 1996, 1997; Chapman *et al.*, 1997). Woods are warmer than the surrounding land at night, and provide superior feeding sites in spring (Jones *et al.*, 1995). This temperature discrepancy may influence insect availability, and is greatest at that time of year. During that period, one of the greater horseshoe bat's key prey, the scarabaeid beetle *Melolontha melolontha* (the maybug), is known to fly between pastures or meadows and woodlands to complete its life cycle. Clearly, using cattle-grazed pastures along the edges of woodland would give greater horseshoe bats access to these insects and to tipulid flies and geotrupid beetles, which also form a major part of the bat's diet in spring.

Of all habitats used by greater horseshoe bats, cattle-grazed pastures, bordered by well-developed field boundaries, provide a more or less ideal habitat for foraging. By browsing the field edges, cattle tend to produce tall archways (1.5-2.1 m high and 0.75-3 m deep) within the vegetation. Many of the resulting bare twigs on the underside of the archway dome, or under the browse line of individual trees, can then be used as hanging places by greater horseshoe bats when perch feeding (Duvergé, 1996). Additionally, cattle dung tends to accumulate along field boundaries, acting as a focal area for insects, which colonise the freshly deposited dung or emigrate from cowpats after mating, oviposition, pupation or hatching.

Nocturnal dung beetles, although most abundant in summer and autumn, fly throughout most of the year (Denholm-Young, 1978), and may therefore be available to greater horseshoe bats throughout the non-hibernation period. Some of those (e.g.

Aphodius rufipes) form an important part of the insect fauna by late July. By that time, they usually appear in large amounts in the diet of greater horseshoe bats, and thereafter dominate the diet until the bats start hibernating in October or November (Jones, 1990; Duvergé, 1996). The readily available and concentrated supplies of insects on pastures may be one of the reasons why greater horseshoe bats are attracted to, and select, cattle-grazed pastures throughout the year.

On a similar note, the use of meadows (primarily along woodland edges or surrounded by thick hedgerows), mixed woods and parkland in summer may reflect attempts to catch moths which abound in those habitats, and which also represent a large proportion of the diet of greater horseshoe bats at that time. Non-migrating insects occur at high concentrations around the crown, and in the lee, of vegetation and shelterbelts. The latter may provide eddies, extending on the leeward side up to four times the height of the shelter, where insects are concentrated (Lewis, 1969a, 1970; Peng, 1991; Peng *et al.*, 1994). When moving further away from such structures, insect densities decrease and are distributed nearer the ground (Lewis, 1969ab, 1970). The observed distance at which bats forage from a field boundary, and differences in flight height, could therefore reflect the distribution of prey items.

Given the evidence presented so far, it appears that the bats may be responding to the availability of prey, while using habitat types rich in their preferred prey. However, by foraging over preferred habitats throughout the year, but still visiting a range of habitats, greater horseshoe bats ensure that they are in a position to take advantage of any prey patches which may appear, at short notice, in less preferred habitats. It is therefore possible, and indeed likely, that greater horseshoe bats fly close to vegetation structures to stay within an area where insects may be concentrated and to be within a range at which they may detect potential prey by using their echolocation calls. Bats also use them as guides along which to travel, and possibly as screens against potential predators.

Threats to greater horseshoe bats on farmland

The well-being of a greater horseshoe colony may be threatened in a number of ways. They may, for example, lose their roosts through dereliction, demolition, change of ownership or conversion of buildings, or lose foraging habitats through changes in land use or physical destruction such as the removal of hedgerows, treelines and the eradication of travel routes, or changes in farming practice. They may also be susceptible to chemical poisoning within roosts (such as by compounds used for remedial timber treatment), as well as outside the roost (through coming into direct contact with vegetation which has been sprayed, or indirectly, through eating insects which have themselves been sprayed or eaten contaminated food).

It is well known that many greater horseshoe bat breeding roosts are found in buildings associated with farms and farming practices. These need to provide quite exacting environmental conditions to enable the bats to breed successfully (Ransome, 1998). However, our studies and those of Jones & Billington (1999) clearly indicate that greater horseshoe bats also depend on a range of buildings within the wider countryside for a number of other uses (e.g. for night roosting or mating). The role of farm buildings for non-breeding activities is important and often overlooked, and there is currently a lack of information relating to these roosts.

Although the loss of roosts and foraging habitats are considered the two most important threats to greater horseshoe bat populations (Stebbings & Griffith, 1986; Stebbings, 1988), indirect chemical poisoning could be important when one considers the use of farmland by these bats. Indeed, Jefferies (1972) showed that there was a possibility that bats could be more heavily contaminated with DDT-type insecticides than either insectivorous or carnivorous birds, whilst Hamon (1985) found high levels of pesticide contaminants in the faeces of rhinolophid bats in eastern France. Additionally, records collated by Brosset *et al.* (1988) showed that population fluctuations of the Mediterranean horseshoe bat *Rhinolophus euryale* in France mirrored those of peregrine falcons *Falco peregrinus* between the years 1950 and 1980 (Monneret, 1987), the period during which DDT was used legally in that country.

More recently, studies have shown that Avermectin worming compounds could cause the death of insects from the cowpat community at both the larval and adult stages, and have sub-lethal effects on coprophagous insects. They may also prevent insect larvae feeding, reduce rates of larval growth, interfere with moulting, inhibit metamorphosis and pupation, and prevent the emergence of adults from the pupal case (Wall & Strong, 1987; Strong, 1993; Strong & James, 1993; Ransome, 1996). Although research carried out on cattle-grazed pastures showed that worming had subtle effects on habitat use by specific bat species and species groups (Chapman *et al.*, 1997), no broad-based impact on the overall levels of bat activity can be quantified at this stage.

CONCLUSIONS AND RECOMMENDATIONS

Studies of habitat use by greater horseshoe bats point to three practical recommendations. First, permanently grazed pasture should not be ploughed up and used for arable crops. Second, as much woodland as possible should be retained, especially if it holds good stocks of deciduous trees. And third, hedgerows should be encouraged to develop within 4 km of a breeding roost. The 'ideal' hedge for a greater horseshoe bat would be in the region of 3-4 m high and 2-3 m wide, with occasional emergent trees. Intensively trimmed hedges (1.5-2 m high by 1-1.5 m wide, with a rectangular cross-section) are of very little use to bats. Broadly speaking, greater horseshoe bats need good cover within the immediate vicinity of their roosts, well-developed field boundaries to commute and forage along, and ancient woodland and cattle-grazed pastures which are easily accessible, for foraging.

Using data collected during our studies, we suggest that habitat conservation for greater horseshoe bats should take into account the flying capabilities and habitat requirements of juveniles, as well as that of adults (Duvergé & Jones, 1994). Hence, a two-tier approach is appropriate. Firstly, particular emphasis should be put on the preservation and enhancement of key habitats within a 1 km radius of the roost, to cater for the foraging ability of recently volant juveniles (the 'young sustenance zone' of Ransome, 1996). Secondly, protected areas should extend to 4 km to take into account most of the foraging areas required by adults within a colony (the 'roost sustenance zone' of Ransome, 1996). Bontadina *et al.* (1997) consider that suitable biotopes should cover at least 30% of the land surface within a 3.5 km radius of a nursery roost, and should be maintained and enhanced if possible (30% being the level at which the potential effects of habitat fragmentation are reduced).

Detailed information now exists regarding improvements which may be carried out to the roosts (Ransome, 1998) and foraging habitats (Duvergé & Jones, 1994; Duvergé, 1996; Ransome, 1996, 1997; Chapman *et al.*, 1997; Anon, 1998; Jones & Billington, 1999) of greater horseshoe bats, as well as to the anthelminthic treatment of grazing stock (Duvergé, 1996; Ransome, 1996; Chapman *et al.*, 1997; Anon, 1998), and management of grazing stock (primarily stock density and timing of grazing) on pastures (Ransome, 1996; Anon, 1998). These should be consulted wherever efforts are being made to conserve this species, and the measures they recommend will also benefit a range of other bats that share the same habitats (Duvergé, 1996; Chapman *et al.*, 1997; Vaughan *et al.,*1997).

Studies of greater horseshoe bats, over the last 40 years, have shown that most greater horseshoe bats move less than 20 km between summer and winter roosts, and therefore spend most of their life in a relatively small geographical area. Continuing human encroachment and the increasing fragmentation of foraging habitats and changes in land use, especially if coupled with the loss of links and corridors between roosts and main habitat blocks, will eventually lead to a reduced number of available feeding areas for these bats. Ultimately, this could prevent their recovery and/or lead to local extinctions. It is, therefore, the duty of all concerned to preserve what we, and the bats, currently have, and to do our utmost to reverse the situation that has prevailed over the last 50 years.

ACKNOWLEDGEMENTS

We would like to thank The Hon. Vincent Weir for funding Dr. Duvergé's work, as well as the following, for the many ways in which they helped our projects during the last decade: Dr R D Ransome, Professor Stephen Harris, Dr Tony Mitchell-Jones, Drs Dean Waters, Kate Barlow, Adam Britton and Nancy Vaughan, and most of the University of Bristol 'bat-lab' members, Dr Henry Schofield, Jacques B Pir, Drs P Robertson & N Aebischer, Christianne Schmidt, Rachel Krusic (née Stevens), Dave Fawcett, Edward Rimmer, Matthew Morton, Tony Sladden, Lyn Jenkins, Robert Haycock, Tom McOwat, Dr Peter Andrews, Annie Poole, John Galvin, Robert Taylor, Geoff Billington (for permission to re-work some of his data), and Dr Fran Tattersall and two anonymous referees for their comments and suggestions.

7

Wood mice in the arable ecosystem

F H Tattersall & D W Macdonald

INTRODUCTION

During the second half of the twentieth century, many British plant and animal species have declined dramatically in abundance as a result of increasingly intensive agricultural practices (Wilson, 1992b; Fuller *et al.*, 1995; Donald, 1998; Chamberlain & Crick, 1999). At the start of the twenty-first century, conservationists and policy-makers are concerned with formulating strategies to reverse these effects (e.g. Johnson & Baker, Chapter 2), but in so doing they require a fuller understanding of the many ways in which farming impacts on wildlife.

Although characteristically a woodland species (e.g. Flowerdew *et al.*, 1985), the wood mouse *Apodemus sylvaticus* is a useful species with which to explore the effects of farming and farmed landscapes on a wild mammal (Macdonald *et al.*, 1993). Wood mice have been intensively studied, on both farmland and elsewhere (Flowerdew & Tattersall, in press), and are sufficiently common and trappable to provide data for analysis. As we will demonstrate below, they are mobile enough to respond at the field scale, but also small enough to respond to variation within fields. Finally, they are midway in the food chain, feeding on plants and invertebrates while themselves being preyed on by, for example, tawny owls *Strix aluco*, foxes *Vulpes vulpes* and weasels *Mustela nivalis* (Green, 1979; Southern & Lowe, 1982; King, 1985; Plesner Jensen, 1993).

Our aims in this chapter are twofold: first, to review the ecology of arable-dwelling wood mice in lowland England, and second, to review the impact on them of specific farming operations. Many of the studies we review used either Longworth live-trapping (Gurnell & Flowerdew, 1990) or radio-telemetry (Tew, 1992). Habitats included in our definition of arable land are cereal and other crops (principally winter wheat, winter barley and oil-seed rape); boundary features such as hedgerows, field margins (grassy strips bordering the crop) and headlands (cropped strips of land at the field edge, where the plough is turned); and small (less than 1 ha) farm woodlots.

THE ECOLOGY OF ARABLE WOOD MICE
Abundance and population ecology

With an estimated pre-breeding population totalling 38 million, the wood mouse is the third most numerous British mammal (Harris *et al.*, 1995). On arable farmland in

Britain, it is one of the commonest small (under 50 g) mammals. Wood mice are the most abundant small mammal in hedgerows (Pollard & Relton, 1970; Kotzageorgis & Mason, 1997), crop fields, including wheat, barley, oats, oil-seed rape, maize, sugar beet and beans (Jefferies *et al.*, 1973; Green, 1979; Pelz, 1989; Macdonald *et al.*, 1993), set-aside (Rogers & Gorman, 1995; Tattersall *et al.*, 1997, 1999ab) and woodlots (Fitzgibbon, 1997).

Until now, however, controlled comparisons of species abundances in different arable habitats were lacking in the literature. We live-trapped small mammals in set-aside, field boundaries (hedgerow, field margin and crop edge), crop and small farm woodlots in Wiltshire and Gloucestershire. While there were differences between these farmland habitats, the wood mouse dominated the small mammal community in each (Figure 7:1), comprising up to 96% of individuals caught in the crop. Common shrews *Sorex araneus*, bank voles *Clethrionomys glareolus* and field voles *Microtus agrestis* were also abundant.

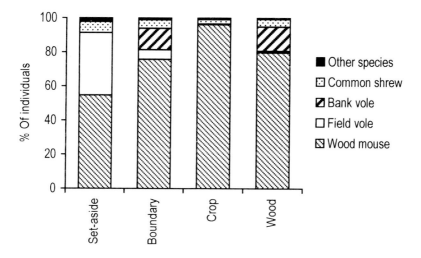

Figure 7:1 Percentage of individuals of each species in small mammal communities live-trapped in set-aside, field boundaries (hedgerow, field margin and crop edge), crop and woodlots (<1 ha). Trapping took place on two Royal Agricultural College farms in Wiltshire and Gloucestershire, with two sites at each farm. Each site consisted of four trapping grids of 0.36 ha, with one grid in each habitat, such that 16 grids were trapped in total, four in each habitat. Trapping occurred every three months throughout 1996 and 1997. 'Other' species were house mice Mus domesticus, yellow-necked mice Apodemus flavicollis, harvest mice Micromys minutus and pygmy shrew Sorex minutus.

Given that, overall, they are proportionally more abundant than any other small mammal on farmland, how do numbers of wood mice differ between these habitats, and how do they change seasonally? In monthly trapping sessions at two sites (each with four trapping grids of 49 traps over 0.36 ha) at Eysey Manor Farm, Wiltshire, wood mice were almost always most abundant in woodlots, and generally least abundant in set-aside; Figure 7:2 presents data from one of these sites. To the best of our knowledge, these data are unique; although there have been many other studies of wood mice on farmland, none have replicated, sequential data collected simultaneously in different habitats over more than a year.

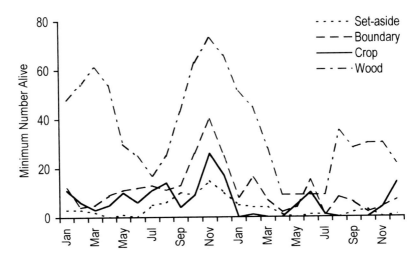

Figure 7:2 Seasonal changes in wood mouse abundance (Minimum Number Alice) in four habitats. Data are for monthly trapping sessions at one Royal Agricultural College farm site in Wiltshire, with four 0.36 ha trapping grids, one in each of set-aside, boundary, crop and woodlot. After Macdonald et al (2000a).

In woodlots at Eysey Manor, abundances showed a spring decline, followed by a period of relatively stable, low population density, and increased again in autumn and winter. There were no clear seasonal trends in abundances in set-aside, boundary or crop. Mice were caught in cropped fields throughout the year, even when there was no vegetative cover, and the presence of burrows and radio-tracking (see below) confirmed that although some probably commuted from the hedgerows, some individuals lived entirely within the fields. Indeed, peak abundances occurred in November 1996, when fields were bare, tilled earth.

In all habitats males began to come into breeding condition in January (1996) or February (1997), and by March almost all had scrotal testes. Breeding females began to appear in February (1997) or March (1996), and all females were breeding by May. Both sexes effectively ceased breeding by November in both years.

The pattern of high winter and low summer abundances that we found in Wiltshire farm woodlots is also characteristic of populations in larger woodlands (Wilson *et al.*, 1993; Mallorie & Flowerdew, 1994), and other habitats as varied as sand dunes (Gorman & Akbar, 1993), sitka spruce plantation (Fernandez *et al.*, 1996) and reed beds (Canova *et al.*, 1994). Kotzageorigis & Mason (1997) reported a pattern of autumn/winter peaks and spring/summer troughs over two years, in eight hedgerows in Essex. Temporal patterns in arable crops and set-aside, however, are more variable: in Wiltshire, north-east Scotland (Rogers & Gorman, 1995) and Berkshire (Tew, 1995) there were no clear seasonal trends, while Green (1979) reported more characteristic woodland-type fluctuations in crops in a Suffolk farm. Conversely, a wood mouse population in cereal fields and hedgerow adjacent to Oxford University's 400 ha Wytham Woods showed a steady increase over the summer, a sharp decline after harvest and low abundance over winter (Tew & Macdonald, 1993; Macdonald et al., 2000d). The authors speculated that after harvest the mice moved into the hedgerows and woodland, and out into the fields again in spring.

In woodland, as in our Wiltshire farmland sites, the main breeding season is from spring to early autumn, but density dependant mechanisms relating to female spatial and reproductive behaviour postpone population increases (Montgomery, 1989ab; Wilson et al., 1993; Mallorie & Flowerdew, 1994). Density dependence in woodland is sufficient to limit peak densities in early winter and promote the recovery of low populations in the summer, although in years with a good tree seed crop, density dependence may be overridden. In cropped fields, the availability of spilt or sown seed might have the potential to play a similar role. Fluctuations of wood mice are synchronous in widely separated woodlands (Mallorie & Flowerdew, 1994), possibly because of synchronised seed crops. The same mechanism could synchronise woodland and hedgerow populations, where hedgerow management allows fruiting.

Densities in our woodlots at Eysey Manor (assuming an effective trapping area of 0.49 ha by adding a notional 5 m border to our 0.36 ha grid: Gurnell & Gipps, 1989) ranged tenfold, from the equivalent of 18-184 per hectare, with a monthly average of 75 per hectare. These estimates are high in comparison with other habitats. For example, in larger mixed deciduous woods, winter increases are usually up to 100-130 per hectare (Gurnell, 1981; Mallorie & Flowerdew, 1994), while in sand dunes a peak of 11.5 per hectare was reported by Gorman & Akbar (1993).

Densities in cropped fields at Eysey Manor ranged from the equivalent of 0-67 per hectare, with a monthly average of 21 per hectare. These are also high in comparison with other long-term studies in arable crops, which, in Britain, have recorded peak densities of only 25 per hectare (Green, 1979; Tew & Macdonald, 1993). However, Rogers & Gorman (1995) recorded a maximum of 33 per hectare in a carrot crop, and Pelz (1989) reported up to 80 per hectare in sugar beet. In set-aside fields, we recorded average densities of 12 per hectare, ranging from 0-102 per hectare, considerably higher than the 0-11 per hectare recorded by Rogers & Gorman (1995) in two-year old whole farm set-aside in Scotland.

With a highly mobile species such as the wood mouse, landscape context is important, but largely neglected, when interpreting abundances in different habitats (e.g. Tattersall et al., 2002). In farm woodlots, abundance of adult wood mice in autumn increases with the length and number of adjoining hedges, a measure of connectedness, while in spring, mice are least abundant in woods surrounded by wheat and rape fields (Fitzgibbon, 1997). Interestingly, Fitzgibbon (1997) found that juvenile abundances

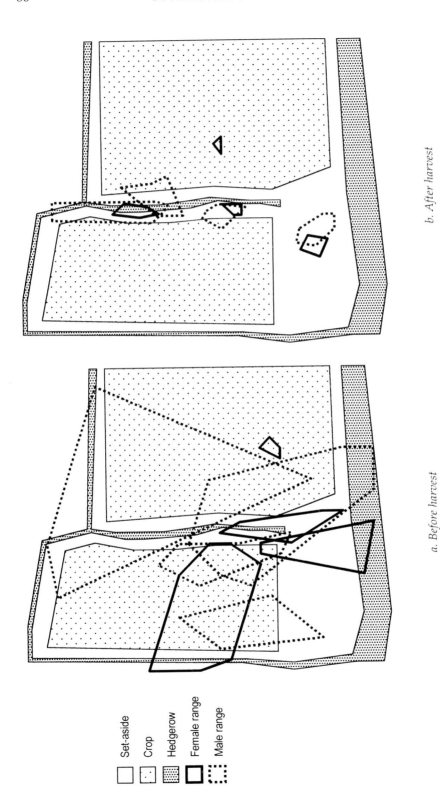

b. After harvest

a. Before harvest

Set-aside
Crop
Hedgerow
Female range
Male range

Figure 7:3 Home ranges of three males and three females before and after harvest at Eysey. After Tattersall et al. (2001).

in autumn are greater in woodlots surrounded by rape, and in isolated woodlots, but it is not clear whether this is due to a crowding effect because the rape is unsuitable to disperse into, or whether on the contrary rape provides good resources and is therefore a source of juvenile colonists. In the boundaries of pastoral fields wood mouse abundances are negatively related to distance from woodland (in winter), and amount of pasture in the surrounding area (in winter and summer) (Montgomery & Dowie, 1993).

Habitat use

Different abundances in different habitats must be mediated by individual decisions about where to live. Studies in Wiltshire and Oxfordshire have looked at how wood mice use cereal fields, hedgerows and field margins before and after harvest.

At Eysey Manor, we radio-tracked nine breeding males and seven breeding females before harvest, and nine animals of each sex after harvest (three of these males, and four of the females, were not breeding), and calculated restricted polygon home ranges. Before harvest (June and July), males had very large home ranges averaging (±SD) 1.87 ha (±0.38) and overlapping several other males and females, while females had smaller, exclusive ranges averaging 0.55 ha (±0.28) (Tattersall et al., 2001). After harvest (September-November), both males and females had small (males: 0.19 ha ±0.05; females: 0.14 ha ±0.07), overlapping ranges that centred on the hedgerow (Figure 7:3). Compositional analysis (Aebisher et al., 1993) revealed that wood mice avoided using the set-aside, and particularly the margins, relative to crop and hedgerow habitats, and this selectivity was statistically significant after harvest and cutting (Figure 7:4). A similar study by Todd et al. (2000) at Wytham Farm, Oxfordshire, compared use of wheat, barley, oil-seed rape and hedgerow, and also found that hedgerow was preferred, significantly so after harvest. In their analysis, wood mice did not consistently distinguish between different crops.

Although cropped fields are superficially homogenous, they contain small, scattered patches of weeds, including plants known to be eaten by arable wood mice, such as wild oats Avena fatua and chickweed Stellaria media (Green, 1979; Plesner Jensen, 1993). In arable fields at Wytham, in spring and summer, Tew et al. (2000) accurately marked the positions of radio-tracked mice with canes to denote two categories of movement: stationary (where mice were active but did not change location for more than ten minutes) and travelling (areas where the mice passed through but did not pause). A third set of canes was placed at random within each mouse's range. At each cane in each category, the team established the plant species and seed bank composition. Places where mice were stationary (where feeding activity was confirmed by piles of discarded seed husks), contained significantly higher abundances of chickweed, wild oats, cleavers Galium aparine and sterile brome Bromus sterilis, and less bare earth, than areas where mice were travelling or random areas, but there were no differences in their seed banks. Thus, patterns of home range utilisation within a field were influenced by the dispersion of weedy patches.

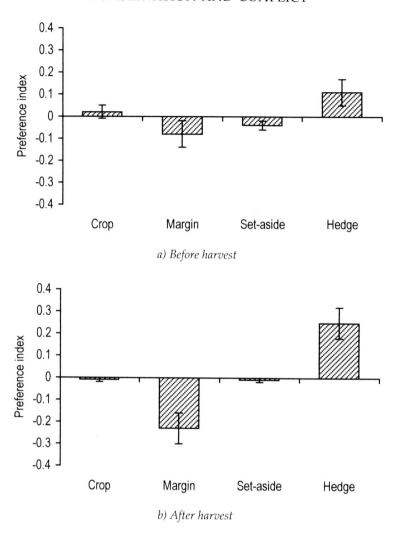

Figure 7:4 Mean (±SE) preference indices for crop, margin set-aside, field set-aside and hedgerow, before and after harvest and topping. A positive value indicates that a habitat is used more than expected (i.e. is preferred) while a negative value indicates that a habitat is used less than expected (i.e. is avoided). From Tattersall et al. (2001).

Diet

In woodland, heathland and sand dunes, varying proportions of seeds and inverte-brates form the bulk of the diet of wood mice (Butet, 1985; Hansson, 1985; Gorman & Zubaid, 1983), and the same is broadly true on arable land. On winter wheat and sugar beet during autumn, winter and early spring, up to 50% of the volume of the diet of wood mice is sown and shed cereal grain. From late spring until harvest, flowers and seeds of weeds such as chickweed *Stellaria media*, goosefoot *Chenopodium* sp. and annual meadow grass *Poa annua* together comprise as much as 80% of their diet (Green, 1979; Pelz, 1989). Invertebrates, of which earthworms *Lumbricus terrestris* are an important component, form at least 10% of the diet, which on ploughed land in late winter and early spring can rise to 65% (Green, 1979), and in sugar beet during spring rises to over 70% (Pelz, 1989). On set-aside in Scotland, wood mouse diet was low in invertebrates and seeds, and high in green leaf plant material, which contributed 65% of the volume of faecal remains.

Arable wood mice feed extensively on weed seeds, such as black grass *Alopecurus myosuroides* and wild oats *Avena* spp., two of the four most common pernicious grass weeds of cereal crops, but also feed on many crop species, such as wheat *Triticum aestivum*, barley *Hordeum vulgare* and oilseed rape *Brassica napus*. When ripe, these crops are preferred to the weed seeds (Plesner Jensen, 1993). As seed predators it is therefore equivocal whether the mice are assets or pests (Povey *et al.*, 1993).

Using a naturalistic field experiment, in which wild sub-adult wood mice were confined in an arena *in situ*, Plesener Jensen (1993) ranked their preferences for different foods and looked at how these changed through the night. Of the 26 foods offered, six (blackberries *Rubus* spp., sweetcorn *Zea mays*, oilseed rape, wild oats, wheat and barley) made up the bulk of the diet in all tests (90-100% in 20 out of 26 trials, and less than 80% in only one trial). The amount eaten peaked early in the night, at around 22:30 hrs, and again at about 04:30 hrs. Meals chosen at the time of the first feeding peak contained significantly more sugars than those at the time of the second feeding peak. The wood mice may have been selecting for sugars early in the night because they are easily absorbed and help raise body temperature quickly when the mice have just emerged from their burrows.

Behavioural ecology

The relative importance of food and cover as resources around which male mice orient, might diminish during the breeding season, when securing a mate is vital. In woodland, food availability sometimes, but not always, explains male dispersion (Wilson *et al.*, 1993), and the factors surrounding the impact of food and females on male behaviour and abundance are likely to be complex.

Using data from arable farmland in Oxfordshire, Tew & Macdonald (1994) looked at the extent to which mice used shared areas. Where overlap between the sexes occurred, large male ranges overlapped several smaller female ranges. They calculated that an average (±SE) restricted polygon male range of 1.44 ha (±0.11; n=104) covered just over half of an average female's range of 0.49 ha (±0.1; n=32), but the area of overlap covered only 17% of the male range. Males shared about 30% of their range with other males, while females shared less than 10% with other females. Males consorted closely with oestrus females for less than four hours, and outside the context of breeding, did not consistently avoid or select areas used by conspecifics of either sex, nor did their movements coincide (Tew & Macdonald, 1994). These results suggest that breeding male wood mice ranged widely to allow them to mate polygynously with promiscuous females.

Home ranges of arable wood mice in Wiltshire and Oxfordshire (see above) were large relative to mean range sizes in woodland, which for males average from 0.03 ha in winter to 1.30 ha in summer, and for females from 0.02-0.40 ha (Wolton & Flowerdew, 1985; Attuquayefio *et al.*, 1986; Corp *et al.*, 1997). However, relative to wood mice in sand dunes, which have an average breeding range of 3.5 ha for males and 1.75 ha for females (Attuquayefio *et al.*, 1986; Corp *et al.*, 1997), arable ranges were small.

Despite variation in range size, the wood mouse's social system is remarkably consistent between habitats. An increase in range size during the breeding season, with large, overlapping, breeding male ranges and smaller, discrete, breeding female ranges, suggesting promiscuous polygyny, have been described for wood mice in a wide variety of habitats (Wolton & Flowerdew, 1985). These include woodland (Wolton, 1985) and sand-dunes (Attuquayefio *et al.*, 1986), as well as arable land (Tattersal *et al.*, 2001, above; Tew & Macdonald, 1994; Rogers & Gorman, 1995).

Ranges of breeding males on Oxfordshire farmland varied considerably, from 0.2 ha to over 3 ha, and individuals with larger ranges clearly had access to more females. Were there any morphological or behavioural characteristics that could be related to range area? Range area correlated statistically significantly with nightly distance moved, average speed and the percentage of fixes where a mouse was moving (Table 7:1), and these measurements were correlated with each other. However, neither range size nor any movement parameter correlated with mouse weight (Tew & Macdonald, 1994). It seems thus that breeding males with large ranges were more mobile, but not heavier, than breeding males with smaller ranges.

Table 7:1 Mean (±SD) measures of mobility, home range size and weight, for 55 breeding male wood mice radio-tracked in arable fields in Oxfordshire between March and August. Regression analysis was used to obtain significance values. ns=not significant; ***=p<0.001. From Tew & Macdonald (1994).

		Statistical significance of relationship with:	
		Weight (g)	Range size (ha)
Weight (g)	23.9 (±2.6)		
Restricted polygon range size (ha)	0.93 (±0.77)	ns	
Min. nightly distance travelled (m)	811.6 (±406.2)	ns	***
Speed (m/min)	1.76 (±0.82)	ns	***
Speed when active (m/min)	2.3 (±0.74)	ns	***
Proportion of time active (%)	76.7 (±18.5)	ns	***

These mobile males were also most likely to be parasitised by the gut nematode *Heligmosomoides polygyrus* (Brown *et al.*, 1994a). Infected male mice travelled at an average moving speed of 3.4 m per minute for 1364 m per night, almost 50% further than uninfected males, which travelled an average of 888 m per night at a speed of 2.5 m per minute. In Oxfordshire, up to 80% of wood mice on arable land were infected and prevalence in adults was much more markedly seasonal than in woodland, possibly in part because ploughing alters transmission of dispersing larvae.

IMPACT OF FARMING OPERATIONS ON ARABLE WOOD MICE

Harvest

To the casual observer, harvest, at the end of the summer, is the most dramatic and noticeable farming operation. The physical process of harvesting itself has little direct effect on wood mice in fields while they remain in their burrows. Of 33 mice radio-tracked by Tew & Macdonald (1993) on Oxfordshire farms during harvest, only one was killed by the combine harvester. The removal of cover, however, greatly increased exposure to predators, and hunting by tawny owls noticeably increased. Of the 32 mice tracked after combining, 17 (53%) disappeared within a week of harvest, and for nine of these predation was observed or inferred from bloodied collars found in weasel nests or lodged in trees. In the trap session following harvest at Wytham there was an 80% decline in the number of individuals caught, presumably the combined result of mortality and emigration. Those individuals that remained in the bare fields after harvest significantly reduced their mobility, possibly in an attempt to lower their risk of being predated. Tew & Macdonald (1993) tracked 12 individuals for at least one full night both before and in the month after harvest. After harvest, wood mice almost halved the total distance they moved per night from an average (±SD) of 927 m (±433) to 568 m (±274). They also moved more slowly — from 1.71 m per minute (±0.68) before harvest to 1.12 m per minute (±0.54) after harvest — and spent 10% less of their time moving. Range sizes could be calculated for only six males immediately before and after harvest (when they averaged 0.76 ha ±0.89 and 0.49 ha ±0.52 respectively), and while five of these showed a decrease, the difference was not statistically significant.

Set-aside management

Set-aside is land removed from arable production for at least one year and sometimes longer (Clarke, 1992; Firbank et al., 1993). While primarily aimed at reducing agricultural surplus, set-aside could also have a major impact on wildlife (Andrews, 1992; Sotherton, 1998). It is an important habitat in European arable ecosystems: 5.6 million hectares were in set-aside in Europe in 1997. A range of management options is available, including how it is established and how often it is relocated, but the land must have a green cover over winter, and this must be destroyed annually with herbicides or by cutting (MAFF, 1998).

Set-aside is cut at around the same time as harvest, but farmers are allowed to leave some areas uncut. To test whether mice preferred cut or uncut margins after harvest and mowing, Tattersall et al. (2001) created an experimental design in which alternate 50 m by 6 m patches, within a standard 20 m-wide set-aside margin, were left uncut next to hedgerow. The remaining 14 m width of margin was cut as normal. After harvest and mowing, compositional analysis showed that radio-tracked wood mice used habitats non-randomly, avoiding cut set-aside and crop, and selecting uncut set-aside and hedge (Figure 7:5). This suggests that the longer vegetation provided mice with cover or food not found in cut set-aside or bare fields. Aerial predators are less successful in habitats with long vegetation than in open habitats (Simonetti, 1989; Longland &

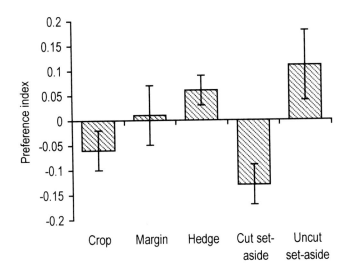

Figure 7:5 Mean (±SE) preference indices for crop, margin, hedgerow, cut set-aside and uncut set-aside after harvest and topping. A positive value indicates that a habitat is used more than expected (i.e. is preferred) while a negative value indicates that a habitat is used less than expected (i.e. is avoided). From Tattersall et al., (2001).

Price, 1991), although this may be less important for wood mice than for other species of British small mammal (Plesner Jensen & Honess, 1995). Invertebrate abundance on field margins is much higher in uncut plots than in plots cut in summer (Feber *et al.*, 1995), and fresh grass seeds remain available in uncut but not cut vegetation.

Cover might also be influenced by establishment method. Tattersall *et al.* (1999b) compared the use made by wood mice of set-aside established by sowing at a high rate with a species-poor mix of agricultural grasses and clover (at Eysey), and at a low rate with a rich mix of wildflowers and grasses (at Jealott's Hill Farm, Berkshire). Vegetation at Eysey was taller and provided more cover than vegetation at Jealott's Hill, but contained fewer species (Table 7:2). Nine males were tracked at each site. At Jealott's Hill, wood mice used the open, species-rich set-aside, crop and other habitats (hedgerow, pasture and fallow) at random, but used the denser, species-poor set-aside at Eysey less than expected in comparison with crop and hedgerow (Table 7:3). The point here is to sound a cautionary note in jumping to conclusions about the impact of farm management practices on wildlife. In this case, although set-aside on both farms was sown, the resulting vegetation, and its attractiveness to wood mice relative to other habitats was different.

Under the Set-aside scheme, farmers are allowed to configure their set-aside as blocks (generally whole fields or large chunks of fields) or ribbons (generally as 20 m wide field margins). Do wood mice differentiate between these configurations? The answer is that they do, but not in the way one might expect. After harvest, wood mice at Eysey avoided set-aside configured as 20 m-wide margins, but not as a 3 ha block

(Figure 7:4; Tattersall *et al.*, 2001). This is surprising because one would expect the mice to forage near to the hedgerow, which is their favoured habitat and provides shelter, and because many mice located their burrows in the margins. One reason for

Table 7:2 *Mean (±SE) vegetation characteristics at Eysey and Jealott's Hill. Significance levels to compare the two sites were obtained with a two-tailed T-test. ***=p<0.001, ns=non-significant. From Tattersall* et al. *(1999b).*

Site	N	Height (m)	% Cover	% Grass	% Herb	Sp. richness
Eysey	34	1.0 (±0.1)	91 (±11)	62 (±25)	31 (±27)	5 (±0.9)
Jealott's Hill	10	0.6 (±0.1)	75 (±11)	45 (±26)	31 (±19)	17 (±3.0)
		***	***	ns	ns	***

Table 7:3 *Mean (±SE) preference indices for crop, set-aside and other habitats before harvest and topping at Eysey and Jealott's Hill. A positive value indicates that a habitat is used more than expected (i.e. is preferred) while a negative value indicates that a habitat is used less than expected. Compositional analysis was used to obtain significance values. ns=not significant; **=p<0.005. From Tattersall* et al. *(1999b).*

Habitat	Eysey	Jealott's Hill
Crop	0.03 (±0.05)	-0.12 (±0.10)
Set-aside	-0.16 (±0.05)	0.12 (±0.10)
Other	0.12 (±0.06)	-0.004 (±0.00)
	**	ns

the distinction between set-aside blocks and margins may be that predators such as stoats *Mustela erminea* and weasels tend to patrol along hedges, making the areas close to them relatively dangerous (King, 1991ab), and mice may have been avoiding the area close to the hedge, rather than margin itself.

Agro-chemicals

Agro-chemicals such as molluscicides, herbicides and fertilisers have pervasive direct and indirect impacts on wood mouse populations and habitat use (see also Shore, Fletcher & Walker, Chapter 4).

Shore *et al.* (1997) monitored wood mouse populations at Wytham on three fields treated with the molluscicide methiocarb in autumn, and two fields treated in spring. (Methiocarb is one of a group of chemicals, known as anticholinesterases, which inhibit the cholinesterase enzymes that are essential for the normal functioning of the nervous system). A large proportion of mice (78% in October and 33% in April) disappeared immediately after methiocarb application. Work elsewhere has shown that populations depressed by methiocarb can rapidly (17-27 days) recover due to immigration of juveniles (Johnson *et al.*, 1991). Although there are no detectable long-term influences of methiocarb on wood mouse populations, sub-lethal exposure to anti-cholinesterase pesticides can cause short-term behavioural changes such as reduced movement, that might make the animals more susceptible to predation and stress (Dell'Omo & Shore, 1996ab).

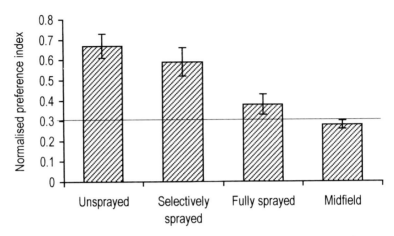

Figure 7:6 Normalised preference indices (mean ±SE) for different habitat types, for 12 wood mice radio-tracked at Wytham. Values above 0.3 indicate selection, values below 0.3 indicate avoidance. After Tew et al. (1992).

Wood mice are capable of differentiating habitats at very small scales, whether patches of weeds in crop or uncut patches of set-aside. Resource-rich areas can also be created by reduction of the pesticides applied to cereal field headlands. Headlands are the edge of the crop where the tractor turns; 'conservation headlands' where herbicides are applied selectively, were developed by the Game Conservancy. Tew et al. (1992) radio-tracked wood mice on 10 m by 20 m experimental headland plots subject to three herbicide spraying regimes: unsprayed, selectively sprayed (conservation headlands) and fully sprayed. They found that mice significantly preferred selectively sprayed and unsprayed headlands over sprayed headlands and mid-field areas (Figure 7:6). When all plots were sprayed, wood mice used mid-field and headland equally. Unsprayed headlands had statistically significantly higher densities of some weeds and invertebrates than sprayed headlands, and it was probably to this that the mice were responding.

Organic and conventional farms represent extreme ends of a continuum of farm management, farming aims and philosophy. Conventionally managed farms have high agrochemical inputs, and intensively manage cropped and non-cropped habitats, while organic farms have limited inputs and less intensive management (Stockdale et al., 2000). One implication of the lack of agro-chemical inputs in organic farming is that farmers cannot easily compensate for excess soil nitrogen losses over winter, and therefore cannot advance their growing season with winter-sown crops. As a result, summer sward height is significantly higher in conventional winter-sown wheat than organic spring-sown wheat. Field (1998) compared wood mouse abundances in 12 pairs of spring-sown organic and winter-sown conventional wheat fields in May and June 1998. Mouse abundances correlated positively with sward height (Figure 7:7), and densities were lower in organic than conventional fields, with an average of 1.3 per hectare in organic and 3.4 per hectare in conventional. Further work is needed to test whether sward height *per se* is important (for example higher swards might provide more protection from predators), or whether it is indicative of some other factor, such as increased food abundance.

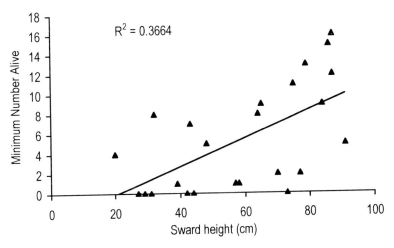

Figure 7:7 Sward height and wood mouse abundance. There was a positive relationship between sward height (cm) and wood mouse abundance (calculated as Minimum Number Alive). The trendline shows linear regression ($F_{1,22}$=12.7, p=0.002) for organic and conventional fields combined. From Macdonald et al. (2000a).

CONCLUSIONS

All habitats used by arable wood mice are strongly influenced by human activity, from crops that are sprayed with pesticides, to hedgerows that are trimmed, to field edges that are mown. Furthermore, farming activities such as harvest, along with crop growth stages, superimpose their own seasonality onto climatic seasons. Changes in farming policy and practice, such as hedgerow removal (Hooper, 1974) or the introduction of set-aside (Clarke, 1992), have potentially long-term effects on wood mice and other arable wildlife because they have altered the landscape itself. However, relatively minor changes in management, such as the timing of chemical inputs and mowing regimes (Smith & Macdonald, 1992; Feber et al., 1995, 1996), can manipulate floral and faunal diversity and abundance, creating subtle short- and long-term variations in habitat suitability.

Farming activities, and the farmed landscape, influence many aspects of wood mouse ecology and behaviour. Wood mice in crop fields show marked differences in behaviour and habitat use before and after harvest. After harvest, they drastically reduce their movements, and focus their activity along hedgerows, rather than ranging widely in the crop as they do before harvest. Range sizes in cropped fields are generally intermediate between woodland and sand dunes, perhaps reflecting food availability. Compared to their woodland counterparts, the population dynamics of farmland wood mice are not well understood, and their abundances are less predictable, both spatially and temporally. In general, however, densities in crop, margins and hedgerows are less than those in woodlots and woodlands. The relationship between habitat fragmentation, dispersal and population dynamics in arable landscapes is a potentially important aspect that remains to be addressed (e.g. Tattersall *et al.*, 2002).

However, some aspects of the wood mouse's behaviour and ecology are the same in arable fields as they are in woodland or sand dunes. Remarkably, their system of large breeding male ranges, overlapping each other and several smaller, discrete, breeding female ranges, persists across woodland, sand dunes and cropped fields, despite a ten-fold difference in range size. The wood mouse is the only small mammal found in any numbers in arable crops, which says much for the species' versatility. Its physical mobility and catholic diet are probably important elements of its ability to cope with the enormous changes to which farmland habitats are subject.

ACKNOWLEDGEMENTS

Parts of this chapter were written while FHT was employed at the Royal Agricultural College, Cirencester, and some of the work reviewed here was carried out on College Farms. We are particularly grateful for the help and support of the farm staff, and other members of the Royal Agricultural College including Will Manley and Barbara Hart. Our work on wood mice has been made possible by grants from MAFF and NERC.

8

The impact of changes in land-use on the Orkney vole

M L Gorman & P Reynolds

INTRODUCTION

The Orkney archipelago lies to the north of mainland Scotland, nearer to Norway than to London. These islands support a mammal that is found nowhere else in the British Isles, the Orkney vole *Microtus arvalis orcadensis*. The main purpose of this chapter is to investigate how populations of this unique rodent may have changed because of past and recent changes in agricultural practice within the Orkneys.

When the glaciers retreated at the end of the last ice age they left behind a barren group of islands, largely devoid of life. Mainland Britain was rapidly colonised by a range of mammals from Europe including humans and the field vole *Microtus agrestis* (Yalden, 1999). The field vole failed to cross the Pentland Firth to reach the Orkneys but, around 5,000 years ago, the common vole *Microtus arvalis* did arrive, probably as a stowaway on the boats of early Neolithic settlers coming from the area we now know as the Low Countries. The long association with man is confirmed by the remains of voles in the lowest levels of middens of the Neolithic village of Skara Brae dated at around 5,500 BP. Today, Orkney voles are to be found on the islands of Mainland, Rousay, Sanday, Westray and South Ronaldsay. Orkney voles are relatively big animals, with large males weighing up to 90 g compared to 35 g for Continental common voles. They are important, not only in their own right as an endemic sub-species, but also because as the only diurnally active rodent in the islands they are a major prey for diurnal raptors.

The Orkney vole has been associated with man from the Stone Age to the space age, and over that long period of time it has witnessed major changes in land-use, many of which may have impacted on its population density. In this chapter, we concentrate on changes on Orkney Mainland from the mid-1800s up to the present time, a period for which changes are well quantified. Over that period the total area used for agriculture has increased from 40% of the land surface to almost 90% while at the same time it has become much more intensive in nature (Table 8:1). Our specific objectives are: (i) to estimate the approximate size of the contemporary population of voles; (ii) to determine how that number might have changed over the last 150 years; and (iii) to speculate on the importance of any such changes, particularly for birds of prey.

METHODS AND DATA SETS

In order to calculate the current size of the vole population we need estimates of densities in different habitat types and of the areas covered by these habitats.

Vole densities

We collected density data for 14 different habitats on the island of Mainland using grids of snap-traps or lines of traps in linear habitats. Each habitat was trapped for five days, four times per year, for three years (Reynolds, 1992). Densities were estimated from an analysis of the decrease in catch per unit effort (Krebs, 1989).

Voles were present in a wide variety of habitats but there were marked differences in density from one habitat to the next. As an example, Figure 8:1 shows August densities averaged over the years 1988 to 1990 inclusive. Maximum densities of 200-500 voles per hectare were found in old peat cuttings, rough grassland and in linear features such as grassy fencelines, drainage ditches and roadside verges. At the other end of the spectrum, voles were completely absent from intensively farmed areas including cereals and root crops and even re-seeded grassland that might appear, at first sight, to be ideal habitat. Within agricultural areas, voles were confined to rough grass habitats along ditches and fence-lines where they were living at very high densities. Radio tracking of voles within agricultural areas confirmed that they rarely, if ever, left the fencelines, verges or ditches in order to forage in adjacent fields (Reynolds, 1992).

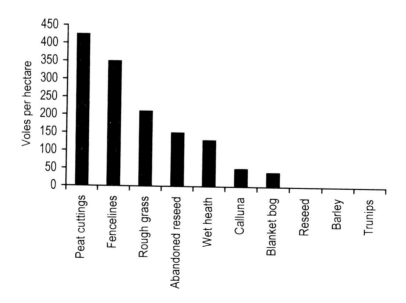

Figure 8:1 Vole densities in various habitats on Orkney Mainland. From Reynolds (1992).

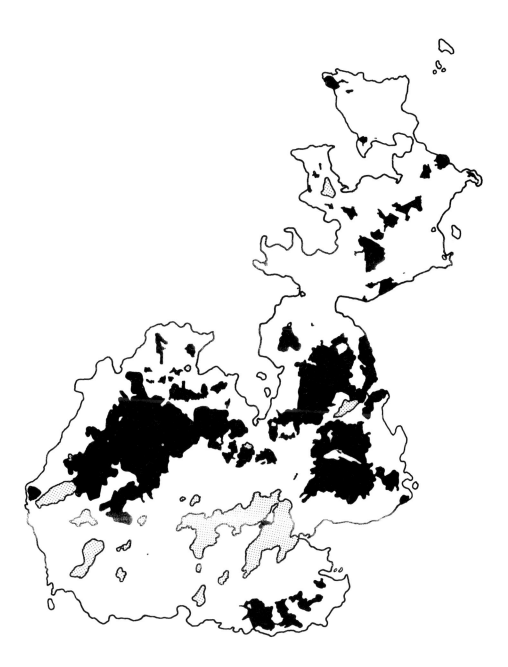

Figuere 8:2 Extent of intensive agriculture (white) on Orkney Mainland. Natural and semi natural habitats are shown in black and open freshwater in grey. Derived from the Land Cover For Scotland (MLURI, 1988).

Changes in land use

We have used three major sources of data on changes in land use. First, we used the Land Cover for Scotland (MLURI, 1988) to calculate the current areas of the Orkney Islands devoted to agriculture and the proportion that is under intensive cropping (Figure 8:2). Second, we used the Scottish Parish Records of Agricultural Statistics. We have used these to calculate the areas of Orkney Mainland devoted to different forms of agriculture in 1866, 1901, 1936 and 1998 (Table 8:1). And third, we used Land-Line Plus 1:2,500 Ordnance Survey maps (as ARCINFO coverages) to measure the total lengths of linear features (ditches, fence-lines and verges) in 35 one kilometre square quadrats randomly chosen from agricultural areas on Mainland. We assumed that, on average, suitable vole habitat extended 0.5 metres to each side of the linear feature. The mean length of linear features per square kilometre was 21.1 km with 95% confidence limits of 19.3-22.8 km.

Table 8:1 Changes in the area (hectares) of Orkney Mainland devoted to agriculture between 1866 and 1998. The areas identified as 'intensive' exclude rough grazing.

	1866	1901	1936	1998
Cereals	6,843	8,244	6,517	2,989
Roots	2,601	3,541	2,718	384
Other	161	81	13	134
Total grass	6,771	10,675	15,284	27,562
Rough grazing	4,000	8,973	12,994	13,046
Intensive	16,376	22,540	24,531	31,069
% Of Mainland	31.7	43.4	47.3	59.9
Total	20,376	31,513	37,525	44,115
% Of Mainland	39.3	60.7	72.3	85.0

VOLE POPULATION DENSITIES

The current population

Putting together the data sets on vole densities in different habitats and land use based on the Land Cover for Scotland we produced an estimate of the size of the contemporary population of voles in August 1990, at the height of the breeding season (Table 8:2). There is no evidence of population cycling in the Orkney vole (Reynolds, 1992) and these estimates for the annual breeding population are probably typical. We have assumed that habitat-specific vole densities are the same on all islands.

Table 8:2 Estimated population sizes of voles on the Orkney Islands for August 1990.
Voles living in linear features are not included.

Mainland	3,025,133
Rousay	457,137
Sanday	110,191
Westray	319,699
South Ronaldsay	217,651
Total	4,129,811

The estimates in Table 8:2 are based on the assumption that voles are totally absent from intensively farmed areas. In fact, although they are absent from much of the intensively farmed areas, voles do thrive at high densities (Figure 8:1) along the extensive and complex network of linear habitats that permeates the agricultural landscape. We estimate that the late summer population living in these linear features on Orkney Mainland could be as high as 250,000 voles, representing some 8% of all the voles on that island.

Historical changes in vole populations

Apart from the important linear habitats, the contemporary agricultural landscape is largely a 'vole-free zone'. Was it always thus? Probably not. In the early years of the twentieth century and as late as the 1940s, voles were present, and probably abundant, in agricultural land. For example, Millais (1904) noted that voles were responsible for considerable crop damage and that many were killed during haymaking. Again, Hewson (1948) found nests and trapped voles in pastures and arable land, and found that cultivated land supported even higher densities than did heather moorland. Although this evidence is anecdotal, it does suggest that voles were once more abundant within the agricultural landscape than they are at present.

It appears, therefore, that there have been major changes in the distribution of voles over the last 50 years, with voles disappearing from cropped areas, including managed grasslands. The reasons for such changes are probably due to changes in cropping regimes, particularly grassland management, and to increases in stocking levels of sheep and cattle. In earlier times, hay meadows and the growing of oats dominated Orkney farming. Today, much of the agricultural land is devoted to silage production, a practice that, several times per year, reduces the sward height to a level that voles cannot tolerate. At the same time, there has been a dramatic increase in stocking densities, of both cattle and sheep (Figure 8:3), with a resultant shortening of the sward to the detriment of voles.

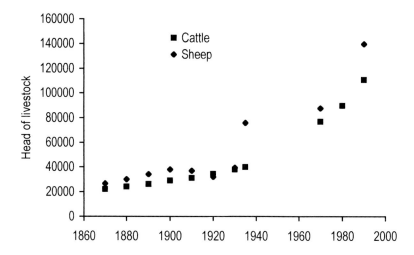

Figure 8:3 Stocking rates of cattle and sheep for Orkney Mainland.
Based on The Scottish Parish Records.

Vole numbers have clearly decreased in size over the last century and a half, but by how much, and when? In trying to quantify the likely losses, two factors need to be considered: the loss of natural habitats to agriculture and the loss of voles from agricultural land.

Reclamation of land has led to considerable losses of natural habitats. For example, between 1932 and 1985, some 45% of the moorland on Mainland was reclaimed with a loss of around 100,000 voles. Similarly, between 1948 and 1968, 11% of wetland habitats were lost. Overall, natural habitats on Mainland have declined from 35,000 ha to less than 20,000 ha since the 1860s, whilst at the same time there has been a corresponding increase in the area devoted to intensive agriculture (Figure 8:4). From the voles'

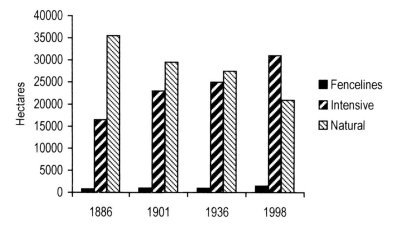

Figure 8:4 Historical changes in the areas of natural habitats, intensive agriculture and fencelines and other linear features on Orkney Mainland. Based on The Scottish Parish Records, Land Cover For Scotland and the Ordnance Survey.

perspective, the only good news is that rough grazing and fencelines and other linear features have increased in tandem with the area under agriculture.

We constructed a simple spreadsheet model to describe how the vole population may have changed on Mainland against this backdrop of changes in the nature of the landscape over the last 150 years. Using the model we can, for example, consider temporal changes in the likely size of the population in August. In doing so we have assumed that:

- The average density in natural habitats, including rough grazing, has remained unchanged at the present value of about 150 voles per ha.

- Voles have always lived at a density of around 350 per hectare along linear features.

- Although voles are now absent from agricultural areas (apart from those associated with linear features), they were present, at one density or another, up to the 1940s. We have run the model with densities in agricultural areas ranging from zero, the current value, to 150 voles per hectare, the average value in contemporary natural habitats.

The predicted losses of voles are shown in Figure 8:5. At one extreme, if voles were formerly present in agricultural areas at a density of 150 per hectare, then the model predicts no losses until the late 1940s. Thereafter, they fell from a population of around 8 million to the current 3.5 million, a 50% loss in 50 years or less. At the other extreme, if voles were already absent from agricultural land in the 1860s then the loss due to reclamation has been more long term and has involved a drop from 5.5 to 3.5 million voles over 150 years. The true scale and temporal pattern of loss probably falls somewhere between these two extreme scenarios.

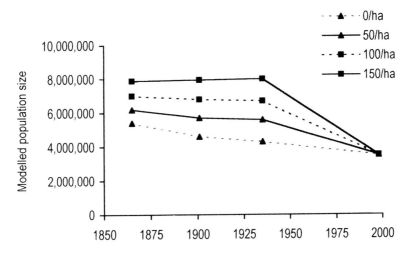

Figure 8:5 Estimated changes in the August population of voles on Orkney mainland as a function of vole densities (voles per hectare) in areas under intensive agriculture. See text for further details.

THE IMPORTANCE OF LOSSES OF ORKNEY VOLES

In terms of survival of the vole as an important endemic member of the fauna of the British Isles, the losses over the last 150 years are probably unimportant; the current population is sufficiently large as to ensure its indefinite survival. However, there is cause for concern when one considers that the vole is the only diurnal rodent on Orkney and of central importance in the diets of nationally important populations of short-eared owls *Asio flammeus*, ground-nesting kestrels *Falco tinnunculus* and hen harriers *Circus cyaneus*. Thus, on Mainland in 1988, voles accounted for 69% of prey items brought to nests by kestrels (one nest), 15% for hen harriers (five nests) and 100% for short-eared owls (19 nests).

Given the importance of voles in the diets of these raptors, is there any evidence that the decline in the vole population has been deleterious for them? The breeding population of the kestrel has certainly decreased in abundance in recent times; for example, an area in west Mainland that supported 19 pairs in the 1950s now supports less than half that number. There is good evidence for a similar decline in hen harriers. For example, the numbers of breeding hen harriers in West Mainland are estimated to have declined by two thirds since their peak in the 1970s (Meek *et al.*, 1998). It is also ominous that breeding output, as measured by the numbers of young fledged per nest, declined from around 1.3 in the 1950s to less than 0.9 by the 1980s (Figure 8:6). During 1996 and 1997, 25% of females which built nests did not lay eggs, while of the nests with eggs 60% failed to produce fledged young (Meek *et al.*, 1998). Unfortunately, the available data on past numbers of short-eared owls, the species most dependent on voles, do not allow an analysis of changes in numbers or productivity.

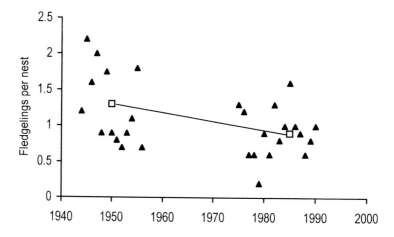

Figure 8:6 Changes in the fledging success of hen harriers on Orkney Mainland. The filled triangles represent the means of the data from the 1940s and 1950s and from the 1970s, 1980s and 1990s. The line is drawn through the two means. Sources of data: 1944 to 1956 (Balfour, 1957); 1976 to 1981 (Picozzi, 1984); 1982 to 1990 (RSPB, personal communication).

CONCLUSIONS AND RECOMMENDATIONS

If one accepts that the vole should be conserved, both in its own right and as a prey item for avian predators, how might this be best achieved? The current vole population is highly fragmented in remaining patches of natural and semi-natural habitat, many of which are small. Within the intensively managed agricultural areas that now dominate, Orkney voles are largely confined to linear habitats consisting of rough-grass tall-herb vegetation associated with fence-lines, ditches and road verges. These linear habitats not only support high densities of voles but they also act as habitat corridors allowing the flow of voles between otherwise isolated patches of natural habitat.

Given this complex situation, site-specific conservation measures such as Site of Special Scientific Interest designation are unlikely to be particularly helpful to the vole. What is required is a more holistic approach to land use and conservation as embodied in the concept of Environmentally Sensitive Farming (RSPB, 1991) in which less intensive management tailored towards the needs of environmental conservation is encouraged and rewarded. All areas of natural and semi-natural vegetation should, as far as possible, be retained. However, of prime importance at the large scale will be the sympathetic management of the complex network of fencelines, ditches and road verges that act as a linear nature reserve within the agricultural heartland.

If we are careful, and creative, in our stewardship of the Orcadian landscape then there is no reason why the ancient association between vole and man should not last for another five millennia.

9

Effects of farming practice and wildlife management on small mustelid carnivores

R A McDonald & J D S Birks

INTRODUCTION

Four species of small mustelid are native to Great Britain: weasels *Mustela nivalis*, stoats *M. erminea*, polecats *M. putorius* and pine martens *Martes martes*. The ecology of the two larger native mustelids, otters *Lutra lutra* and badgers *Meles meles*, are sufficiently distinct from the species considered here that they are dealt with separately in this book (Strachan, Moorhouse & Macdonald, Chapter 5; Woodroffe *et al.*, Chapter 11; Harris & White, Chapter 12; Delahay *et al.*, Chapter 13). Feral American mink *Mustela vison* are also common and widespread in Britain but are not included in this chapter because they are not native and represent a conservation threat in most situations. The four native species of small mustelid have much in common. They are all predators of small mammalian prey. They are sexually dimorphic and all have the elongated body shape characteristic of the Mustelidae and its associated high energetic demands (Brown & Lasiewski, 1972). They are highly mobile and generally live at low density relative to their body size. They also have conflicts with certain human interest groups, most notably with game conservation (Tapper, 1992; McDonald & Murphy, 2000).

While these common aspects of their biology and ecology have similar implications for their conservation and management, there are several contrasts between them at an ecological level and in the effects of human actions on their status. Stoats and weasels are abundant and widespread, while polecats and pine martens are relatively rare and restricted in range. Stoats and weasels are not protected under conservation legislation while pine martens are subject to protection under Schedule V and polecats under Schedule VI of the Wildlife and Countryside Act (1981) and subsequent amendments. All four species, to some extent, have been the focus of popular conservation concern. However, none of them is currently in steep decline and there is no current threat of extinction in their main ranges. Because of these comparisons and contrasts within the group and the range of problems they face, mustelids provide a useful set of case studies for conservation and management of British mammals on farmland.

In this chapter, we consider first the status of small mustelids in Great Britain. As their ranges change, each species will encounter different habitats and land management regimes, and we outline the likely effects of these range alterations. We examine the various ways in which management of farmland may affect mustelid populations.

Conservation & Conflict
ISBN 1 84103 001 5

In particular, we describe the effects of wildlife management and the perception of mustelids as pests and efforts towards their control. The control of other farmland mammals, notably rats *Rattus norvegicus* (see Cowan & Quy, Chapter 14) and rabbits *Oryctolagus cuniculus* (Dendy *et al.*, Chapter 17; Trout, Chapter 18) may also have an impact on their mustelid predators. Throughout this chapter, we outline potential concerns for the future status of mustelids in Britain and highlight areas for future work on these species.

CHANGING STATUS AND DISTRIBUTION

Based upon their status and distribution, the four native small mustelids in Great Britain fall into two categories. Stoats and weasels are common and widespread, and population estimates suggest that they are Britain's most abundant native carnivores (Harris *et al.*, 1995; Table 9:1). Conversely, pine martens and polecats have restricted ranges and relatively small populations (Langley & Yalden, 1977; Harris *et al.*, 1995).

Table 9:1 Estimated pre-breeding population sizes of small mustelids in Great Britain. Foxes and badgers are included for comparison. Data are from Harris et al. *(1995). Range is number of 10 km squares in which species has been recorded since 1960 (after Arnold, 1993). Reliability rating is on a scale of 1=most confident to 5=least confident. Changing status: + suggestions of increase; – suggestions of decline.*

Species	England	Wales	Scotland	Total	Range	Status	Reliability
Weasel	308,000	36,000	106,000	450,000	1,490	–	4
Stoat	245,000	37,000	180,000	462,000	1,514	–	4
Polecat	2,500	12,500	0	15,000	235	+	3
Pine marten	< 100	50	3,500	3,650	262	+	2
Badger	195,000	25,000	35,000	250,000	1,800	+	2
Fox	195,000	23,000	22,000	240,000	1,744	+	4

Changes in the status of weasels and stoats

Stoats and weasels are small, fast-moving and spend much of their time hunting underground (King, 1989). As such, they are hard to see and they leave no signs that can be used for indirect monitoring. Therefore, they are greatly under-recorded by recording schemes based on sightings (Arnold, 1993) and there has been no national survey of their populations. Macdonald *et al.* (1998) highlighted the problems with monitoring these species and recommended the use of live trapping and camera traps to estimate population densities and trends. However, these methods are expensive and have not yet been tested in the field, hence they suggested that trapping records maintained by gamekeepers were the best existing method of monitoring populations.

To date, the only information on trends in stoat and weasel populations is based on these trapping records (Tapper, 1992, 1999; Harris *et al.*, 1995). Tapper (1992) reported a continuous downward trend in weasel bags from the initiation of the National Game Bag Census in 1961 and in stoat bags since 1975. He believed this to be the result of a reversal of an earlier increase of weasel populations during myxomatosis (Sumption & Flowerdew, 1985) together with a general decrease in the amount of trapping effort made by gamekeepers (Tapper, 1992). McDonald & Harris (1999) found that differences in trapping effort between gamekeepers concentrating on wild and reared game, and a gradual increase in the importance of reared game, could have led to the apparent decline in stoat and weasel records. McDonald & Harris (1999) recommended that game bags could be used to monitor stoat and weasel populations, but only if effort was taken into account when examining trends. Their recommendations have yet to be implemented and so, in the meantime, we have considered that there are no national trends in stoat and weasel populations, but have discussed how management may affect populations at a local or temporary level.

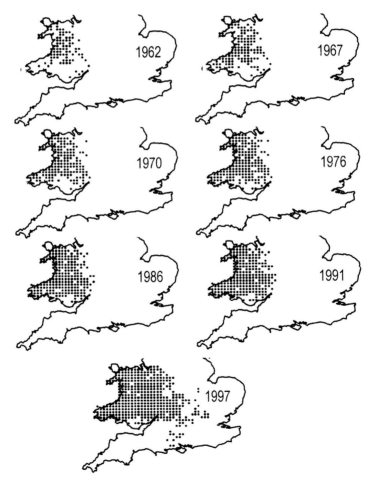

Figure 9:1 Changing distribution of polecats in western Britain, 1962 to 1997. Key to surveys: 1962 (Walton, 1964), 1967 (Walton, 1968), 1970 (Corbet, 1971), 1976 (Arnold, 1978), 1986 (Blandford, 1987), 1991 (Blandford, 1987; Arnold, 1993) and 1997 (Birks & Kitchener, 1999).

Changes in the status of polecats

At their nadir in the early twentieth century, polecats occupied an area of just 70 km radius centred on Aberdovey in north-west Wales (Langley & Yalden, 1977; Harris *et al.*, 1995). This area is characterised by sparsely populated marginal uplands dominated by sheep farming. Significant reductions in persecution pressure following the First World War (1914 to 1918) led to a recovery in polecat numbers and range, which has continued to the present day (Birks & Kitchener, 1999). This range expansion has been mapped since the first structured survey in 1959 to 1962 (Walton, 1964) and has shown the consolidation of the core Welsh range followed by recolonisation of the English Midlands (Figure 9:1). Unrecorded reintroductions of polecats to northern and southern England and western Scotland, together with confusion between polecats and polecat-ferret hybrids, have probably contributed to conservative distribution mapping in the past. Nevertheless, there is evidence of a consistent increase in the rate of range expansion to 4.3 km per year between the mid-1980s and mid-1990s (Birks & Kitchener, 1999). This can be attributed either to the recovery in numbers of rabbits, following the effects of myxomatosis, or to improved performance of polecat populations in the more productive lowland landscapes east of the Welsh border (Birks & Kitchener, 1999).

New conservation issues may arise as the polecat's range expands from a sparsely populated marginal upland refuge into lowland landscapes where agriculture is more intensive (Birks, 2000). Using the Institute of Terrestrial Ecology's (ITE, now Centre for Ecology and Hydrology CEH) land classification scheme, it is possible to track the initial increase in polecat occupation of pastoral landscapes as it expanded its range within Wales up to 1991 (Figure 9:2). Subsequent expansion through the English Midlands is reflected in increased occupation of lowland landscapes dominated by arable farming. The limited occurrence of upland landscapes, appearing for the first

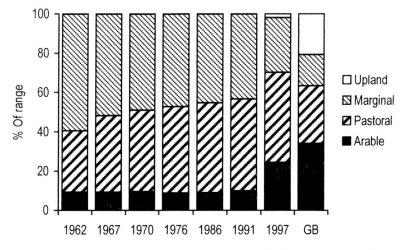

Figure 9:2 Changes in relative importance of land class groups (landscape types) as polecat range has expanded. The last column represents the landscape composition of the whole of Great Britain. For a key to the surveys, see Figure 9:1.

time in the 1997 distribution, coincides with the description of previously unmapped polecat populations in Cumbria and Argyll. The general pattern, however, is one of intensive agriculture occupying an increasingly large part of the polecat's British range.

Changes in the status of pine martens

Like polecats, pine martens suffered a catastrophic decline due to persecution by humans in the twentieth century. By the 1914 to 1918 war, pine martens survived mainly in a stronghold in the north-west of Scotland, with smaller, isolated populations persisting in the remoter parts of northern England and Wales (Langley & Yalden, 1977). By the 1940s, there was some evidence of recovery in Scotland, and subsequent decades revealed a slow expansion within and beyond the North-west Highlands (Harris *et al.*, 1995). Some early distribution maps undoubtedly underestimated the pine marten's Scottish range (e.g. Corbet, 1971). Since the early 1980s, however, distribution mapping has tracked their range expansion more accurately (Figure 9:3). The latest Scottish survey, completed in 1994, confirmed that pine martens are no longer confined to the Highlands (Balharry *et al.*, 1996). They are now also present in parts of Grampian, Tayside, Central and Strathclyde regions. In addition, animals released into Galloway Forest Park in the early 1980s have re-established a small population in south-west Scotland (Shaw & Livingstone, 1992).

Unlike polecats, pine martens have not yet expanded their range into lowland landscapes (Figure 9:4), despite a reputation for long distance dispersal movements. This may be an effect of greater persecution in the lowlands (Balharry *et al.*, 1996), or a consequence of habitat constraints. The densely populated Scottish central lowland belt is also likely to be a significant barrier to dispersal of pine martens to southern Scotland. There is little, if any, evidence of recovery among the fragmented pine marten populations in England and Wales. Evidence of their persistence is sparse and difficult to gather, and recent surveys have drawn conflicting conclusions about their status (Bright & Harris, 1994; McDonald *et al.*, 1994; Strachan *et al.*, 1996; Messenger *et al.*, 1997). Future surveys and studies of pine marten ecology at the fringes of their range, especially studies of the small numbers persisting outside of Scotland, will prove useful in assessing how pine martens react to encounters with more intensive management regimes.

AGRICULTURAL INTENSIFICATION AND HABITAT LOSS

Given the available information about the current ranges of the four species, what recent changes in land use and habitat may have had an impact upon them? Using the data collected during the Countryside Surveys of 1984 and 1990, processed using the Countryside Information System (CIS; from CEH, Monks Wood, Huntingdon, Cambridgeshire, UK), we examined the largest changes in habitat in each of the species' ranges between the two surveys. Overall, there were surprisingly few major changes in landscape cover (Table 9:2). The most noticeable process in this period was a loss of hedgerows, stone walls and banks, which are valuable to farmland wildlife as sources

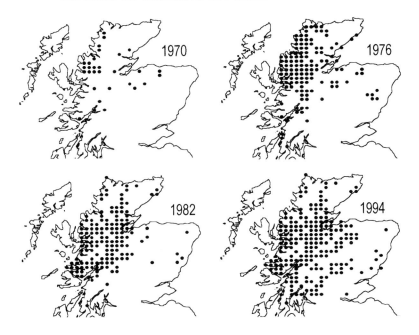

Figure 9:3 The changing distribution of pine martens in northern Scotland, 1970 to 1994.
Key to surveys: 1970 (Corbet, 1971), 1976 and 1982 (Velander, 1983), 1994 (Balharry et al., 1996,
added to Velander, 1983).

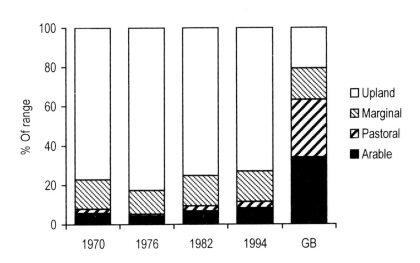

Figure 9:4 Changes in relative importance of land class groups (landscape types) as pine marten range
has expanded.The last column represents the landscape composition of the whole of Great Britain.
For a key to the surveys see Figure 9:3.

of food and cover. Linear features that are rich in food and cover have been largely replaced by fencelines and relict hedges, which provide little cover for either predator or prey.

Table 9:2 Major changes in the estimated percentage cover of habitat types and length of linear boundaries in the ranges of small mustelids in Great Britain between the Countryside Surveys of 1984 and 1990.

Species	Increases		Decreases	
Weasel and Stoat:	Fence	5.9	Barley	-2.8
	Weedy swards	2.1	Hedge	-2.6
	Non-cropped arable	1.2	Fenced hedge	-2.1
	Fenced relict hedge	1.1	Managed grass	-1.9
	Open canopy heath	0.7	Recently sown grass	-1.6
	Oilseed rape	0.7	Unclassified	-1.0
	Pure rye-grass	0.6	Other root crops	-0.9
	Potatoes	0.6	Wall	-0.7
	Grass strip	0.5	Fenced wall	-0.6
	Conifer woodland	0.5	Hedge and bank	-0.6
Polecat:	Fence	6.3	Managed grass	-4.3
	Weedy swards	3.4	Recently sown grass	-2.6
	Fenced relict hedge	1.5	Fenced hedge	-2.4
	Open canopy heath	1.3	Fenced hedge& bank	-1.9
	Pure rye-grass	0.9	Hedge	-1.9
	Oilseed rape	0.8	Barley	-1.6
	Potatoes	0.6	Wall	-1.5
	Fenced wall	0.6	Unclassified bound.	-1.0
	Wheat	0.6	Other root crops	-0.8
	Non-cropped arable	0.6	Upland grass	-0.6
Pine marten:	Fence	9.0	Fenced wall	-2.7
	Fenced relict hedge	1.3	Fenced hedge	-2.2
	Wet heaths and bogs	0.8	Fenced hedge & bank	-1.3
	Weedy swards	0.7	Barley	-1.2
	Open canopy heath	0.7	Hedge	-1.2
	Wheat	0.6	Fenced bank	-0.8
	Wetland	0.3	Recently sown grass	-0.7
	Felled woodland	0.3	Moorland grass	-0.7
	Mixed woodland	0.3	Wall	-0.7
	Grass strip	0.2	Dense bracken	-0.6

Habitat changes, such as the loss of linear features, can have a direct impact on mustelid populations by reducing food supply. Small mustelids have evolved as predators of small mammalian prey (King, 1989), and in intensively farmed landscapes, hedgerows and stone walls form refuges of semi-natural habitat for small rodents and rabbits. Agricultural intensification may also lead to greater spatial and temporal heterogeneity in prey populations, because of reduced dispersal between sources and sinks. The low energetic reserves of mustelids suggest that increased heterogeneity in food supply may lead to reduced survival and productivity (King, 1989). Behavioural mechanisms, such as prey switching (Stephens & Krebs, 1986), surplus killing and prey caching (Oksanen *et al.,* 1985), can buffer individuals against unpredictable food supplies. However, while prey switching may facilitate survival of individuals, in many cases it does not appear to provide sufficient energy for successful breeding. For example, weasels fail to breed if field vole *Microtus agrestis* densities fall below 10-15 per ha, even if the availability of alternative prey remains high (Tapper, 1979; King, 1980a). Weasels are the shortest-lived and most specialist feeders of this group, and this may explain the pattern of recurring extinction and recolonisation in many weasel populations (King, 1980b). In contrast to weasels, pine martens have a more varied diet (Balharry, 1993; Halliwell, 1997) and are longer-lived. Thus, individual pine martens can switch prey to improve survival, whilst delaying breeding until adequate food supplies become available.

Less intuitively, the loss of hedgerows and walls will reduce cover and safe dispersal routes, and may increase mustelids' vulnerability to predation by larger predators (King, 1989). In this case, increased predation will reduce rates of recolonisation following local extinctions in stoat and weasel populations and reduce the rate of range expansion in polecat and pine marten populations. The loss of semi-natural habitats such as hedgerows is also part of a wider process of habitat fragmentation arising from agricultural intensification. Bright (1993) provided a framework for considering the sensitivity of British mammals to fragmentation, based on a Principal Component Analysis of life history variables. He suggested that pine martens were among the most sensitive because of their low rate of population increase, low density, high range, intolerance of matrix habitats and specific habitat requirements. By comparison, polecats, stoats and weasels were relatively insensitive to habitat fragmentation because of high rates of population increase, tolerance of matrix habitats and general habitat requirements. While Bright (1993) provided a unique consideration of the potential effects of habitat fragmentation, his description of each species was somewhat subjective and his estimates of mustelid population density were inaccurate by a factor of 100. Furthermore, while stoats and weasels have indeed been recorded in a wide range of habitats, breeding is not likely to occur every year in many of these, since food supplies are low in many habitat types. Given the importance of recolonisation to fluctuating considerable stoat and weasel populations, and the current range expansion in pine martens and polecats, habitat fragmentation may have considerable effects on mustelid populations.

FARMING AND GAME MANAGEMENT INTERESTS

The management of farmland for game brings both costs and benefits for small mustelids. The habitat characteristics and management systems of game estates compare favourably with those on farms not managed for shooting (Macdonald & Johnson, 2000; Macdonald & Johnson, Chapter 1). Conversely, farms with a local game shooting interest are more likely to cull small mustelids on their land (Packer & Birks, 1999; and see below). Interest in maintaining a rich network of cover and feeding opportunities for game provides a direct incentive to farmers and landowners to conserve and create hedgerows, woodlands and wetlands. Moreover, intensive farming practices may be modified to benefit game, by reducing chemical inputs to crop edges ('conservation headlands'), by creating 'beetle banks' and by selecting set-aside options. In respect of such sympathetic habitat management and reduced chemical inputs, 'best-practice' game estates may be regarded as being similar to organic farms (Lampkin, 1990; Wilson *et al.*, 1997; Chamberlain *et al.*, 1999). The benefits for a range of farmland wildlife of measures taken to enhance game populations have been demonstrated through the Game Conservancy Trust's Loddington project (Tapper, 1999; see also Oliver-Bellasis & Sotherton, Chapter 3).

Fox control

The control of foxes *Vulpes vulpes* is presently a fundamental part of the practical management of game in Britain (Tapper, 1992, 1999). A 1997 survey of 215 members of the National Gamekeepers Organisation in Great Britain, described in detail by McDonald & Harris (1999, 2000), found that 99% practised some form of fox control. Foxes are also controlled on farmland for the protection of lambs and are hunted with dogs in other areas (Macdonald & Johnson, 1996). The effectiveness of fox control is a controversial point (see Macdonald & Johnson, Chapter 1; Macdonald *et al.*, Chapter 18), but considered in isolation there are two reasons to suggest that large scale fox control may have beneficial side effects for small mustelids.

First, small mustelids are vulnerable to predation from other carnivores (Palomares & Caro, 1999). Foxes have been found to kill or influence populations of weasels, stoats and polecats (Latham, 1952; Mulder, 1990) and particularly pine martens (Storch *et al.*, 1990; Lindström *et al.*, 1995; Helldin, 1998; Bright, 2000; Webster, 2001). However, a recent study has suggested that there is no reciprocity in the distribution of foxes and pine martens and that increased levels of habitat fragmentation have not augmented levels of fox predation on pine martens (Kurki *et al.*, 1998). Cause and effect between fox control and mustelid abundance has not been tested experimentally. However, any effect of fox control on mustelid abundance would require large sample sizes in order to be distinguished statistically from other factors, such as improved habitat quality, that are characteristic of farmland managed for game. Furthermore, even if the removal of foxes could be demonstrated to have an effect on mustelid abundance, management of one carnivore to enhance populations of others is unlikely to be a socially acceptable tool for conservation.

Second, reduction of fox abundance may enhance populations of prey for other predators by decreasing competition for food. Rabbit abundance tends to be higher in areas with high levels of fox control (Trout & Tittensor, 1989). Similarly in Australia, fox control can increase rabbit population growth rates , particularly when populations have been reduced by other factors such as drought (Pech *et al.*, 1992; Banks *et al.*, 1998). The impact of foxes on small rodent populations is far from clear (Lloyd, 1980). Although, Furlong (1999) detected no effect of fox exclusion on small mammal populations, dietary studies of foxes indicate that the impact of fox predation may be substantial (Baker & Harris, Chapter 10). As long as the ecological impact of fox predation on populations of rabbits and rodents is unclear, the effect of competitive release on mustelids remains a further level of complexity to resolve.

Perception of small mustelids as pests

Lethal predator control is an integral component of modern game management (Tapper, 1992, 1999) that may counter any benefits to small mustelids on farms where game is a main interest. Historically, predator culling by gamekeepers had a major and long-lasting effect upon the abundance and distribution of pine martens, polecats and wild cats *Felis silvestris* in Britain (Langley & Yalden, 1977). Presently, reduced numbers of gamekeepers, legal constraints and the emphasis on predation control rather than predator eradication (Tapper, 1992, 1999), mean that pressure on mustelid populations is lower (see below). Nevertheless, recent surveys have revealed that most gamekeepers still perceive small mustelids as pests, albeit less serious than foxes, because of the threat they pose to game (Balharry *et al.*, 1996; McDonald, 1998; Packer & Birks, 1999; Figure 9:5). A survey of farmers' attitudes towards polecats revealed some recognition that they contribute to the control of rodent pests, but this view was outweighed by concerns about their impact upon game on the farm (Packer & Birks, 1999).

While stoats and weasels are legal targets for gamekeepers, polecats and pine martens are protected against deliberate capture or killing without a licence. However, the traps set for stoats and weasels in tunnels, usually Fenn Mark IV or Mark VI steel spring traps (A. Fenn & Co., Redditch), are also suitable for capturing polecats and pine martens. Therefore, polecats and, to a lesser extent, pine martens are likely to be occasionally caught in tunnel traps. This has led to calls for the development of grilles that can be placed in front of tunnels in order to exclude the two larger species (Short & Reynolds, 2001). Although the Wildlife and Countryside Act 1981 requires trappers to take reasonable precautions to prevent trapping protected species, accidental capture is widely regarded as a defence for taking them (Birks & Kitchener 1999, Packer & Birks 1999). However, this defence has not been tested in law and farmers and game-keepers may be vulnerable to prosecution if the precautions they took were deemed inadequate. Presently the mandatory use of an excluder grille by trappers is viewed as unacceptable because of an apparent reduction in trap success rate with large rats and squirrels *Sciurus carolinensis* and larger male stoats (Game Conservancy Trust, 2000). Non-lethal alternatives or additions to trapping programmes have been suggested. Electric fencing and closing pheasant access holes at night can effectively exclude pine martens and polecats from pheasant rearing pens (Balharry, 1998; Balharry & Macdonald, 1999; Packer & Birks, 1999). In a recent survey, rearing a greater number of birds to compensate for losses to predators would be adopted by only one of 203 English gamekeepers (McDonald & Harris, 1999), hence this is apparently not a viable alternative to predation control.

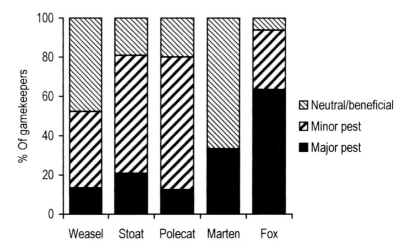

Figure 9:5 Perceptions of small mustelids and foxes as pests of game interests. Because of survey design the black column for pine martens also includes those who viewed pine martens as a minor pest. Sources: Weasels and stoats (McDonald & Harris, 1999), polecats (Packer & Birks, 1999) pine martens (Balharry et al., 1996) and foxes (National Gamekeepers' Organisation survey, R McDonald unpublished data)

Tunnel trapping is still part of a gamekeeper's job, though there are now clear differences in the trapping effort made by keepers working to protect wild game and those dependent on rearing and releasing birds. As rearing becomes more prevalent, fewer farmers and landowners require their gamekeepers to catch large numbers of stoats and weasels (McDonald & Harris, 1999). In England in 1997, a typical gamekeeper on a wild game estate ran 90 traps for nine months in order to catch 34 stoats and 28 weasels, while on a reared game estate a typical gamekeeper ran 40 traps for six months to catch 19 stoats and 11 weasels (McDonald & Harris, 1999). In 1996, without taking into account trapping effort, the Game Conservancy Trust's National Game Bag Census recorded a national average bag of 1.75 stoats and 0.86 weasels per square kilometre (S. Tapper, personal communication). Because of their protected status, polecats and pine martens are probably under-recorded by the National Game Bag Census. Packer & Birks (1999) reported that in 1995 a mean of 2.2 polecats were trapped on 22 farms and that over 90% of gamekeepers caught polecats, though only a few of these (11%) caught more than ten. Because of the pine marten's status as a fully protected Schedule V species, people are unlikely to admit to catching any. Nevertheless, any tunnel traps set in areas where pine martens are living are liable to capture them, unless excluder grilles are fitted.

While gamekeepers' bags of some mustelids can be substantial, the impact of trapping on the population growth rate of small mustelids is unclear. Langley & Yalden (1977) attributed the decline of polecats and pine martens in the nineteenth century to the control efforts made by gamekeepers. Packer & Birks (1999) considered that current levels of polecat trapping were not likely to reduce population growth, where it took place on isolated estates surrounded by untrapped areas. This is now the common pattern of game management, at least in the western part of Britain, but as polecats extend their range eastwards, game interest becomes more pronounced and trapping is conducted over larger contiguous regions (Tapper, 1992). Packer & Birks (1999) suggested that in conjunction with other risks inherent in modern agriculture, trapping might play a significant part in slowing the rate of expansion in polecat populations. In contrast, weasels and stoats have suffered no reduction in their ranges because of gamekeepers' efforts and this has been interpreted as evidence for their tolerance of trapping (King & Moors, 1979; King, 1980b; Tapper *et al.*, 1982).

McDonald & Harris (2002) took an analytical approach to a preliminary assessment of the impact of trapping and shooting on stoat and weasel populations using Leslie matrix models to simulate growth rates in closed populations. This analysis indicated that model stoat populations would decline under an average trapping regime, as a result of concerted trapping effort during spring when juveniles were still dependent on their mother's survival. In contrast, weasels maintained their population growth rate as the result of high productivity when sufficient prey was available. As the result of high rates of immigration, gamekeepers' control efforts were not thought to be capable of causing long term declines in real stoat and weasel populations (McDonald & Harris, 2002). Bright & Harris (1994) drew on demographic data from American marten *Martes americana* populations to develop a stochastic model of the effect of changes in vital demographic rates on the growth of hypothetically reintroduced pine marten populations. They found that growth was highly sensitive to adult survival and concluded that pine marten populations would be strongly affected by anthropogenic mortality (Bright & Harris, 1994). These analytical approaches to understanding the effects of anthropogenic mortality should be expanded, now that more accurate demographic data and analytical tools are readily available, and will certainly provide clearer guidance for management and conservation of small mustelids.

CONTROL OF FARMLAND PESTS

Although small mustelids are often considered pests in their own right, control directed towards other mammals has the potential to impact more significantly upon their populations than control directed at mustelids themselves. Anticoagulant rodenticides (Pelfrene, 1991) are used to control rodents on approximately 80% of farms in Britain (De'Ath *et al.*, 1999; Garthwaite *et al.*, 1999). Walton (1970) first identified the role that accidental poisoning with rodenticides might play in the mortality of polecats feeding on rats in farmyards, and subsequent studies confirmed that secondary exposure is common in this species (Shore *et al.*, 1996; Birks, 1998; Newton *et al.*, 1999; Shore *et al.*, 1999; see also Shore, Fletcher & Walker, Chapter 4). Most rodenticides in Britain are used in and around buildings, leading to assumptions that farm buildings are the main areas of risk for mustelids in agricultural landscapes (Birks, 1998). However, a significant proportion of rodenticides are also used in fields and farm woodlands, and there is evidence of outdoor misuse of those chemicals licensed only for use inside buildings (De'Ath *et al.*, 1999; Garthwaite *et al.*, 1999). This raises the possibility that a broader range of prey species might be contaminated with rodenticides over wider areas of farmland. McDonald *et al.* (1998) confirmed this scenario by finding that secondary exposure to rodenticides was commonplace in stoats and weasels in lowland Britain. This was despite the fact that stoats and weasels rarely eat target species such as rats, squirrels or house mice *Mus domesticus* but mainly eat non-target species such as birds, rabbits, wood mice *Apodemus sylvaticus* and voles (Day, 1968; McDonald *et al.*, 2000).

The toxicity of anticoagulant rodenticides has been demonstrated for weasels (Harradine, 1976; Townsend *et al.*, 1984), stoats (Grolleau *et al.*, 1989; Alterio, 1996; Alterio *et al.*, 1997; Brown *et al.*, 1998; Murphy *et al.*, 1998a), polecats (Alterio, 1996; Shore *et al.*, 1996; Birks, 1998) and pine martens (Lund & Rasmussen, 1986; Berny *et al.*, 1997). Despite this, no studies have yet provided any indication of the impact of widespread exposure to rodenticides on mustelid populations. Any such impact is likely to be greatest on intensively farmed land where the range of prey available to small mustelids is limited, and where they are more dependent upon the higher concentrations of rodents targeted by farmers.

Because of the cost of the damage they cause to agricultural crops and property (Edwards, 1997; Dendy *et al.*, Chapter 17), rabbits are widely controlled on farms and non-agricultural land (Trout, 1994; Trout, Chapter 16). Some control methods are likely to impact directly or indirectly upon small mustelids because of their dependence upon rabbits as prey and their burrows as resting sites. In Ireland, 21% of resting sites used by radio-tracked stoats were in rabbit burrows (Sleeman, 1990), while in England, radio-tracked polecats foraged frequently within rabbit warrens, and 80% of daytime resting sites were in warrens (Birks & Kitchener, 1999). Such heavy use of warrens presents a serious risk when rabbits are controlled by mechanical warren ripping, flooding or fumigation. Burrow fumigation is one of the most widely used methods of rabbit control (Trout, 1994; McDonald & Harris, 2000). Its consequences for small mustelid populations are difficult to assess because any subterranean mortality cannot be monitored. One of the oldest methods of rabbit control, ferreting, may also lead to an increase in the frequency of hybridisation between ferrets and wild polecats (Davison *et al.*, 1999).

CONCLUSIONS AND RECOMMENDATIONS

Through its profound effects upon wildlife habitats and sources of prey, agriculture inevitably exerts a powerful indirect influence upon small mustelid populations. Given the major changes in agricultural landscapes that have occurred since the mid-twentieth century, it is encouraging to note that two of the species considered in this chapter remain widespread, and one is currently expanding its range in farmed landscapes. However, there is little information on the ways in which small mustelid population density, survival, reproductive performance and dispersal are affected by changes in farm management. This seriously limits our understanding of how these species perform on farmland. Similarly, despite some advances through recent research, our understanding of the more direct effects of agricultural management, particularly pest control, upon small mustelid populations remains limited. This shortfall in knowledge, which owes much to the special challenges involved in studying these small and inconspicuous mammals, hinders attempts to design sound management and conservation programmes for them at both local and national scales.

In order to improve our knowledge of the status of small mustelid populations on farmland we recommend five areas as priorities for action. First, assessment of variations in small mustelid population densities in farmed landscapes should be improved; for example, McDonald & Harris's (1999) recommendations for recording trapping effort should be adopted. Second, how individuals use farmland and the factors critical to their successful dispersal, reproduction and survival, should be investigated using standard techniques of telemetry and survey, in conjunction with the application of Geographic Information Systems technology (Gough & Rushton, 2000). A third area for investigation is the interaction between small mustelids and foxes; this is contingent on the application of accurate survey methods for both foxes and mustelids, though in principle simulation and experimental methods could be applied to this question. Fourth, monitoring the exposure of individual mustelids and populations to pesticides, especially to rodenticides and fumigants, is necessary for assessing the population effects of lethal and sub-lethal exposure. Finally, analytical approaches to understanding the effects of anthropogenic mortality on small mustelid populations should be expanded and, where necessary, trapping practice should be modified to reduce the risk of accidental capture of protected species.

ACKNOWLEDGEMENTS

Robbie McDonald was supported by a scholarship from The Wingate Foundation. We are grateful to Will Manley and Fran Tattersall for inviting us to participate in the Farming & Mammals Symposium, to the National Gamekeepers' Organisation and the British Association for Shooting and Conservation for supporting questionnaire surveys of their members, to Liz Balharry for data from pine marten surveys and to James Packer for information and for the loan of slides. The maps were produced using DMAP.

10

A review of the diet of foxes in rural Britain and a preliminary assessment of their impact as a predator

P J Baker and S Harris

INTRODUCTION

The red fox *Vulpes vulpes* has arguably the greatest ecological and economic impact of all British carnivores. This is because it is widespread and abundant, it is large, implying a substantial daily food requirement, and it has no specialised dietary requirements so is able to exploit a diverse array of prey. Foxes are also able to respond to declines in the abundance of different prey species by switching to alternative prey, and therefore have the potential to reduce the abundance of individual prey species with no con-current reduction in their own numbers. This can bring them into conflict with people where their prey is of economic or conservation concern. Conversely, foxes may be beneficial to agricultural interests where prey species are pests. In the mid-1980s rabbits *Oryctolagus cuniculus* were estimated to cause over £100 million of damage annually (Mills, 1986). Since then rabbit numbers have approximately doubled (Harris *et al.*, 1995) suggesting that current levels of damage may be substantially higher (see also Trout, Chapter 16). A recent analysis showed that predator control was inversely related to rabbit abundance (Trout *et al.*, 2000).

Thus, the diet of foxes in different habitats, and the impact of foxes on their prey, are of considerable applied interest. Two inter-linked questions need to be considered: how important is the prey to the predator, and how important is the predator to the prey? The first assesses the extent of predation on a prey species, and is related to predator diet composition and the effect of prey abundance on predator numbers. The second relates to the impact of predation on the prey species, particularly the level of mortality exerted by the predator and whether this is additive to other forms of mortality. In this chapter, we synthesise information in the available literature to address these issues.

Experimental manipulations are the best way to quantify the impact of a predator but are costly and difficult to perform over large areas. They can also be confounded by several factors (Côté & Sutherland, 1997). A more equivocal approach is to assess the effect of an observed level of predation on the dynamics of a prey population of a given size. This requires data on the abundance and diet of the predator in relation to the abundance and dynamics of its prey, that is, fox abundance, fox diet and prey abundance should be recorded simultaneously. However, to date, this has only been done in a single study in Britain (Reynolds & Tapper, 1995b). Our approach, therefore, was to collate existing information on the patterns of prey consumption by rural foxes and to relate these to the estimated annual prey requirements for a typical fox group. We then assess the potential impact of fox predation on selected prey groups.

COLLATING AND ANALYSING THE DIETARY DATA

Twenty-seven published and one unpublished study listed the diet of rural adult foxes; one study, using feeding remains at dens (Hewson, 1985) was excluded. One study dealt only with cub diet (Kolb & Hewson, 1980a). For each study we recorded: the number of sites and diets; the materials used; sample sizes; the seasons in which data were gathered; whether the data were expressed as frequency of occurrence, percentage volume or percentage mass ingested; if authors had analysed faeces for earthworm chaetae; and whether authors had analysed the diet of cubs.

Many authors analysed their data according to the location from which samples were taken; the data for each location have been treated separately and are referred to as a 'diet'. Locations equivalent to a county or larger are termed 'regions'; smaller areas are termed 'sites'. As some sites were used in several studies, we refer to these diets as 'non-independent'. To avoid over-dependence on individual locations, data from the same site and using the same material were averaged. Diets treated in this way are referred to as 'independent'. Four landscape types were recognised: arable, pastoral, marginal upland and upland (Bunce et al., 1996). Each diet was assigned to one landscape.

Data were initally collated into 16 prey groups: small mammals (less than 0.1 kg), medium-sized mammals (0.1-5.0 kg), large mammals (more than 5 kg), galliformes, passerines, other birds, bird eggs, reptiles, amphibians, fish, fruit and berries, non-digestible plant remains, insects, earthworms, other invertebrates and miscellaneous. For analysis, eight prey groups were used: small mammals, medium-sized mammals, large mammals, birds, bird eggs, other vertebrates, digestible plant remains and invertebrates. Non-food items (non-digestible vegetation and miscellaneous categories) and earthworms were removed. For those diets that listed only 'total vegetation' we have assumed that this was non-digestible plant remains. Miscellaneous items were typically non-digestible remains such as earth, soil or litter, although some authors — e.g. Hewson & Leitch (1983) and Hewson (1990) — combined amphibians, insects and vegetation in their 'other' category. But these occurred infrequently (2-3%) and any errors are likely to be minimal.

The frequency of occurrence of earthworms was a major problem; only 19 out of 31 diets were analysed for the presence of earthworms. Consequently, we excluded them to increase comparability between studies. On average, earthworms accounted for only 5% of occurrences, although four studies recorded frequencies of 12-35%. Therefore, this is potentially a major source of error and may significantly underestimate the importance of invertebrates as a food source. It is also likely to reduce the frequency of occurrence of bird remains (Reynolds & Aebischer, 1991).

Three types of material were used: scats, stomach contents and feeding remains. Diets derived from more than one type of material are termed 'mixed samples'. Diets derived from the different material types are not directly comparable and so we have presented data derived from each type of material separately. Data for adults and cubs were treated separately. Six different measures were used: frequency of occurrence as a percentage of scats or stomachs examined; frequency of occurrence as a percentage of all occurrences of all prey items; the percentage frequency with which the prey type was the major component in each scat or stomach; percentage volume; percentage mass of prey in total mass of remains; or percentage mass ingested.

We used two standardised measures of diet composition (Fedriani & Travaini, 2000). The frequency of occurrence of each prey type was standardised as the percentage of all occurrences of all prey types. This enables dietary breadth to be calculated. However, it can be complicated by the accuracy to which authors analysed separate prey groups, and so we standardised the level to which authors analysed their data. For birds, mammals and other vertebrates, we combined the occurrences of all species or families included in that group (for example, large mammal occurrences would equal the sum of occurrences of deer and sheep). This assumes that foxes were unlikely to have eaten more than one species from any group in any meal. For invertebrates, we used the frequency of the most commonly recorded item; this may underestimate the true frequency of occurrence of invertebrate prey.

The second standardised measure of diet composition was based on the percentage mass of prey ingested. For diets expressed as the percentage volume of remains in scats, we divided a standard 100 cm³ of prey remains by the density figures given by Reynolds & Aebischer (1991). The figures were then multiplied by the conversion factors listed by Reynolds & Tapper (1995b) to give the mass of prey ingested. For diets listed as percentage mass of prey in scats, we multiplied a standard 100 g of prey by the same conversion factors. Diets listed as the percentage volume in stomach contents were divided by the density figures in Reynolds & Aebischer (1991). For those studies that presented data as mass of prey remains in stomachs, we have taken these data to indicate percentage mass consumed.

Seasons were defined as spring (March-May), summer (June-August), autumn (September-November) and winter (December-February). For studies that presented data by season, the overall diet was calculated as the average for the seasons listed: we also present data for the seasonal variation in adult diet within each landscape. Where only a total diet was given for each study, this has been used: these are potentially confounded by differences in the number of samples collected in different seasons. The diet of cubs was calculated for April to June when the cubs are dependent on the adults for food. Cub diets expressed as frequency of occurrence were treated as for adults. For those expressed as the bulk of prey consumed, the data were treated as for adults but with different conversion factors (Reynolds & Tapper, 1995b).

Dietary breadth was calculated using the standardised Levin's index and used to compare the diet of adult foxes between landscapes, between seasons and to analyse temporal changes. At least two major changes in prey abundance have occurred: an increase in the biomass of rabbits following myxomatosis (Harris *et al.*, 1995) and an increase in the number of pheasants *Phasianus colchicus* released for sport shooting (Tapper, 1992). If these changes had affected fox diet, we would expect to see a decrease in dietary breadth.

Sources of error

Pooling data across studies is problematical because of methodological differences, non-random sampling — particularly with small sample sizes and data not collected throughout the year — errors in conversion factors and differences in diets between sites. Each of these errors will be additive, and accounting for all types of error would generate extremely large and unrealistic confidence limits for the contribution of each prey type to the average diet for each landscape. The most parsimonious approach would be to use the single source of error that accounts for most variation. However, there are no data on the magnitude of most of these sources of error, so we assumed that dietary differences between sites reflect the greatest source of variation, and have used these to obtain maximum and minimum values for each prey category in each landscape.

THE DIET OF RURAL FOXES

Scope of the diet data

Twenty-seven published and one unpublished study listed the diet of rural adult foxes. The 28 studies listed 36 non-independent diets (Table 10:1): 20 were derived from scats, six from stomach contents, four from the separate analysis of scats and stomach contents, five from the combined analysis of scats and stomach contents, and one from the combined analysis of scats and feeding remains. On average, diets were derived from 288 samples collected over three seasons.

Three non-independent diets (8%) were derived from samples collected throughout Britain, 22 (61%) from Scotland, ten (28%) from England and one (3%) from Wales. Two diets (6%) used material collected prior to the outbreak of myxomatosis in 1953, nine (25%) between 1953 and 1970, 18 (50%) between 1970 and 1980 and seven (19%) since 1980 (Table 10:1). The diets from Britain were collected before or soon after the outbreak of myxomatosis. Most samples from Scotland (55%) and England (50%) were collected between 1970 and 1980. In Scotland, the majority of the remaining material was collected before 1970 (32%), and in England after 1980 (40%). Therefore, most studies were conducted in a period when rabbits were less abundant than they are currently.

The 36 non-independent diets comprised 28 independent diets from 18 sites and ten regions. Of these, seven were classified as arable (three in England, three in Scotland, one in Britain), four as pastoral (all in England), eight as marginal upland (five in Scotland, one in Wales, two in Britain) and nine as upland (one in England, eight in Scotland). In each country, at least one landscape had not been sampled; upland and

Table 10.1 Summary of the studies used to analyse the diet of adult foxes. Regions were defined as areas equivalent to a county or larger, while sites refer to smaller, more specific areas; sites are denoted by numbers, regions by letters. Habitats were: A=arable, P=Pastoral, M=marginal upland, U=upland. Materials were: S=scats, G=stomachs, F=feeding remains. Material separated by a comma denotes that the author analysed each material separately, material joined by an ampersand denotes that the author combined materials for analysis. Seasons were: Spring (SP), summer (SU), autumn (AU), winter (WI). Results were expressed as: F=frequency of occurrence in scats/stomachs, F2=frequency of occurrence as main item in scats/stomachs, O=percentage of all occurrences, I=percentage of mass ingested, M=percentage mass in total mass of remains.

Reference	Diet ID	Year	County	Country	Site or region label	Habitat	Material	Sample size	Seasons data collected	Results expressed as:
Baker et al. (unpubl. data)	1	1996-1997	Wiltshire	England	1	A	S	885	SP-WI	F, O, I
Burrows (1968)	2	1963-1966	Gloucestershire	England	2	P	S	?	SP-WI	F, V
Butler (1980)	3	1974-1978	Aberdeenshire	Scotland	3	A	S, G	225, 62	SP-WI	F, V, I
Douglas (1965)	4	1959-1961	Aberdeenshire	Scotland	4	U	G	15	SP	F
Frank (1979)	5	1973-1974	Aberdeenshire	Scotland	3	A	S & G	117+1	SP-WI	F, I
Hewson (1983)	6	1979	Highlands	Scotland	5	U	S	82	WI, SP	O
Hewson (1984)	7	1977-1978	Highlands	Scotland	5	U	S	309	SP-WI	F
	8	1976-1977	Highlands	Scotland	6	U	S	818	SP-WI	F
Hewson (1990)	9	1987-1990	Highlands	Scotland	7	U	S	209	SP-WI	F2
Hewson et al. (1975)	10	?	Aberdeenshire	Scotland	8	M	S	137	SP-WI	F
	11	?	Argyll	Scotland	9	M	S	386	SP-WI	F
Hewson & Leitch (1983)	12	1977-1978	Argyll	Scotland	10	M	S	295	SP-WI	F, F2

Reference	No.	Years	Locality	Country				n	Season	
	13	1977-1978	Highlands	Scotland	5	U	S	684	SP-WI	
Howes (1974)	14	1973	E. Yorkshire	England	11	P	S	41	AU	F
Howes (1978)	15	1975	Lincolnshire	England	12	P	S	111	AU	F
Howes (1980)	16	?	E. Yorkshire	England	11	P	S	165	SP, SU, AU	F
Kolb & Hewson (1979)	17	1971-1976	North-east	Scotland	A	U	S, G	617, 272	SP-WI	F, M
	18	?	West	Scotland	B	M	S, G	1168, 137	SP-WI	F, M
	19	?	Lochaber	Scotland	C	M	S, G	34, 36	SP, SU	F, M
Leckie et al. (1998)	20	1992-1996	Dumfries	Scotland	13	A	S	340	SP-WI	F
Lever (1959)	21	1955-1958	-	Britain	D	A	S & G	277+95	SP-WI	F
	22	1955-1958	-	Britain	E	M	S & G	78+28	SP-WI	F
Lloyd (1980)	23	?	Kent	England	F	A	G	64	WI, SP	F
	24	?	-	Wales	G	M	G	67	WI, SP	F
Lockie (1956)	25	1955-1956	Borders	Scotland	H	A	S & G	19	WI, SP	F
	26	1955-1956	Aberdeenshire	Scotland	I	U	S & G	7	SP	F
Lockie (1963)	27	1956, 1959	Highlands	Scotland	14	U	S	178	SU, WI	-
Lockie (1964)	28	1956, 1959	Highlands	Scotland	14	U	S	178	SU, WI	-
O'Mahony et al. (1999)	29	1996-1998	Northumberland	England	15	U	S	176	WI	-
Reynolds & Aebischer (1991)	30	1985-1987	Dorset	England	16	A	S	393	SP-WI	F, V, I
Reynolds & Tapper (1995b)	31	1985-1987	Dorset	England	16	A	S	633	SP-WI	-
Richards (1977)	32	1974-1975	Devon	England	17	P	S & F	186+64	SP-WI	F
Southern & Watson (1941)	33	1939-1940	-	Britain	J	M	G	40	WI, SP, SU	F
Watson (1955)	34	1951-1952	Highlands	Scotland	18	U	G	6	AU,WI, SP	F
Watson (1976)	35	1963-1964	Highlands	Scotland	18	U	G	219	SP-WI	O
Wilson (1990)	36	1988	Aberdeenshire	Scotland	3	A	S	31	SP, SU	-

marginal upland were over-represented, arable and pastoral landscapes under-represented (Table 10:2).

Table 10:2 Landscape composition and stratification of adult fox diet studies in Britain. Figures are the percentage of independent diets in each country; figures in brackets are the percentage land area in each landscape type.

Habitat	England	Scotland	Wales	Britain	Total
Arable	37.5 (50.2)	18.8 (18.1)	0.0 (4.2)	33.3 (35.0)	25.0 (35.0)
Pastoral	50.0 (37.7)	0.0 (10.4)	0.0 (46.5)	0.0 (29.1)	14.3 (29.1)
Marginal upland	0.0 (8.7)	31.3 (18.9)	100.0 (49.1)	66.7 (15.9)	28.6 (15.9)
Upland	12.5 (3.3)	50.0 (52.6)	0.0 (0.2)	0.0 (20.0)	32.1 (20.0)
N diets	8	16	1	3	28

Thirty-one non-independent diets were expressed as a measure of frequency of occurrence; 27 (87%) were expressed as the frequency of occurrence in scats/stomachs, two (6%) as the frequency of all occurrences and two (6%) as the frequency of occurrence as the main item in scats/stomachs. Thirteen (42%) were expressed as the average across the seasons for which material was collected: 18 (58%) were expressed as the total for the study, and are potentially confounded by different sample sizes from each season.

All 31 diets indicated the frequency of occurrence of the three mammal groupings. Seventeen (55%) provided data on the three bird groups, the remainder the frequency of occurrence of bird remains as a whole. Vegetation was subdivided into edible and non-edible components in 26 diets (84%). We were able to determine the frequency of occurrence of insects and other invertebrates in all 31 diets. However, only 19 (61%) analysed samples for earthworms.

Adult diet composition in each landscape

Standardised data on the occurrence of prey remains are illustrated in Figure 10:1. Combining diets from the same location and type of material, there were 29 separate diets based on measures of frequency of occurrence.

The average occurrence of medium-sized and large mammals was not consistently higher in scats than stomach contents (Figure 10:1). However, there was large variation within all landscapes in the frequency of occurrence of all prey types. As the figures derived from each type of material in each landscape were generally similar, we averaged the results to derive a single diet for each landscape.

Medium-sized mammals were the most frequently recorded prey group in arable and pastoral areas, averaging 33% and 28% of occurrences respectively. In both landscapes, small mammals averaged 22% of occurrences. Large mammals averaged 8% in arable and 4% in pastoral areas. Birds were more common in arable (25%) than pastoral areas (17%); the reverse was true for invertebrates (8% versus 23% respectively).

In marginal uplands, the dominant prey type was small mammals (34%), followed by medium-sized mammals (19%), large mammals (18%) and birds (18%). In uplands, large mammals were most frequent (36%), followed by small mammals (22%), medium-sized mammals (20%) and birds (18%).

Eleven studies (14 non-independent diets) measured the relative bulk of prey consumed. Seven diets were expressed as percentage mass ingested, two as percentage volume and five as percentage mass in remains. Standardised figures (for nine independent diets) are illustrated in Figure 10:2. In arable areas, medium-sized mammals (63%) and birds (21%) were the most important prey types, with all other prey groups comprising less than 10% of mass ingested. Small mammals, medium-sized mammals, birds, plants and invertebrates contributed approximately equal amounts to the mass of prey ingested for the single diet from pastoral areas. In marginal uplands, large mammals were the major component (56%), followed by small (24%) and medium-sized (13%) mammals, with birds comprising only 4%. Similarly, upland diets were dominated by large (38%), medium-sized (28%) and small mammals (26%), with birds accounting for just 7% of mass ingested.

Dietary breadth

Average dietary breadth indices based on mass ingested were lower than those based on frequency of occurrence in arable (0.351, n=6 versus 0.382, n=8), pastoral (0.390, n=1 versus 0.406, n=5) and marginal upland (0.311, n=3 versus 0.347, n=8) landscapes, but higher in upland areas (0.355, n=4 versus 0.302, n=10). Diets expressed as mass ingested tend to indicate greater dietary specialisation because of the increased importance of medium-sized and large mammals in arable and marginal upland areas respectively; the pattern for pastoral areas is less clear as only a single diet was available.

To compare between landscapes, indices for the same location derived from more than one material type or expressed as frequency of occurrence and percentage mass ingested were averaged. Dietary breadth was similar for arable (mean: 0.393; range: 0.259-0.533) and pastoral landscapes (0.391; 0.236-0.477), lower in marginal upland (0.343; 0.209-0.479) and lowest in upland landscapes (0.308; 0.234-0.457). In all landscapes, the range of values was similar. There was no evidence for an increase in dietary specialisation in any landscape following the outbreak of myxomatosis (Figure 10:3).

Seasonal variation in adult diet composition

Eleven studies calculated seasonal diet composition of adult foxes using measures of frequency of occurrence (Figure 10:4). In arable and pastoral landscapes, birds and bird eggs were most frequently recorded during summer, invertebrates during summer and autumn, and edible vegetation during autumn and winter. The frequency of occurrence of both small and medium-sized mammals tended to decline from spring through to autumn, although variation between diets was large. Studies in marginal upland and upland landscapes only presented data for spring/summer and autumn/

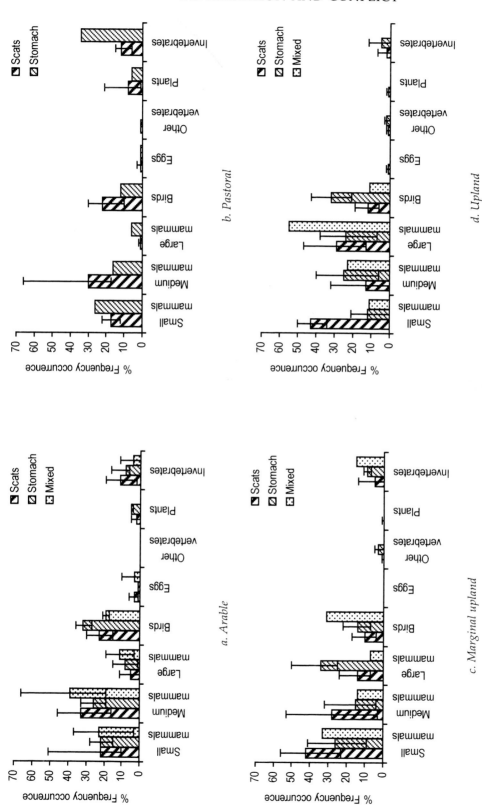

Figure 10:1 Composition of the diet of adult foxes, based on measures of frequency of occurrence, in (a) arable landscapes, (b) pastoral landscapes, (c) marginal upland landscapes, and (d) upland landscapes. Diets derived from different materials (scats, stomach contents and mixed samples) are presented separately. Figures are the percentage frequency of all occurrences. Error bars are minimum and maximum values: these were derived as the lowest and highest frequency of occurrence from any diet in that landscape.

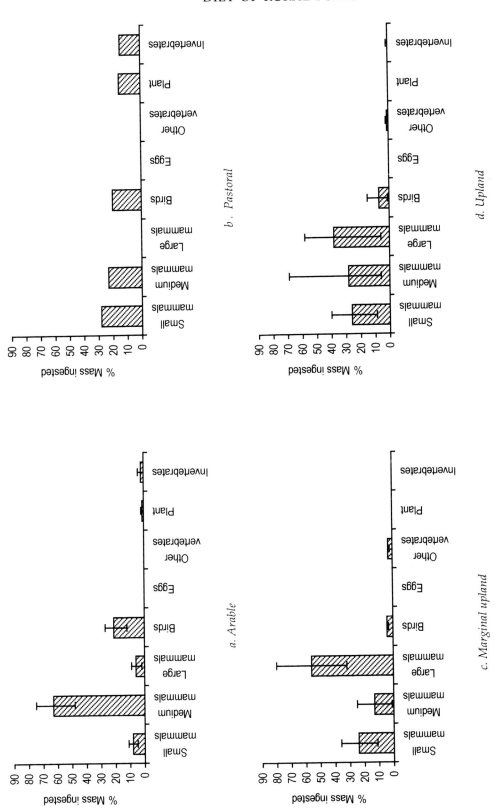

Figure 10:2 Composition of the diet of adult foxes, based on mass of prey ingested in (a) arable landscapes, (b) pastoral landscapes, (c) marginal upland landscapes, and (d) upland landscapes. Error bars are minimum and maximum values: these were derived as the lowest and highest percentage from any diet in that landscape.

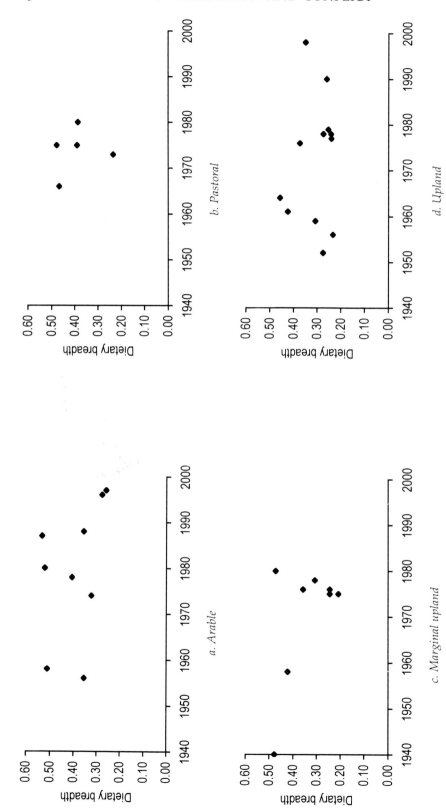

Figure 10:3 Temporal trends in adult dietary breadth in (a) arable landscapes, (b) pastoral landscapes, (c) marginal upland landscapes, and (d) upland landscapes. Each point refers to one of the non-independent diets listed in Table 10:1.

winter. In upland regions, there was very little seasonal variation for the average figures, although there were often pronounced differences within studies, particularly in the occurrence of small and large mammals. The one diet from marginal uplands showed a large inverse seasonal variation in the occurrence of small and large mammals.

Seasonal variation in dietary breadth was compared using the average dietary breadth of all diets in that landscape (Table 10:3). In arable areas, dietary breadth was lowest in spring and winter, reflecting the increased specialisation on birds and mammals, and highest in summer and autumn when other prey groups were more common. The pattern for pastoral landscapes was very different to arable areas but was complicated by the large variation in the two diets analysed. The pattern of dietary specialisation in the marginal upland diet was the same as for upland landscapes, with higher dietary breadth in spring/summer and increased dietary specialisation in autumn/winter. For upland landscapes, there was considerable variation in the values from individual diets.

Table 10:3 Seasonal dietary breadth (Levin's standardised niche breadth index) of adult foxes based on frequency of occurrence.

Habitat	N diets	Seasons	Dietary breadth	
			Average	Range (minimum-maximum)
Arable	3	Spring	0.331	0.300 - 0.387
		Summer	0.357	0.346 - 0.367
		Autumn	0.340	0.333 - 0.347
		Winter	0.338	0.321 – 0.347
Pastoral	2	Spring	0.375	0.275 - 0.476
		Summer	0.333	0.329 - 0.337
		Autumn	0.285	0.251 - 0.392
		Winter	0.548	-
Marginal upland	1	Spring/summer	0.193	-
		Autumn/winter	0.114	-
Upland	5	Spring/summer	0.245	0.217 - 0.288
		Autumn/winter	0.215	0.139 - 0.341

Cub diet

Six of the studies in Table 10:1 and one further study (Kolb & Hewson, 1980a) described nine cub diets. Six were based on measures of frequency of occurrence (Figure 10:5) and three, all from arable areas, on mass ingested. In arable areas, the relative distribution of the different prey types (Figure 10:5) was generally similar to that of adults (Figure 10:1), although the absolute frequency of occurrence of medium-sized mammals and birds was higher in the diet of cubs; invertebrates were also taken more

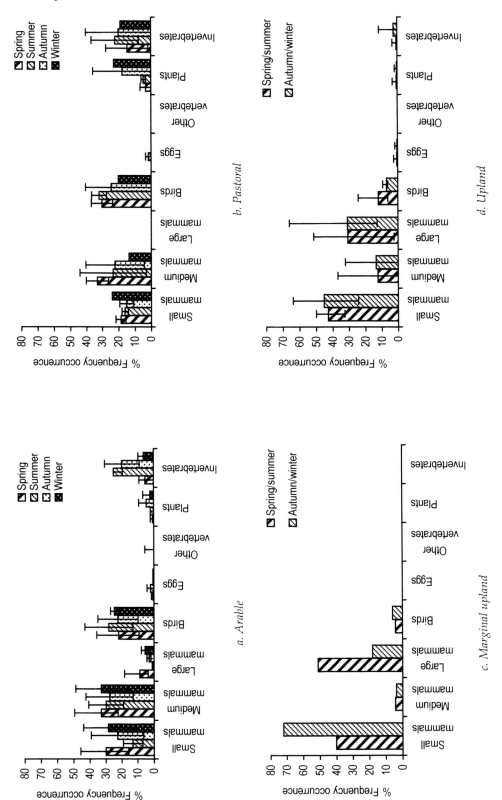

Figure 10:4 Seasonal composition of the diet of adult foxes based on the frequency of occurrence of prey groups, in (a) arable landscapes, (b) pastoral landscapes, (c) marginal upland landscapes and (d) upland landscapes. Diets derived from different materials have been averaged. Error bars are minimum and maximum values: these were derived as the lowest and highest frequency of occurrence from any diet in that landscape from any material.

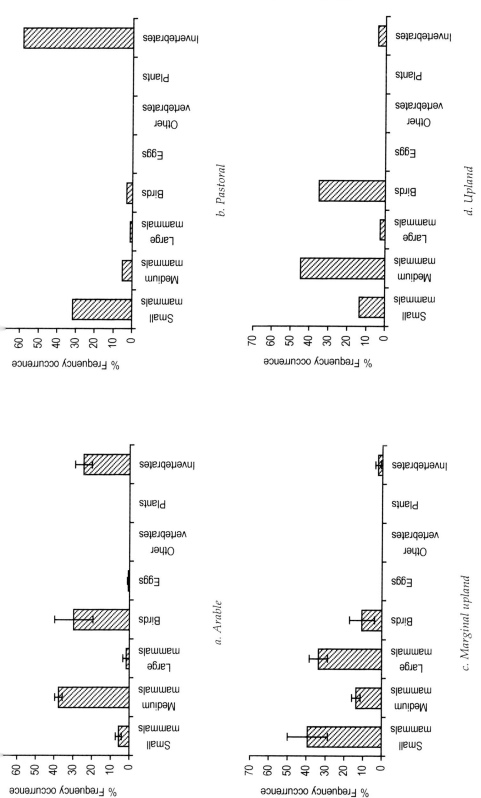

Figure 10·5 The diet of cubs during the denning period based on frequency of occurrence in (a) arable landscapes, (b) pastoral landscapes, (c) marginal upland landscapes, and (d) upland landscapes. Error bars are minimum and maximum values. Diets derived from different materials have been averaged. Diets derived from any diet in that landscape from any material. These were derived as the lowest and highest frequency of occurrence from any diet in that landscape from any material.

frequently by cubs. In terms of biomass, birds contributed more to the diet of cubs than adults, while medium-sized mammals were equally important to both. In marginal uplands, large mammals were much more frequent in the diet of cubs than adults. The frequency of occurrence of all other prey types was broadly similar, except invertebrates, which were more common in the diet of adults. For both pastoral and upland landscapes, there was only one diet for cubs and it is unclear whether either is representative.

The dietary breadth of cubs in arable habitats (mean: 0.323, range: 0.290-0.356, n=2) was similar to that of adults in spring (Table 10:3). In pastoral areas, the single diet for cubs indicated a more specialised diet (0.174) than adults, while those in marginal upland (0.530, n=2 cub diets) and upland areas (0.275, n=1 cub diet) indicated the contrary. So, measures of dietary breadth did not indicate a more specialised cub diet during the denning period, although there is a tendency for cubs to consume larger prey types.

ANNUAL PREY REQUIREMENTS

The total mass of each prey group ingested annually by a standardised fox group was estimated by extrapolating from a captive study (Sargeant, 1978), using body weights of 6.5 kg for males and 5.5 kg for females (Harris & Lloyd, 1991). Prey consumption rates for foxes in Scotland will be higher, given their larger body mass.

All cubs were assumed to have been born on April 1st. During weeks 1-4, the cubs were assumed to consume no prey; the additional prey burden that lactation placed on the female was calculated from Sargeant (1978). During weeks 5-13, the diets for cubs and adults were calculated separately. From July 1st, cub diet was assumed to be the same as for adults. Cub prey consumption was assumed to peak 28 weeks after birth, at 1.3 times the adult requirement, and then to decline to the adult rate at 52 weeks (Reynolds & Tapper, 1995a). Cubs were assumed to attain a final mass at 52 weeks of 6 kg. We calculated group prey consumption under two scenarios. 'No-dispersal' assumes that no juveniles disperse and they are resident from October 1st onwards: this indicates the maximum amount of prey consumed. 'Dispersal' assumes that no juveniles remain after October 1st: this indicates the minimum amount of prey consumed.

Fox group size was 2.5 adults (1 male to 1.5 females) in all habitat types. This may slightly over-estimate prey consumption in upland areas where more than two adults are rare, and under-estimate prey consumption in lowland areas where larger groups may occur. Litter size was four cubs: this is smaller than birth litter size and reflects pre-emergent mortality. Only one female bred.

The total number of prey consumed annually was estimated by dividing the total mass of each prey group ingested by the standard fox group by the typical body mass of the most commonly occurring species in that prey class. For mammals, minimum values were obtained by dividing by the average weight of an adult; maximum values were obtained by dividing by the average weight of a juvenile. The following species were used: field vole *Microtus agrestis*, rabbit and brown hare *Lepus europaeus*. Large mammals were not included because these are generally taken as carrion and the impact of predation is likely to be minimal. The same approach was taken for birds, using pheasants as an example and male and female weights instead of adult and juvenile weights.

Annual consumption of prey

The annual prey requirements for a standardised fox group are summarised in Table 10:4. In the presence and absence of juvenile dispersal, the group would require 594 kg per year and 928 kg per year respectively. The annual mass of prey ingested in the four landscapes is summarised in Table 10:5. We have had to assume that the diet for cubs in pastoral, marginal upland and upland areas was the same as for adults, although this is unlikely to be the case.

Relating these figures to consumption rates for individual species is inherently difficult. First, the prey groupings are composites of several species and in most instances it is not possible to identify species. This was particularly true for birds, which can only be identified to Order. Therefore, it is only possible to estimate the total number of animals killed for prey groups dominated by one species. Second, total figures will be the sum of different age and sex classes that differ in their unit mass. Minimum values can be obtained by assuming the minimum mass ingested relates solely to adult prey, maximum values by assuming that the mass ingested relates solely to juvenile prey. This will indicate the typical range of animals killed on a fox territory characteristic of that landscape (Table 10:6).

Table 10:4 Annual prey requirements (kg) of a standardised fox group (1 male, 1.5 females, 4 cubs) based on mass ingested.

	Dominant male	Dominant female	Subordinate female	Cubs
N foxes	1	1	0.5	4
Body weight (kg)	6.50	5.50	5.50	6.00
Mass specific requirement (kg prey/kg/week)	0.44	0.46	0.46	-
Weekly requirement (kg prey)	2.86	2.53	1.27	-
Total lactation requirement (kg prey)	0.00	10.56	0.00	-
Total requirement during weeks 1-4 (kg prey)	11.44	20.68	5.08	0.00
Total requirement during weeks 5-13 (kg prey)	25.74	22.77	11.43	72.96
Total requirement during weeks 14-26 (kg prey)	37.18	32.89	16.51	164.64
Total requirement during weeks 27-52 (kg prey)	74.36	65.78	33.02	333.72
Annual requirement (kg prey)	148.72	142.12	66.04	571.32

Table 10:5 Annual mass (kg) of each prey group consumed by a standardised fox group (1 male, 1.5 females, 4 cubs) in each landscape. Minimum values were derived from the minimum percentage of the diet (mass ingested, Figure 10:2) with dispersal: maximum values were derived from the maximum percentage of the diet without dispersal.

	Arable	Pastoral	Marginal upland	Upland
Small mammals	26-94	166-260	65-334	54-371
Medium-sized mammals	285-696	137-213	6-232	36-640
Large mammals	11-84	0	190-743	36-538
Birds	71-273	119-186	17-37	6-139
Bird eggs	0	0	0	0
Other vertebrates	0	0	11-29	0-19
Plants	0-17	89-139	0	0-4
Invertebrates	0-34	13-83	0	0-9

Table 10:6 Number of prey consumed annually by a standardised fox group (1 male, 1.5 females, 4 cubs). The minimum number of prey animals consumed was derived from the minimum mass of prey ingested in Table 10:5 and assuming all prey were adults: the maximum number of prey animals consumed was derived from the maximum mass of prey ingested in Table 10:5 and assuming all prey were juveniles.

	Small mammals: Field vole	Medium-sized mammals: Rabbit	Birds: Pheasant
	Juvenile=0.01 kg Adult=0.03 kg	Juvenile=0.60 kg Adult=1.50 kg	Female=1.10 kg Male=1.45 kg
Arable	867-9,400	190-1160	49-248
Pastoral	5,533-26,000	91-355	82-169
Marginal upland	2,166-33,400	4-387	12-34
Upland	1,800-37,100	24-1067	4-126

THE IMPACT OF FOX PREDATION

Ideally, to indicate the impact of fox predation it is necessary to relate the number of animals killed (Table 10:6) to the abundance, productivity and mortality of the prey. In the absence of such data, we review the available literature on the impact of fox predation on each prey group.

It is often difficult to interpret the significance of fox predation, since the impact on a prey population is inevitably influenced by how that species is utilised or viewed by people. For example, sport shooting in Britain is typically reliant on large numbers of driven birds and requires much higher densities of game than would normally be present. These are generally achieved by intensive management regimes. Consequently, a level of predation that exerts little effect at typical prey densities may be unacceptable at an economic level. For species of conservation concern, even very low levels of predation may be deemed significant. Therefore, we have interpreted the significance of fox predation in accordance with the views of the original authors. For example, for game species, the term limiting implies the effects of fox predation on the maintenance of enhanced prey densities.

Impact on small mammals

Three species of small mammals are commonly eaten by foxes: field vole, bank vole *Clethrionomys glareolus* and wood mouse *Apodemus sylvaticus*. However, it is unlikely that foxes exert any limiting effect. For example, field voles can attain densities of 100 animals per hectare (Harris *et al.*, 1995), and are able to produce 19 young in a year, based on an 183-day breeding season, 20-day pregnancy and 28-day lactation, producing 3.8 litters averaging five young (Gipps & Alibhai, 1991). Assuming a spring density of just 5 per hectare (500 per square kilometre) on a fox territory of 2.5 km^2, there would be 1,250 adults. With a 1:1 sex ratio, this population could produce more than 11,800 young a year. So, with high levels of productivity, foxes alone are unlikely to affect small mammal populations.

Impact on medium-sized mammals

Fox predation can limit the abundance of rabbits in Australia under certain circumstances (e.g. Banks, 2000) and may also affect rabbit abundance in Britain (Trout *et al.*, 2000; Macdonald *et al.*, Chapter 18). But rabbit abundance in Britain may generally be higher than can be limited by fox predation, particularly as rabbit numbers are increasing post-myxomatosis and there is no evidence of increased dietary specialisation in any landscape (Figure 10:3). However, given the economic impact of rabbits, rabbit predation may be beneficial to farmers: this depends, in part, on the degree to which levels of predation are compensated for by increased productivity and survival of the remaining individuals. There are no quantified data on this.

The extent of predation on brown hares is essentially unknown, as only one study has subdivided lagomorph remains into rabbits and brown hares (Reynolds & Tapper, 1995b). In that study, hares comprised 23% of lagomorphs consumed. These authors also recorded hare density and concluded that predation on some fox territories was sufficient to reduce the standing crop, but not on others. In a modelling approach, Reynolds & Tapper (1995a) used parameter values derived from field studies where prey density was approximately 15 hares per square kilometre. They concluded that annually fox predation would be significant relative to productivity. An experimental reduction in predation pressure showed that hare abundance increased following the onset of predator control, but this also occurred in the absence of control (Tapper *et al.*, 1991). In East Anglia, rapidly increasing fox numbers since 1970 have not had an impact on hare numbers (Tapper, 1992). Therefore, the effect of fox predation on brown hare populations is equivocal. No data are currently available for the impact of fox predation on mountain hares *Lepus timidus* in Britain.

Impact on birds

Fox predation on birds is of concern since many species, including some game birds, are currently declining or considered threatened. The major game bird in Britain is the pheasant, and large numbers are reared and released annually for sport shooting (Tapper, 1992). Predation rates on released birds can be substantial, and predator control is an important component of pheasant management (Reynolds & Tapper, 1996). Because of the poor breeding performance of pheasants in Britain (Robertson, 1991), it is likely that sustainable shoots based on wild-bred populations of this introduced species can only be generated by extensive alteration of the countryside and predator control (Mayot *et al.*, 1998).

The impact of foxes on grey partridge *Perdix perdix* can be substantial, as the density of these birds has declined as a consequence of habitat degradation (Potts, 1980) and possibly the increased number of pheasants released (Tompkins *et al.*, 2000). The impact of predators, including foxes, on high-density partridge populations maintained for sport shooting has been demonstrated experimentally by Tapper *et al.* (1996). It is not clear whether foxes affect populations at lower densities.

The impact of foxes on red grouse *Lagopus lagopus scoticus* in upland and marginal upland landscapes is also unclear. Jenkins *et al.* (1964) concluded that foxes tended to predate non-breeding birds and the effect on population growth was minimal. More recent studies have refuted this conclusion. Hudson (1992) observed a negative relationship between fox and red grouse numbers on three game estates. He also found that birds with heavy parasite burdens were more susceptible to predation by foxes and low levels of predation may increase the size of the grouse population by removing parasitised birds. Other ground-nesting species, such as some terns, are also vulnerable (Forster, 1975). However, the impact of foxes on hen harriers *Circus cyaneus* was negligible in comparison with other causes of breeding failure (Green & Etheridge, 1999).

So, there is a mounting body of evidence that fox predation can severely affect the size and growth of populations of some ground-nesting birds, but that this is not true for all species or even for the same species at different sites. In most circumstances, current levels of predation by foxes may only be problematical because of declines due to other factors such as habitat structure, habitat loss or environmental pollution. In a recent review, predation was listed as a possible cause of recent declines for only one species of farmland bird (Evans & Wilson, 2001).

CONCLUSIONS AND RECOMMENDATIONS

Two general conclusions can be drawn. First, there is a general paucity of data concerning the diet of foxes in Britain. Second, there is substantial variation in the relative importance of any prey group within any landscape. The lack of data is evident when the data are stratified: the maximum number of diets available in any of the four landscapes was nine, spanning a period of approximately 50 years. In addition, there is very little information on the diet of cubs and on seasonal variation in the diet of adult foxes.

The available data are also limited by a lack of direct comparability, principally because of differences in the way data have been analysed and presented. One major limitation is that a significant proportion of diets only analysed the macro-remains in faecal samples. Although most diets were expressed using measures of frequency of occurrence, direct comparisons between these were limited by differences in: (i) the materials used to estimate diet composition; (ii) the accuracy to which prey groups were identified; (iii) the grouping of species into categories; (iv) whether the data were expressed as a total for all samples, or as a mean across seasons; and (v) whether the data were simply expressed as the frequency of occurrence as the main item in faecal remains or stomach contents. So, few studies are directly comparable, and some of the assumptions we have had to make will affect diets from some landscapes more than others.

Despite the small number of studies, and the inherent problems of comparing between them, it is abundantly clear that the over-riding characteristic of the diet of foxes in Britain is substantial variation in the importance of different prey groups both within and between landscapes. This presumably reflects differences in the abundance of prey at different sites and also prey selectivity. But because almost no studies have simultaneously measured diet composition and prey abundance, it is not possible to assess which prey species or prey groups are preferentially selected, nor how foxes respond to variation in these prey types. Variation between sites also makes it difficult to describe a typical fox diet in any of the four landscapes.

Relatively little is known about the impact of foxes on different prey species in Britain. Although it is possible to estimate the total mass of prey ingested using conversion factors, this does not provide any information on the consumption rates of different age and sex classes that may have fundamental impacts on the dynamics of the prey population. To date, only one study in Britain has manipulated the abundance of predators but there are no specific data on foxes. Ground-nesting birds are most likely to be limited by fox predation, but this is not true for all species and the impact of foxes will vary between sites and will be affected by the availability of other prey and the abundance of other predators. The impact of foxes may also be augmented by other factors such as habitat loss or pesticides. Lagomorphs may be limited by fox predation under some circumstances. However, for all three species further research is needed to determine under what circumstances fox predation is limiting. The majority of small and large mammals are unlikely to be limited by foxes.

Since there is limited information on the extent of fox predation, and the available data indicate marked differences between landscapes, we currently know very little about the impact of fox predation on both wild and domestic species. Below, we identify where further information is needed to determine the impact of fox predation with recommendations for methodologies where appropriate.

There is an obvious need for further studies of the diet of adult foxes (including seasonal variation) and especially cubs during the denning period across the whole of Britain. Data must be collected, analysed and presented according to a standardised format. For data collection, such studies should be stratified using the land class system (Bunce *et al.*, 1996). For data analysis, we recommend the procedures of Reynolds & Aebischer (1991). In particular, studies based on faecal material must analyse the micro-fragment to ascertain the importance of earthworms and birds. We recommend that mammals are analysed to species and then grouped into small (less than 0.1 kg), medium (0.1-5.0 kg) and large (over 5 kg) categories. Bird remains should be analysed to order but must also include a total figure for avian prey. Because of their limited importance, other vertebrates can be classified simply as reptiles, amphibians and fish. Invertebrate remains can also be classified simply as insects, earthworms and 'other'. Suggested other prey classes are bird eggs and edible vegetation. Lastly, data should be presented as both frequency of occurrence and percentage mass ingested (Fedriani & Travaini, 2000), and analysed by seasons, with study totals expressed as the average across seasons rather than for all samples analysed.

Carefully designed predator manipulation experiments are needed to assess the impact of foxes on their prey species. Similarly, experiments are needed to assess the effect of habitat structure in promoting predation. Information is required on how foxes respond – functionally and numerically – to temporal and spatial variations in the abundance of particular prey species. Consequently, studies should concurrently measure diet composition, fox abundance and the abundance of different prey groups. Techniques that allow the identification of age and sex classes from faecal remains or stomach contents are required to determine age and sex classes preferentially predated by foxes (see Calzada & Palomares, 1996). Finally, although a great deal of attention has been paid to the possible impact of foxes on their prey in Britain, relatively little is known about how the abundance of prey affects fox numbers and reproductive output (but see Kolb & Hewson, 1980b).

III
Economic
Impacts

11

Towards a sustainable strategy to control bovine tuberculosis in cattle

R Woodroffe, F J Bourne, C A Donnelly, D R Cox, G Gettinby, J P McInerney & W I Morrison
(The Independent Scientific Group on Cattle TB)

INTRODUCTION

In the past, the control of bovine tuberculosis (TB) has presented serious difficulties for the coexistence of livestock with wildlife in the agricultural environment. *Mycobacterium bovis*, the pathogen involved, is usually seen as an infection of cattle, but it has a wide range of mammalian hosts (including man). Badgers *Meles meles*, in particular, appear able to sustain infection within their populations and can transmit the disease to cattle under experimental conditions (Little *et al.*, 1982; Cheeseman *et al.*, 1989). To reduce the risks of such transmission occurring in the wild, the British Ministry of Agriculture, Fisheries and Food (MAFF) adopted a series of badger culling policies between 1972 and 1996. On 8th June 2001, MAFF was subsumed into the Departmennt of Environment Food and Rural Affairs (DEFRA).

Badger culling was extremely unpopular with the conservation lobby and the general public. This, combined with culling's apparent failure to prevent the incidence of cattle TB from rising steadily through the 1990s, led the Government to commission an Independent Scientific Review of the problem in 1996 (Krebs *et al.*, 1997). This review recognised that no simple solution could be formulated on the basis of available data, and suggested instead that a broad array of research be put in place to develop a range of policy options (Krebs *et al.*, 1997). This chapter outlines the major research initiatives that have subsequently been established, under the development and direction of a second committee of independent experts (Bourne *et al.*, 1998, 1999).

History of the Problem

Bovine tuberculosis was once a major human health problem in the UK. In the 1930s, there were 50,000 new human cases each year and an annual death toll in excess of 2,500 people. These human cases were mainly caused by drinking unpasteurised milk from the national dairy herd, where nearly half in Britain carried the infection (Krebs *et al.*, 1997). A concerted strategy to slaughter cattle shown to be infectious by the skin tuberculin test 1979 commenced in the 1950s and by 1979 less than 0.1% of the herds that were tested showed evidence of infection. At that time it appeared that control of the disease had been achieved, although foci of infection remained in the south west of England where incidence never fell below 0.61% of herds tested. Since the end of

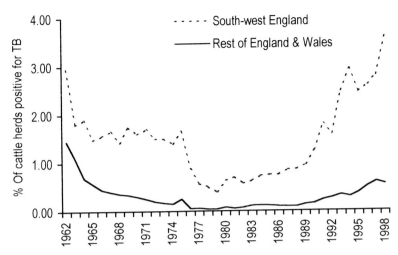

Figure 11:1 Percentage of cattle herds confirmed to have been infected with Mycobacterium bovis *from 1962 to 1999. The figure for 1999 is provisional. Data from MAFF (2000c).*

the 1980s, however, the rate of infection in cattle began to rise and it has continued to do so ever since (Figure 11:1). There were 872 new herd incidents in 1999, an increase of 18% on the previous year, involving the slaughter of 6,083 cattle (MAFF, 2000c). Of great concern is that these new incidents are occurring not only in the chronically affected areas in the south-west of England, but also in previously unrecorded counties such as Cheshire (MAFF, 2000c).

The presence of a TB reservoir in wildlife was first suspected in the early 1970s, when the incidence of TB in cattle in the south-west of England had been roughly constant for ten years (Figure 11:1; earlier declines in TB incidence were likely to have been caused by improvements in diagnostic techniques, Bourne *et al.* 1999), despite the repeated testing of cattle and slaughter of animals showing evidence of infection. Since the test-and-slaughter approach had successfully eradicated TB from other countries, this suggested that cattle were being re-infected from some outside source. The potential role of badgers was first recognised in 1971, when a badger was found dead from generalised tuberculosis on a farm in Gloucestershire where bovine TB had recently been confirmed in cattle. Further studies produced additional circumstantial evidence of badgers' involvement in transmitting TB to cattle, and badger culling, in various forms, was adopted as MAFF policy from 1972 (Krebs *et al.*, 1997). Harris & White (Chapter 12) provide an assessment of the efficacy of these culling regimes and a more detailed historical analysis, and, together with Delahay *et al.* (Chapter 13), explore some of the reasons why culling may not be an effective method for controlling TB in either badgers or cattle.

The Krebs review and the Independent Scientific Group on Cattle TB (ISG)

The resurgence of bovine TB showed that past policy had been inadequate, and highlighted the need for a new strategy to control cattle TB. To base future strategy on a sound scientific footing, the Government in 1996 appointed an independent scientific committee, chaired by Professor Sir John Krebs, to review the problem. Krebs' report concluded that the evidence of badgers' involvement in transmitting TB to cattle, while largely circumstantial, was compelling (Krebs *et al.*, 1997). It pointed out, however, that badgers' contribution to the problem had not been quantified, and that the effectiveness of badger culling in controlling the problem was unknown. Krebs recommended that a large-scale field experiment be carried out to address these two questions, as part of a broad based research programme investigating a variety of policy options, including vaccination of cattle. He advised that MAFF halt culling of badgers, except in areas covered by the experiment.

A major recommendation of the 'Krebs report' was that the field trial be designed and directed by a group of independent scientists. In response, the government appointed the Independent Scientific Group on Cattle TB, with a remit that was widened to include overseeing (and, where necessary, expanding) all of the research required to underpin future control policies (Bourne *et al.*, 1998). The Group's ultimate objective is to present Ministers with a range of scientifically based policy options that will be technically, environmentally, socially and economically acceptable. A future policy will aim to ensure that cattle and badgers can co-exist harmoniously: Ministers have made it clear that the widespread elimination of badgers is not an option for future policy (Bourne *et al.*, 1998).

Limitations imposed by diagnostic techniques

Accurate diagnosis of infection is the mainstay of both epidemiological research and test-and-slaughter approaches to disease control. Limitations in diagnostic techniques have two vitally important implications. First, there is currently no reliable test for diagnosing TB infection in live wildlife, which means that accurate diagnosis of TB in wildlife can currently be achieved only by *post mortem* examination. This hinders attempts to control TB by removing only infected badgers (Woodroffe *et al.*, 1999), and is also a severe limitation of longitudinal studies of TB epidemiology in wildlife (but see Delahay *et al.*, Chapter 13).

Accurate diagnosis of TB-infected animals is arguably more crucial still in cattle. While the tuberculin test has been used successfully to eradicate TB from cattle in several other countries, a small proportion of infected — and infectious — animals are wrongly diagnosed as negative (reviewed in Morrison *et al.*, 2000). Recent analyses raise the possibility that these 'false negatives' may have a profound impact on the persistence of the infection in cattle populations (Morrison *et al.*, 2000).

FORMULATING THE QUESTIONS

The development of a scientifically based sustainable strategy to control cattle TB demands a better understanding of the epidemiology of the infection in both cattle and wildlife. It is frustrating that, despite efforts to control cattle TB over the past three decades, many crucial questions remain unanswered. The majority of these are addressed by the ongoing research programme, including:

- What proportion of TB cases in cattle is caused by (i) other cattle, (ii) badgers or (iii) other wildlife species?

- What risk factors predispose cattle herds to TB outbreaks? Various factors have been proposed to contribute to the risk of such outbreaks, but no rigorous analyses have been carried out. Risk factors might include (i) farm husbandry (e.g. livestock management, stocking density, source and purchases of stock, and cropping practices); (ii) presence of undetected TB-infected cattle; (iii) TB outbreaks in nearby herds; (iv) climate; (v) geographical features; (vi) presence of infection in badgers and other wildlife; (vii) different strains of *M. bovis*.

- What is the distribution and dynamics of TB infection in badgers and other wildlife? The distribution of TB in badgers and other wildlife is poorly known at both national and local levels. Likewise, the factors that influence the persistence of infection in wildlife populations, and how this may vary locally, are largely unknown. Factors affecting the epidemiology of TB in badgers and other wildlife might include (i) abundance and social structure, including disturbance from previous culling operations (Smith *et al.*, 1995; White & Harris, 1995; Swinton *et al.*, 1997; see also Delahay *et al.*, Chapter 13); (ii) population dynamics; (iii) pathology and virulence of infection; (iv) different strains of *M. bovis*.

- What are the possible and most probable routes of transmission of infection to and within cattle herds and between species? Possibilities include (i) urine, (ii) faeces, and (iii) respiratory discharge.

ADDRESSING THE QUESTIONS

Answering the questions raised above demands a multi-facetted approach. Some studies can address more than one question, and some questions need to be tackled from more than one perspective. We, the Independent Scientific Group on Cattle TB, have established a wide-ranging programme of research which includes: (i) a field trial; (ii) use of epidemiological questionnaires; (iii) investigations of TB pathogenesis and diagnosis; (iv) a survey of TB infection status of badgers killed in road traffic accidents; (v) analysis and re-analysis of existing data, as well as (vi) other related research. We shall outline the various approaches in turn.

The field trial

The field trial is designed to address three objectives: (i) to quantify the proportion of TB outbreaks in cattle that can be prevented by culling badgers; (ii) to enable assessment of the cost-effectiveness of badger culling as a strategy to control cattle TB; and (iii) to provide data on the underlying epidemiology of TB in cattle and badgers.

The trial design is detailed in our first report to Ministers (Bourne *et al.*, 1998), and summarised in Table 11:1. It consists of three experimental treatments: proactive culling (in which badgers are culled throughout an entire trial area), reactive culling (in which badgers are removed from prescribed locations around farms experiencing outbreaks

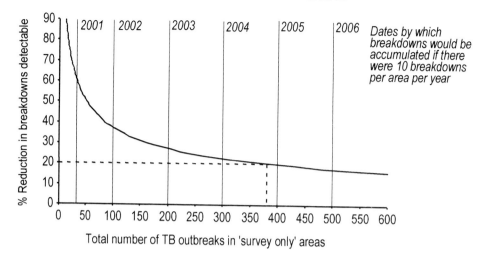

Figure 11:2 Representation of the statistical power of the field trial. The curve shows the percentage reduction in cattle TB incidence that would be detectable by the experiment (with a probability of 90%) as a function of the underlying incidence of cattle TB measured in the survey only (experimental control) areas. Dates indicate the time span over which data may be accumulated, assuming ten TB outbreaks per survey-only area, per year. For example, a 20% reduction in TB incidence ought to be detectable after 385 breakdowns have occurred in the survey only areas; this is projected to occur, under these assumptions, in the later part of 2004. Lower underlying incidences of TB could prolong the experiment; higher levels would give results more rapidly. It must be borne in mind, however, that each triplet must be observed for a minimum of two years, to establish the stability of effects, to give cattle testing timetables sufficient opportunity to reveal any effects, and to allow for any lag in the impact of badger culling. This places a lower limit on the duration of the trial.

of cattle TB) and survey only (an experimental control where no badger removal takes place). Comparing the culling treatments with the 'survey only' control will quantify the capacity of badger culling to reduce TB incidence in cattle. The proactive culls will also provide in-depth data on TB epidemiology in cattle and badgers.

The trial is being carried out in the areas of Britain suffering the highest incidence of TB in cattle. A total of thirty trial areas have been recruited, each measuring approximately 100 km². These areas are assembled into ten sets of matched triplets; trial areas within a triplet are then randomly allocated to the three treatments. Calculations of statistical power (detailed in Bourne et al., 1999) indicate that this design has the capacity to detect differences in cattle TB between treatments within a reasonable time-frame, and to take account of factors such as non-compliance by landowners (see also Harris & White, Chapter 12). The exact number of years required to detect such differences depends on the strength of the effects, as well as the underlying incidence of TB in cattle. The power calculations show, for example, that if breakdowns were occurring at a rate of 100 per year in the ten survey-only areas (an average of ten per area per year), a reduction in TB incidence as low as 20% in either of the treatment areas would be detectable in 50 'triplet-years' (i.e. an average of five years for each of the ten triplets; Figure 11:2). This means that, according to current schedules, results may be expected towards the end of 2005 (Bourne *et al.*, 1999). Greater treatment effects, or higher underlying TB incidence, would give results sooner; large-scale non-compliance could delay results.

All trial areas are surveyed for signs of badger activity before they are randomly allocated to experimental treatments. This provides indices of badger abundance that may be used to explain variation within and between triplets. Surveys are repeated periodically to maintain up-to-date information on badger numbers, and to assess the impact of culling on badger activity (Table 11:1).

Badger culling operations have been designed to reduce animal suffering to a practical minimum. Badgers are caught in cage traps (welfare concerns led to rejection of the Krebs Report's recommendation that snares be used) and despatched by gunshot. Despatch is subject to both external and internal audit. Animals other than badgers are released. To avoid dependent cubs, that might be left below ground, suffering if their dams were killed, culling operations are restricted to the period May-January (Bourne *et al.*, 1998).

All badgers captured in the course of the trial are subject to full *post mortem*, including culture from lymph nodes, body fluids, and any lesions found, to determine whether they are infected with TB. Blood samples are also collected for serology. DNA profiling is carried out both on badgers themselves, and on *M. bovis* isolates cultured from them (*M. bovis* isolates from cattle are subject to the same DNA profiling as a matter of routine). These genetic analyses should reveal relationships between TB epidemiology and badger population structure and dispersal. This will be especially valuable in evaluating hypothesised impacts of past badger culling on badger social organisation and, hence, on TB transmission (Swinton *et al.*, 1997; Rogers *et al.*, 1998).

Evaluation of badger culling as a measure to control cattle TB requires information on its environmental impact and economic cost, as well as its technical effectiveness. For this reason, we ensured that MAFF commissioned a full environmental impact assessment (Bourne *et al.*, 1998) as well as economic evaluations of the cost of badger culling (to farmers, Government, and wider society; Bourne *et al.*, 1999). The results of these studies will be taken into account in developing future policy options.

Epidemiological questionnaires: investigating risk factors

Several risk factors, particularly in the area of cattle husbandry, have been proposed as predisposing some farms to TB outbreaks (e.g. Griffin *et al.*, 1996,). Such risk factors are not amenable to experimentation, primarily because of the abundance of potential factors, and the relative paucity (in statistical terms) of TB outbreaks. We have therefore adopted an alternative approach, using questionnaires to compare farms that have experienced TB breakdowns in recent years with unaffected herds. This 'case control' approach is well established in the fields of medical and veterinary epidemiology (e.g. Doll & Hill, 1950; Griffin *et al.*, 1996).

Table 11:1 Summary of treatments in the field trial. Each treatment will be applied across ten areas, each measuring approximately 100 km². Trial areas are clustered into 'triplets'. The trial is expected to last approximately 50 'triplet-years' in total.

	Proactive culling	Reactive culling	Survey only
Aim of treatment	(i) To quantify the proportion of TB outbreaks in cattle that can be prevented by culling badgers (through comparison with the survey-only area) (ii) To provide data on the underlying distribution of TB in badgers for comparison with data on cattle	(i) To determine the effectiveness of culling as a measure to reduce TB incidence in cattle (through comparison with the survey-only area)	(i) To quantify the proportion of TB outbreaks in cattle that can be prevented by culling badgers (through comparison with the survey-only area) (ii) To determine the effectiveness of culling as a measure to reduce TB incidence in cattle (through comparison with the reactive area)
Objective of culling	To remove as high a proportion of badgers as possible within the constraints of welfare concerns, and to maintain very low badger densities throughout the trial	To remove badgers associated with particular TB incidents in cattle, minimising disturbance to other badgers	No culling
Culling tactics	(i) Initial cull carried out across entire area (ii) First follow-up cull 6-9 months after initial cull (iii) Annual follow-up culls thereafter	(i) Culling in response to confirmed TB incidents in cattle, at all setts within estimated home ranges that contain the breakdown land (ii) No follow-up culling	No culling
Survey data	(i) Initial survey of entire area (ii) Follow-up surveys in 3rd and 5th years after proactive cull (iii) Annual sett-checks prior to follow-up culling	(i) Initial survey of entire area (ii) Follow-up surveys in 3rd and 5th years after proactive cull (iii) Small-scale surveys prior to reactive culls	(i) Initial survey of entire area (ii) Follow-up surveys in 3rd and 5th years after proactive cull (iii) Quarterly sett-checks to seek evidence of illegal killing

We have been involved in the design of two such questionnaires. The first, designated 'TB99', collects detailed information on herd composition, cattle movements, farm enterprises, management practices and the occurrence of other cattle diseases, as well as steps taken to avoid contact between cattle and badgers. TB99 is administered to all farms experiencing TB outbreaks in Britain ('cases'). Where cases occur within trial areas, a similar questionnaire is administered to three matched 'control' farms that have not experienced TB outbreaks.

The second questionnaire, the 'enrolment questionnaire', is administered postally within proactive areas only. This questionnaire is designed to ensure that factors such as cattle husbandry can be directly compared with badger data. The presence of infected badgers may influence the magnitude of other risk factors — for example, measures to keep badgers and cattle apart may be effective only where badgers are infected. Because accurate diagnosis of TB in badgers requires *post mortem* examination, data on prevalence from control farms will be available only in the proactive areas, when culls are carried out. Since a relatively small proportion of TB99 forms may be administered in these areas at these times, we have developed the enrolment questionnaire — a greatly simpli—fied version of TB99 — to ensure that best use is made of the badger data.

Studies of TB pathogenesis and diagnostics

We attach a high priority to determining the relative importance of cattle-to-cattle transmission of TB. We consider that this has not been adequately addressed in the past, and suggest that the currently used diagnostic tests might limit control of the disease (Morrison *et al.*, 2000). Long-established procedures for TB control — such as restrictions on cattle movement following disclosure of disease — are based on the assumption that cattle-to-cattle transmission is of critical importance. Such measures, combined with regular testing and slaughter of reactor cattle, would be expected to eliminate infection in the absence of an external disease reservoir. However, high risks of infection, irrespective of the source of disease, may place greater demands on the testing programme. Of particular concern are (i) the observation that the tuberculin test misdiagnoses *circa* 10% of infected animals as uninfected; (ii) the finding that newly-infected animals may be infectious before they develop an immune reaction to the tuberculin test (both reviewed in Morrison *et al.*, 2000); and (iii) the fact that farm-based testing regimes, combined with movement of cattle between farms, means that a proportion of cattle are never subjected to TB testing.

Given these difficulties, it is clear that a sustainable TB control policy demands a better understanding of the pathogenesis of TB in cattle, at what stage of infection cattle can transmit disease to others, and how this relates to reactions to the tuberculin test and other diagnostic tests. It is our intention that studies of cattle pathogenesis be cross-linked with the expanded programme of vaccine development that has been established since publication of the Krebs report (Bourne *et al.*, 1999).

While detailed information on TB epidemiology and pathogenesis must be amassed over several years, it is possible that modifications of cattle testing regimes will contribute to the control of the disease in the shorter term.

Survey of TB prevalence in badgers killed in road traffic accidents

Future cattle TB control policies may require information on the prevalence of TB in badgers across the country (for example, some husbandry recommendations might be appropriate only where badgers are infected). Trial data on badger TB prevalence must therefore be supplemented by information from areas outside the trial with both high and low levels of TB in cattle. This in turn will allow an extension of the analysis of the relationship between TB in badger populations and in cattle. Since no reliable live test is available for use in badgers, we have proposed that badgers killed in road traffic accidents be collected from the counties of Cornwall, Devon, Gloucestershire, Herefordshire, Worcestershire, Shropshire and Dorset, and subjected to detailed *post mortem*. These data will initially be used to estimate regional TB prevalence and later, as sample sizes increase, to estimate local prevalence. Overlap between areas recruited into the field trial and into the road accident survey will allow validation of the use of road-killed animals to obtain measures of TB prevalence in badgers.

Analysis of old data

DEFRA holds a large body of data collected on past TB breakdowns, including information on TB testing histories, farming practices, past badger culls and *M. bovis* strain types. Analysis and re-analysis of these data is being carried out under the auspices of the Independent Scientific Group, on the assumption that these analyses will help answer some of the questions posed, and complement other ongoing research activity.

Other related research

In addition to the work carried out directly by MAFF/DEFRA and the Independent Scientific Group, a large number of other research projects have been funded by MAFF/DEFRA to address scientific issues with direct or indirect importance to the development of future policy.

Ecological studies form a large component of the complementary research. These include practical issues, such as the development of techniques for measuring badger density from field signs, as well as analysis of the ecological characteristics of high TB-risk areas, and investigations of the capacity of wildlife other than badgers to maintain and transmit *M. bovis* infection. Further studies are investigating the impact of badger culling on badger social and spatial organisation and, hence, on TB epidemiology.

Vaccine development is another important component of the TB research initiative. Vaccines for either cattle or badgers are regarded by some as a panacea for TB control (reviewed in Krebs *et al.*, 1997). Much effort is being directed towards this area of research, which is effectively linked to an extensive international programme developing vaccines for human use. The completion of the project to sequence the *M.*

bovis genome is likely to accelerate vaccine development, but the difficulties in developing a successful vaccine remain enormous. Vaccines are unlikely to be available within the next decade. Moreover, the development of a successful vaccination campaign, aimed at cattle, badgers or other wildlife, would depend absolutely on the epidemiological insights generated by the broader programme of research.

Recent developments in the genetic fingerprinting and strain typing of *M. bovis* will also allow more discrete tracking of the tubercle bacillus through animal populations — cattle, badgers and others — and complement the epidemiological study.

Finally, we expect that additional research projects will be established in response to the recommendations of a recent review of the effect of husbandry practices on TB in cattle.

CONCLUSIONS

The role of the Independent Scientific Group on Cattle TB is to provide a scientific basis for future policy to control tuberculosis in cattle. From the outset, we have adopted a holistic approach, recognising that sustainable control policies (which exclude the eradication of badgers from large tracts of the countryside) could only be achieved through a better understanding of TB epidemiology in cattle and wildlife.

Accurate and sensitive diagnosis of disease forms the cornerstone of both TB control and epidemiological studies. In the absence of a reliable live test for use in badgers and other wildlife, epidemiological questions can currently (and unfortunately) be answered only through killing wild animals. However, it must be borne in mind that similar restrictions apply to TB diagnosis in cattle: the possibility remains that the cattle population itself contains undiagnosed reservoirs of infection. Research must therefore be directed at both wildlife and cattle.

Our research initiative has attracted controversy — particularly concerning the field trial, on which media attention has focussed. Stakeholders often have entrenched opinions on tuberculosis control: many farmers (encouraged by past MAFF policies) are convinced that culling badgers is the only way to protect their cattle, while many members of the conservation and welfare lobby maintain that badger culling is inappropriate and unnecessary. Both parties have argued that parts of our research programme are unnecessary because 'the answer is obvious'. Clearly, these viewpoints cannot both be right. In the absence of solid data on the questions raised in this chapter, neither viewpoint can form the basis of an effective, sustainable policy to control cattle TB.

It will take time for research to feed through into policy. Nevertheless, if a lasting answer to the TB problem is the goal — as it must be — the outcome of our epidemiological investigation, including the field trial, should not be pre-judged and must be allowed to run its course without interference. Thus far, our studies are running to plan, and we look forward to producing a stream of advice to farmers, conservationists and Ministers over the next few years.

12

Bovine tuberculosis in cattle in Britain: Can badger culling ever solve the problem?

S Harris & P C L White

INTRODUCTION: BOVINE TB IN CATTLE AND HUMANS

In the 1930s, bovine tuberculosis *Mycobacterium bovis* in cattle in Britain was at a high level, with at least 40% of dairy herds being infected. Like *M. tuberculosis* in humans, *M. bovis* in cattle normally infects the lungs initially. However, as the infection progresses, it can travel to other organs. If it reaches the udder, the cow's milk can become infected. During the 1930s, there were no controls on bovine tuberculosis (TB) in cattle, and as a result there were approximately 50,000 cases of bovine TB each year in the human population in England and Wales. These were responsible for over 2,000 deaths annually (Kayne, 1937; Bryder, 1988).

The introduction of skin-testing to detect infected cattle in the 1950s, the slaughter of infected animals and the pasteurisation of milk, all contributed to reducing the transmission of TB from cattle to humans, and also to a decline in the overall level of TB in cattle. By the early 1960s, the proportion of cattle herds infected with TB in England and Wales had been reduced to 1.5%, although it remained at around 3% in south-west England (Krebs *et al.*, 1997). The level of TB in cattle continued to be reduced until the late 1970s. It reached its lowest level in 1979, when around 0.05% of herds were infected in England and Wales overall, and around 0.4% in south-west England (Krebs *et al.*, 1997).

Since that time, however, TB in cattle has been increasing. In 1990, around 0.2% of herds in England and Wales, and 1.2% in the south-west, were infected. In 1998, these figures had risen to 0.5% and 3.6% respectively (MAFF, 2000c). Bovine TB remains rare in the human population, and in 1995, only 32 human cases of bovine TB were recorded in the UK (Krebs *et al.*, 1997). The risk to human health in the UK is therefore currently very small, although it has been argued that bovine TB retains the potential to become more of a threat in the future (Krebs *et al.*, 1997).

THE HISTORY OF BADGER CULLING TO REDUCE TB IN CATTLE IN BRITAIN

In 1971, a dead badger infected with bovine TB was found on a farm on Gloucestershire where TB had been found in the cattle. Subsequent investigations by the Ministry of Agriculture, Fisheries and Food (MAFF) showed that up to 20% of badgers could be infected with TB in areas where TB occurred in cattle. Other wildlife species were also found to be infected, but at a much lower prevalence. On this basis, MAFF concluded that badgers were the principal wildlife threat in terms of providing a reservoir for TB in cattle.

From 1973, farmers who had TB-infected cattle were allowed to kill badgers on their land. However, because of the animal welfare concerns raised by this action, MAFF took over all badger control in 1975. The history of official badger control from that date until 1994 has been reviewed by White & Harris (1995a). A summary of TB control strategies and the various reviews into TB in cattle are summarised in Table 12:1. Badgers were initially culled by gassing in setts with hydrogen cyanide. Gassing ceased temporarily between October 1979 and October 1980 while the first review into bovine tuberculosis was undertaken (Zuckerman, 1980). Following the recommendations of this review, gassing recommenced in October 1980. However, the results of investigations into the humaneness of gassing led to the 'gassing' strategy being stopped in July 1982.

Table 12:1 History of badger control and TB reviews in Britain.

Date	Event
1971	TB first found in badgers
1973	Farmers with TB-infected cattle allowed to kill badgers on their land
1975	MAFF takes over official badger control, using 'gassing' strategy
October 1979 - October 1980	Gassing operations temporarily halted for Zuckerman review
July 1982	Trapping and shooting replace gassing as form of control Control based around 'clean ring' strategy
April 1986	Dunnet review published Clean ring strategy replaced by 'interim' strategy
November 1994	Interim strategy modified to include 'live test' strategy in certain areas
April 1996	'Live test' component suspended for Krebs review
December 1997	All badger control halted following publication of Krebs review Krebs recommends large-scale field-based 'experimental' comparison of control strategies
August 1998	Krebs experiment endorsed by Independent Scientific Group in form of a randomised culling trial First trial areas announced No culling policy remains in areas outside trial areas
February 2001- May 2002	Culling trial suspended due to Foot and Mouth Disease outbreak

Gassing was replaced by cage trapping and shooting. Between August 1982 and April 1986, trapping was concentrated around the setts and badger groups were trapped in a centrifugal manner until no more infected animals were caught. This was the 'clean ring' strategy. From April 1986, following the recommendations of the second government review into bovine tuberculosis (Dunnet *et al.*, 1986), an 'interim' strategy was introduced whereby trapping was limited to that part of a farm where cattle were believed to have been infected by badgers. From November 1994, this strategy was modified to incorporate the use of a new test for the diagnosis of TB in live badgers in certain areas. This 'live test' strategy remained in place for 18 months and was suspended in April 1996 when the third review of bovine tuberculosis in cattle and badgers (Krebs *et al.*, 1997) was started. All badger control operations under the interim strategy stopped in December 1997 when the Krebs *et al.* (1997) report was published.

From 1978 up to and including 1997, MAFF examined 20,075 badgers post-mortem for signs of TB. Of these, an average of 17.5% were found to be positive for TB (MAFF, 2000c). Over the same period, 28,366 cattle were compulsorily slaughtered due to TB infection (MAFF, 2000c). Badger control cost £1.48 million in 1997 (MAFF, 2000c), which gives an average cost per badger killed of £605.

Krebs *et al.* (1997) recommended that a large-scale field experiment comprising at least thirty 10 km by 10 km TB 'hot-spot' areas should be established to quantify the contribution made by badgers to TB in cattle. In August 1998, this experimental approach was endorsed by the Independent Scientific Group on Cattle TB in the form of a randomised culling trial (Bourne *et al.*, 1998; see also Woodroffe *et al.*, Chapter 11). The treatments to be compared in the trial were: (i) proactive culling, where all badgers within a treatment area are culled; (ii) reactive culling, where only badgers from social groups associated with a cattle herd breakdown are culled; and (iii) no culling. Bourne *et al.* (1998) recommended that the treatments should be grouped geographically into triplets, with each 100 km² area within each triplet being assigned to one of the three treatments. This effectively reduced the number of different geographical locations involved in the trial from 30 to 10, but had the benefit of reducing other potential differences between the areas, thereby increasing the sensitivity of the trial to the effects of badger culling *per se*. The total area of land involved in the trial would therefore be 3,000 km², 1,000 km² for each of the three treatments. Krebs *et al.* (1997) originally envisaged that the field experiment should be started in all treatment areas concurrently. Bourne *et al.* (1998) recognised that this would not be feasible due to insufficient resources, but identified two triplets, one on the Gloucestershire/Herefordshire border and one on the Devon/Cornwall border, where work should begin immediately. They further recommended that field work for the remaining eight triplets should start in 1999. However, the complications involved in identifying treatment areas and undertaking surveys and badger control on the ground meant that this target was over-optimistic; only four of the remaining eight triplets had been identified by the end of 1999; another two were announced in March 2000 and the final two in October 2000. Delays of the culling trial due to Foot and Mouth Disease have meant that trapping has not yet started in these two final areas, although it is due to start there by the end of 2002.

THE EFFECTS OF BADGER CULLING ON TB IN CATTLE

Previous work by MAFF in Gloucestershire, Dorset and Devon has shown that the large-scale removal of badgers can dramatically reduce the incidence of TB in cattle (Wilesmith *et al.*, 1982; Clifton-Hadley *et al.*, 1995b; Krebs *et al.*, 1997). However, as Krebs *et al.* (1997) pointed out, none of the badger culling strategies used to date has been assessed in a properly designed experiment. They argued that because of other, possibly confounding, factors that may influence the level of TB in cattle, it was not possible to separate out the effects of badger culling. From a strictly experimental viewpoint, this may be true. However, few real-world environmental problems can be analysed experimentally, and comparative approaches are much more commonly used to investigate environmental problems.

White & Harris (1995a) used a simulation modelling approach to compare the relative effectiveness of the gassing, clean ring and interim strategies in reducing TB in badgers. They found that the gassing and the clean ring strategies were the most effective, and that the interim strategy offered only marginal benefits over a no culling strategy (Table 12:2). They also analysed the likely effectiveness of the live test, which was only just being introduced at the time. Their results suggested that the live test would not work, and that although it would be more effective than the interim strategy, it would not be so effective as either the clean ring or the gassing strategies. Since the White & Harris (1995a) paper was published, data on badger removal operations conducted under the live test strategy have become available from MAFF's Wildlife Unit. These have now been analysed by Woodroffe *et al.* (1999). On the basis of their analysis, Woodroffe *et al.* (1999) concurred with the predictions of White & Harris (1995a), concluding that the live test would be unlikely to reduce the overall prevalence of TB in badgers.

Table 12:2 Relative effectiveness of the gassing, clean ring, interim and no culling strategies in controlling localised TB in badger populations. The figures in the table represent the probability of successful disease eradication in nine social groups (forming a 3 x 3 square) centrally located within a main grid of 10 x 10 social groups. The results are based on 50 simulations of the spatial stochastic model of White & Harris (1995b) with a disease-free equilibrium group size of eight adults and yearlings. Modified from White & Harris (1995a).

Control strategy	Initial prevalence of TB in the nine groups (%)					
	10	20	30	40	50	60
Gassing	0.96	0.94	0.84	0.90	0.88	0.72
Clean ring	0.70	0.96	0.88	0.96	0.96	0.98
Interim	0.58	0.24	0.12	0.02	0.06	0.08
No culling	0.38	0.10	0.12	0.06	0.02	0.04

Whilst the above studies have analysed the effectiveness of badger culling in reducing TB in badgers, the core problem does not concern TB in badgers, but TB in cattle. Therefore, the most important analysis is the effect of culling badgers on the level of TB in cattle. Because this linkage is not a direct one, and because of the number of other factors involved, this is a more difficult proposition, certainly from a modelling perspective. McInerney (1987) used a simple econometric model to test whether the introduction of badger control in 1975 had led to a reduction in the incidence of TB in cattle. The results showed no significant downward time trend in the data, implying that badger control had not been successful in progressively reducing TB in cattle. More recently, Cox et al. (1999) modelled the relationship econometrically, to include the period of the interim strategy. They carried out a partial cost-benefit analysis of the different control strategies, and found that neither the gassing, clean ring nor interim strategies offered net financial benefits to the nation (Table 12:3). However, the results of their econometric model also showed that gassing, and then the clean ring strategy, had been the most effective strategies in terms of their effect on TB in cattle, and that the interim strategy may in fact have made the disease problem in cattle worse. This finding corresponds with earlier suggestions that relatively non-intensive culling of badgers can disturb the social hierarchy, hence leading to an increased potential for disease spread between badgers, and therefore a potentially greater transmission risk to cattle (Brown et al., 1994b; see also Delahay et al., Chapter 13).

Table 12:3 Results from a partial cost-benefit analysis of the gassing, clean ring and interim strategies. The costs and benefits (1994 values) are expressed (a) per badger culled and (b) per badger removal operation under the different control strategies. The costs are based on MAFF estimates and those made by Power & Watts (1987). The range of benefits is based on 95% confidence limits for the marginal effects of the strategies. The historical values have been inflated to 1994 values using the Consumer Price Index. Modified from Cox et al. (1999).

Control strategy	Costs (£)	Benefits (£)	Net benefits (£)
Gassing	660	37 to 35	-623 to -625
Clean ring	2,576	141 to 135	-2,435 to -2,411
Interim	1,470	-1,057 to -1,063	-2,527 to -2,533

a. Cost per badger culled.

Control strategy	Costs (£)	Benefits (£)	Net benefits (£)
Gassing	42,646	2,277 to 2,135	-40,369 to -40,511
Clean ring	115,921	6,810 to 6,562	-109,111 to -109, 359
Interim	8,866	-6,935 to -7,060	-15,801 to -15,926

b. Cost per badger removal operation.

THE EFFECTS OF CULLING WILDLIFE VECTORS ON TB IN CATTLE IN OTHER COUNTRIES

In a review of bovine TB in developed countries, Clifton-Hadley & Wilesmith (1995) defined five broad groupings of countries with respect to their progress towards eradicating TB in cattle, Group 1 being countries where the greatest, and Group 5 being countries where the least, progress had been made. Britain was categorised in Group 4, defined by a country where significant progress has been achieved in eradicating TB from cattle, but where the complete eradication is hindered by the presence of significant TB in wildlife. The other country in this group was New Zealand, where the principal wildlife reservoir species for TB is the brush-tailed possum *Trichosurus vulpecula*.

TB in cattle is also a significant problem in both Northern Ireland and the Republic of Ireland. Clifton-Hadley & Wilesmith (1995) categorised both these countries in Group 5. This group represented countries where progress towards eradication has been hindered by a wildlife reservoir (principally badgers in both cases), but more importantly, where significant cattle-to-cattle transmission still occurred, and therefore were different from Britain and New Zealand. Although some cattle-to-cattle transmission probably does still occur in Britain and New Zealand, it has hitherto been considered of relatively little importance compared with the wildlife-cattle transmission route, although this has been questioned more recently (Bourne *et al.*, 1999). There is no official badger control policy in Northern Ireland, yet despite this major difference, King *et al.* (1999) demonstrated a close similarity between the annual pattern of herd breakdowns in Northern Ireland and south-west England between 1975 and 1993. These authors used this as evidence to demonstrate the failure of badger culling strategies in Britain, and to draw attention to the fact that factors other than those relating to badgers may be influencing the incidence of TB in cattle.

As in Britain, the main means of attempting to eradicate TB from cattle herds in New Zealand has been a test and slaughter campaign for cattle, in conjunction with a wildlife reservoir culling programme. Modelling work has shown that the control of TB in possums can significantly reduce the levels of TB in cattle (Barlow, 1991ab), and these models have been used as the basis for possum control operations to reduce TB in cattle in New Zealand. A significant difference between bovine TB in New Zealand compared with Britain is that in New Zealand there are quantified data to show that TB in cattle is directly linked with the level of TB in possums. Barlow (1991a) demonstrated a significant linear regression relationship between the log of cattle reactor rates and the log of TB prevalence in possums, using data from three different sites. More recently, an analysis by Kean *et al.* (1999) based on data from Hohotaka in New Zealand showed that possum control between 1988 and 1994 resulted in a reduction in TB-infected possums by 87.5%, with a subsequent 83% decline in annual TB incidence in local cattle herds. However, even for TB in New Zealand, such data are few.

Possum control in New Zealand began at about the same time as badger control in Britain. The trend in cattle TB incidence and expenditure on wildlife control operations in the two countries up to 1995 is quite similar (Figure 12:1). In both countries, the

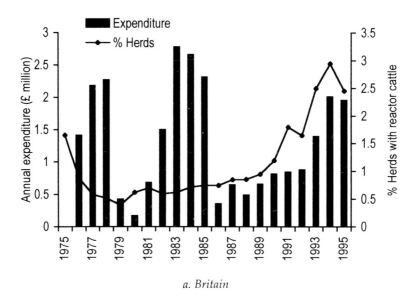

a. Britain

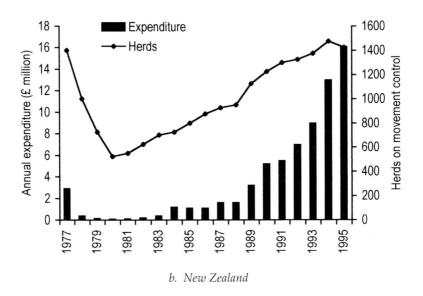

b. New Zealand

Figure 12:1 Pattern of annual expenditure on control of wildlife vectors in relation to the incidence of bovine TB in cattle in (a) Britain and (b) New Zealand for the period 1975 to 1995 inclusive. The figures for Britain are for the south-west only, where the vast majority of herd breakdowns and vector control operations occurred during this period. Data on expenditure on badger control in the UK are taken from Cox et al. (1999), and data on the incidence of TB in cattle in the UK are taken from Krebs et al. (1997). All the data for TB in New Zealand are taken from O'Neill & Pharo (1995). Data for herds in New Zealand are for numbers of herds.

overall trend suggests that the control of wildlife had some effect in decreasing TB in cattle in the late 1970s, although this period also coincided with improvements in the tuberculin test itself. Since then, increased expenditure on wildlife control has had no apparent effect. In New Zealand, this period coincided with a dramatic reduction in government funding for possum control due to the government believing that the possum-related TB problem had effectively been solved (O'Neill & Pharo, 1995). In Britain, this coincided with the change from gassing to a trapping-based control strategy. In both countries, the net effect was a dramatic reduction in the number of wildlife reservoir individuals killed. Since 1995, the incidence of TB in cattle in New Zealand has fallen (Animal Health Board, 1997), whereas that in Britain has continued to rise at an increasing rate.

Despite the lack of an overall relationship between TB in cattle and possum control in New Zealand, possum control has nevertheless been successful in controlling TB in cattle in specific areas (Animal Health Board, 1997). In New Zealand, possum control operations are undertaken proactively in Vector Risk Areas where TB is known to occur in the wildlife. The aim of initial possum control operations in New Zealand was to reduce the population by 75%. Since 1996, a Residual Trap Catch Index (RTCI) measure has been used, and all operations must now have a RTCI of less than 5%. This approximates to a population reduction in excess of 80%. It appears that this proactive strategy, with a very high initial vector population reduction followed by maintenance control, is finally achieving success in controlling TB in cattle in some areas in New Zealand. These results are also in accordance with the predictions of earlier simulation models (Barlow, 1991ab), which suggested an initial cull of 70% would be sufficient to virtually eradicate TB in possum populations.

IMPLICATIONS FOR THE CONTROL OF TB IN CATTLE IN BRITAIN

Various models have been produced that address the control of TB in cattle in Britain in a similar way to those of Barlow for TB in New Zealand. White & Harris (1995ab) and White et al. (1997) showed that endemic TB could be brought to very low levels in badgers by an initial cull of 80% or more of the population (Table 12:4). However, for a cull of 90% or less, it took 10-15 years for the disease to be eradicated, and even with a cull of 95%, the disease was never eradicated within less than six years. These predictions were broadly in accordance with those from other, independent modelling studies by Smith et al. (1997) and Swinton et al. (1997). Generally, the disease could only be eradicated completely by either a very high initial cull (in excess of 90%) or repeated high intensity control efforts.

Of the badger control strategies employed to date in Britain, White & Harris (1995a) estimated that only gassing and the clean ring strategy would have achieved culling rates at or above 80%. However, for the clean ring strategy, their simulation model assumed a perfect ring-based strategy, which included re-visiting previously trapped areas. Complications in the field would have meant this was not achieved in practice, so efficiencies of culling would be reduced, possibly significantly.

Table 12:4 Frequency distribution of the percentage of simulations in which successful control of endemic TB in badgers was achieved within specific time periods following a single culling operation of varying efficiency. The results are based on 50 simulations of the spatial stochastic model of White & Harris (1995b) with a disease-free equilibrium group size of eight adults and yearlings. Modified from White & Harris (1995a).

Culling efficiency	Time taken for successful control of endemic disease (years)				
	<5	5-9	10-14	15-19	≥20
50%	0	0	0	0	0
60%	0	0	0	0	0
70%	0	0	0	0	0
80%	0	0	6	2	0
90%	6	28	22	8	10
95%	22	44	12	14	4

Basic disease epidemiology informs us that the greater the reduction in the host population, the greater the likelihood that a disease will be unable to persist in that population. None of the badger culling-based strategies used to date have been effective in reducing TB in badgers. The results of the models discussed above suggest that a major reason for this might be that insufficient badgers were killed to reduce the level of TB in the badger populations significantly. Indeed, the repeated occurrence of herd breakdowns in many areas where badgers had been culled reinforces this view.

On the basis of their modelling work, White & Harris (1995a) concluded that no reactive strategy based on badger culling would provide a long-term solution to the problem of bovine tuberculosis in badgers. They recommended that culling directed proactively in areas with a recent history of TB in cattle would have a far greater chance of success. This is similar to the strategy that has since been adopted, with indications of early success, in New Zealand. In Britain, a proactive culling strategy was finally included as part of the culling trial, four years on from these initial recommendations.

THE PUBLIC ACCEPTABILITY OF BADGER CULLING

So far, we have drawn comparisons between the TB situation in Britain and New Zealand. Many aspects of the disease situation are similar. In both countries, the eradication of TB in cattle has been complicated by the existence of a wildlife reservoir. It is also widely argued that to reduce TB in cattle further, significant numbers of this reservoir need to be culled. However, this is where the TB situations in the two countries diverge. In New Zealand, possums are a major pest. They are an introduced species, and as well as serving as a reservoir for TB, they also pose a major threat to the native podocarp forests.

Studies of public attitudes to possums and TB control in New Zealand have been carried out by the Ministry of Agriculture and Forestry and Landcare Research

(Ministry of Agriculture and Forestry, 1999). They questioned 1,127 members of the New Zealand public and found that most had unfavourable attitudes to possums. Most respondents were aware that possums threatened the country's wildlife and carried TB, and only 10% believed that possums were native to New Zealand and harmless. The effect of possums on the country's native wildlife was considered much more important than their role in carrying TB. Not controlling possums at all was seen as unacceptable, and therefore the study concentrated on the public acceptability of different forms of control. Of those forms of control considered, shooting and trapping were the most acceptable, and poisoning — the current strategy — was the least acceptable, due to concerns about the effects on other species and the human population.

White & Whiting (2000) carried out a preliminary analysis of public attitudes to badger culling to control TB in Britain. They considered three alternative management strategies: widespread culling, the current culling trial and no culling. They interviewed 100 people selected opportunistically in York and Glastonbury and asked the interviewees to give each of the three treatments a preference rating on a scale of 1 (least preferred) to 10 (most preferred). These preference ratings were based on summary information provided on the cost of badger culling, the effect on TB in cattle and the effect on badger populations of each of the three treatments. The preference ratings are summarised in Figure 12:2. The single most preferred treatment was no culling, and the least preferred was the widespread cull. Analysis of the responses using logistic regression showed that those people favouring the widespread cull or the culling trial cited the level of TB in cattle as the primary factor guiding their

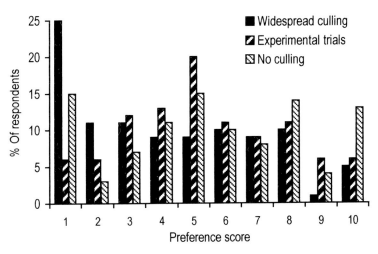

Figure 12:2 Frequency distributions of preference scores for three TB management strategies (widespread badger culling, the culling trials and no badger culling) based on interview questionnaires of 100 Glastonbury and York residents. Data from respondents from the two towns were not significantly different, so only overall totals are shown for clarity. A preference rating of 10 indicates the highest, and 1 the lowest, preference. Data are taken from White & Whiting (2000).

preferences, whilst those who favoured no culling cited conservation and welfare impacts on badgers as the most important factors. Only 30% of respondents rated widespread culling as their preferred treatment, and 47% of respondents gave it a preference rating of 3 or less. These results suggest that proactive, widespread culling of badgers would be an extremely unpopular strategy for the control of TB in Britain.

CONCLUSIONS AND RECOMMENDATIONS

The culling trial currently underway aims to compare the effectiveness of proactive culling, reactive culling and no culling of badgers in controlling TB in cattle. The trial has numerous problems associated with it. These include access restrictions imposed by landowners, thereby preventing MAFF/DEFRA from killing badgers in certain areas and so blurring the differences between treatments, the illegal killing of badgers by farmers in no-control areas (hence invalidating the experimental controls), and possible edge effects from neighbouring areas with different treatments. There are also severe doubts as to whether the original aim of removing all badgers from an area is achievable.

For example, MAFF removed 238 badgers from the Putford area of the triplet on the Devon/Cornwall border in their proactive trapping operation in December 1998. This proactive removal area was surveyed independently for badger activity between March and June 1999, three months after the MAFF badger culling operation. The survey followed the methods outlined in Wilson et al. (1997). The area was subdivided into one-kilometre squares and all the linear features were walked within each square. The location and type (main, disused main, annexe, subsidiary or outlying) of all setts, the number of well-used, partially-used and disused badger holes within each sett and any signs of sett disturbance by humans were recorded. Access was gained to 94% of the 100 km^2 study area. Of the 332 setts found in total, 80 were main setts and 11 were disused main setts. Extrapolating to the total area of the site, this gives a total of 85 main setts. Using the regression equation originally formulated by Wilson (1998), and updated by Sadlier & Harris (unpublished data), to estimate adult badger group size (G_A), based on the number of active holes found per main sett (h_a), $G_A=0.52 * h_a + 1.66$, it was estimated that the average group size in the Putford area was 5.5 (±3.7) adult badgers (95% prediction intervals). This extrapolates to a total adult badger population of 468 (±315). Based on reproductive data from Cresswell et al. (1992), the total number of litters in the area would be 77 (±52) with an average of 2.7 cubs per litter. This gives an approximate total number of cubs born in spring 1998 of 208 (±140). Thus, in spring 1998, the estimated total badger population for the removal area was 676 (±455). The total badger population would have declined during the year between spring 1998 and December 1998 when the culling operation took place. The exact level of this decline is unknown but there is likely to have been a minimum of 19% mortality overall (White & Harris, 1995b), resulting in an estimated population in December 1998 of 548 (±368). The 238 badgers initially removed by MAFF falls just within the lower 95% prediction interval from this relationship, but is nevertheless likely to be well short of total removal intended in the trial and also short of the *circa* 80% efficiency required for the control of bovine TB from a one-off culling operation (White *et al.*, 1997). The relative inefficiency of the initial trapping operation is also indicated by the results of follow-up trapping by MAFF in November 1999, which removed a further 85 badgers. It is likely that further annual follow-up control operations by MAFF will remove more badgers in due course.

Notwithstanding these practical complications, there is no indication of how the trial will relate to future badger control strategies. The trial has no clearly defined goal and no recommendations in terms of what should be done in relation to the findings. It is likely to show that proactive control is the most effective approach, and that killing a very high proportion of the badgers in an area will reduce TB in cattle in that area. If this is the case, it would confirm the findings of all the modelling studies that have been done to date, reinforce the conclusions from MAFF's early badger removal studies, and also be in accordance with the results from studies in County Offaly in Ireland (O'Mairtin *et al.*, 1998). However, in terms of policy, it is unclear where this would lead. If badgers are indeed the principle source of TB in cattle, and other wildlife species such as deer are unable to maintain the disease in the absence of badgers, then removing all badgers will remove TB in cattle. However, removing badgers more or less completely from all areas where TB in cattle occurs would not be acceptable as a long-term solution, and the data from the Devon/Cornwall border suggest that it would also not be easily achievable.

The most pertinent policy questions are in fact more subtle. For example, what level of reduction of TB in cattle should be aimed for, and what is the best way of doing this? If killing badgers is thought to be the best way forward in the short-term, what proportion of badgers need to be killed? What is the relationship between badger population reduction and cattle TB reduction, what form does this relationship take, and what are the marginal effects of badger removal on the level of TB in cattle? Although the original design of the experiment as envisaged by Krebs *et al.* (1997) would not have addressed these questions, the very fact that there will be differences in removal efficiencies between areas under the present culling trial will potentially allow them to be tackled. The major stumbling block to this work will be deriving reliable estimates of badger density both before and after badger removal; good density estimates are essential to help refine the current epidemiological models. Until accurate methods exist for deriving absolute badger densities, it will not be possible to make reliable quantified comparisons in terms of the relative effectiveness of the different badger removal strategies being considered in the culling trial.

From a theoretical perspective, if badgers are the sole wildlife reservoir for TB, badger culling does have the potential to considerably reduce the TB problem in cattle. The evidence from New Zealand suggests that intensive culling of the wildlife host over wide areas can significantly reduce TB in cattle. However, possums are a major pest in New Zealand, and the acceptability of their control is unquestioned. In contrast, badgers are a species with considerable conservation interest in Britain, and large scale culling, which would effectively remove badgers from certain areas completely, is considered as unacceptable by the general public. Some proactive culling of badgers would be beneficial in the worst affected areas, and limited proactive culling of badgers might be publicly acceptable. However, badger culling alone will never solve the TB problem in Britain, even if badgers to cattle transmission is the root cause of the problem, because the scale of culling required would be neither acceptable nor achievable.

A vaccine remains the best long-term option for the control of TB in cattle in Britain (Newell & Hewinson, 1995; Krebs *et al.*, 1997). However, despite recent progress towards overcoming some of the practical problems associated with such a vaccine (e.g. Vordermeier *et al.*, 1999), this goal remains a long way off, and interim strategies are required. Whilst proactive culling of badgers would probably be useful in the worst affected areas, other responses such as changes in husbandry and alternative land use also have the potential to be of considerable benefit.

The long-term solution to the problem of TB in cattle in Britain undoubtedly lies in the hands of scientists working on vaccine development, but the short- and medium-term responses are more questions of policy, husbandry and land use. Previous work has suggested that husbandry can have a significant impact on the potential for transmission of TB from badgers to cattle (Hutchings & Harris, 1997), and a case-control study is currently underway which will include a consideration of the role of husbandry, together with environmental and climatic factors, in the incidence of TB in cattle. If significant linkages were established between TB incidence and specific cattle husbandry practices, this would be a major step forward in terms of identifying practical management that could be carried out to reduce the risk of the disease in cattle in the short-term. In the medium-term, especially if the problem in cattle continues to worsen, we may need to reassess our farming practices in terms of land use in certain areas. For example, it may not be economically worthwhile to continue to farm dairy cattle in TB hotspot areas until an effective means of controlling TB in cattle is found. Such a course of action would undoubtedly be unpopular with the farmers in the affected areas, and would in no way be a long-term solution. Nevertheless, it would act to reduce the immediate problem of TB levels in cattle. This would maximise the number of herds qualifying for official tuberculosis-free status according to the EU directive 64/432/EEC and would, therefore, be of longer-term benefit to the farming community as a whole.

13

Bovine tuberculosis in badgers: Can culling control the disease?

R J Delahay, G Wilson, L M Rogers & C L Cheeseman

INTRODUCTION

Bovine tuberculosis (TB) caused by *Mycobacterium bovis* is a serious disease in British cattle, which can also affect a wide range of other mammals including humans. In the early 1970s, the persistence of herd breakdowns in the south-west of England was associated with a relatively high prevalence of infection in badgers *Meles meles* (Muirhead *et al.*, 1974). Cattle are thought to be at risk of infection from environmental contamination with urine, sputum and faeces from infectious badgers (Muirhead *et al.*, 1974; MAFF, 1979), and possibly also via direct contact with dead and moribund badgers (Dolan, 1993; Flanagan, 1993).

The removal of infectious individuals by lethal control is a principal tool of disease management. However, where infectious individuals cannot be accurately identified from a wildlife reservoir, then local reduction in the size of the reservoir population may be considered. In the absence of a rapid and accurate means of detecting *M. bovis* infection in live badgers, the latter has essentially been the approach of successive UK government policies to control the transmission of bovine TB from badgers to cattle. However, despite approximately 27 years of attempting to control herd breakdown rates by culling badgers, the incidence of disease in British cattle has continued to rise. The only period of substantial decline in UK herd breakdown rates occurred from 1975 to 1979 during the first four years of the badger gassing strategy. It is tempting to interpret this as evidence that gassing was the only strategy that was successful in reducing herd breakdown rates. However, direct comparisons of the various strategies employed are not scientifically valid as they took place at different times. Consequently, the impact of badger culling on the trend in cattle herd breakdowns is potentially confounded by a range of other factors that may have co-varied over this period. (Woodroffe *et al.*, Chapter 11 and Harris & White, Chapter 12 provide further details of past and present culling strategies and their efficacy).

In 1975, a long-term study of the ecology of a naturally infected badger population began at Woodchester Park in Gloucestershire. Since then, the Central Science Laboratory has carried out ecological and epidemiological investigations of this and other badger populations. Such intensive, long-term studies of mammal populations are rare — notable exceptions include the studies of red deer *Cervus elaphus* on Rum (e.g.

Clutton-Brock *et al.*, 1997), and Soay sheep *Ovis aries* on St Kilda (e.g. Clutton-Brock *et al.*, 1996) — but they provide unique opportunities to study basic ecological and epidemiological concepts, and to explore potential options for the management of complex ecological problems. The purpose of this chapter is to highlight research findings from the Woodchester Park study that have particular implications for badger culling as a strategy for the control of TB in cattle. Our review of these data indicates several lines of evidence that raise serious questions about the likely efficacy of badger culling as a method of controlling bovine TB in cattle, and may go some way towards explaining the lack of success of previous policies to control this wildlife disease reservoir.

THE WOODCHESTER PARK STUDY

Woodchester Park lies on the Cotswold escarpment within an area of high cattle herd breakdown risk. The landscape is dominated by a mixture of fragmented deciduous woodland, coniferous plantations and pasture. Dairy and beef cattle farming predominat, with smaller areas of arable and sheep grazing. A local abundance of Cotteswold sandstone and well drained soils provide ideal conditions for the construction of badger setts, which are concentrated under woodland cover, whilst the proximity of managed grasslands provides excellent conditions for foraging on invertebrates, particularly earthworms. The high local density of badgers (Rogers *et al.*, 1997) is likely to have been facilitated largely by the combination of a local abundance of optimum habitat and an absence of persecution.

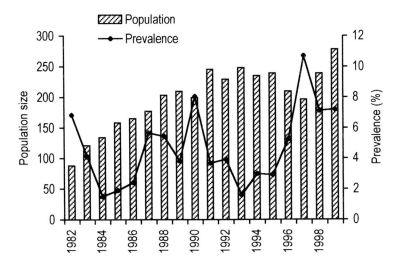

Figure 13:1 Population size (expressed as the number of adults caught in the core of the study area) and prevalence of infectious (culture positive) badgers at Woodchester Park (1981 to 1999).

The Woodchester Park study area covers approximately 11 km², which has in-corporated up to 36 badger social groups. Within this is an intensively studied core area of approximately 7 km², which has contained 21-25 social groups since 1981 (Rogers *et al.*, 2000). The configuration of badger social group territories is determined annually by bait marking (Delahay *et al.*, 2000a) each spring.

From 1977 to the present, badgers have been routinely captured in cage traps placed next to active setts and baited with peanuts. Individual badgers are captured on average twice each year. Captured badgers are anaesthetised by an intramuscular injection of ketamine hydrochloride, and permanently marked with a unique tattoo (Cheeseman & Harris, 1982). The infection status of captured badgers is determined by bacterial culture of faeces, urine, sputum, pus from abscesses and bite wound swabs (Clifton-Hadley *et al.*, 1993), and a serological test for the presence of antibodies to *M. bovis* (indirect ELISA; Goodger *et al.*, 1994). The ELISA test is not sufficiently sensitive to be used in isolation to identify infection in live badgers for disease management purposes (Clifton-Hadley *et al.*, 1995a; Woodroffe *et al.*, 1999), but in conjunction with sequential sampling and culture results it provides a useful research tool. After a period of recovery, all badgers are released at the point of capture.

BADGERS AND TB AT WOODCHESTER PARK

Badger density and disease prevalence

Rogers *et al.* (1997) described a steady increase in badger population density from 7.8 per square kilometre in 1978 to 25.3 per square kilometre in 1993 in the core of the Woodchester Park study area. The period of most rapid population growth was from 1980 to 1988, with a levelling off in numbers during recent years (Rogers *et al.*, 1999). However, the prevalence of infectious badgers detected in the population has fluctuated widely (Figure 13:1) and bears no linear relationship to badger density (Spearman Rank Correlation r=0.004, n=18, p=0.98).

There appears, therefore, to be no simple linear relationship between the preva-lence of infectious badgers and the density of the population. Under this scenario, the removal of a given fraction of the local badger population (achieved by a given culling effort) will not necessarily yield a proportionate reduction in the number of infectious badgers present and, by extrapolation, may fail to provide a concomitant reduction in the disease risks to cattle.

The aggregation of disease

A second barrier to effective culling lies in the highly aggregated pattern of infection in the population, which means that the target for efficient disease control may be relatively small. The spatio-temporal distribution of infected individuals in the Woodchester Park population was investigated for the period 1982 to 1996 (Delahay *et al.*, 2000b). Throughout the study period, infection remained concentrated in social groups at the west end of the study area, although temporal trends in disease were not synchronised amongst them. Infection persisted for several years within certain groups.

However, variations in the proportion of new infections in badger social groups were not statistically significantly related to territory size, the number of latrines therein, the number of adults or cubs, the adult sex ratio, group density or group size (logistic regressions, all p>0.1).

Evidence from Woodchester Park and other populations (Cheeseman *et al.*, 1981; Cheeseman *et al.*, 1985) shows that infected social groups may be found in close proximity to apparently disease-free groups. Unfortunately, as the diagnostic tools required to target only infected groups are currently unavailable, the geographical extent of culling operations must be determined by other means, which may at best be epidemiologically inaccurate, or at worst irrelevant.

Movement and disease incidence

The third line of evidence from the Woodchester Park study to suggest that culling is unlikely to successfully control TB in badgers relates to what has been referred to as the 'perturbation effect' (Swinton *et al.*, 1997; Tuyttens *et al.*, 2000). Typically, in undisturbed badger populations, individuals live in social groups, each of which occupy and defend a territory (e.g. Kruuk, 1978; Cheeseman *et al.*, 1981; Kruuk & Parish, 1982; Cheeseman *et al.*, 1985). This segregation of the population is likely to reduce the opportunities for disease transfer between groups. However, perturbation (disruption) of this social organisation could result in increased mixing of individuals, and enhanced rates of disease transmission. This contention is supported by evidence from the Woodchester Park population, where years of increased rates of movement between social groups were followed by rises in the incidence of new infections.

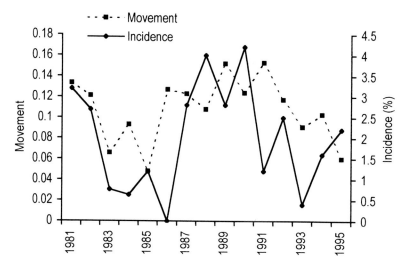

Figure 13:2 Rates of movement between badger social groups (proportion of the population that moved) and the incidence (proportion of the population that became infected) of disease at Woodchester Park (1981 to 1995).

During the period 1978 to 1995, the movement of badgers between social groups was investigated by analysis of trapping records (Rogers *et al.*, 1998). Nearly half (44%) of the 475 badgers for which detailed sequential records were available, had made either temporary or permanent movements between social groups. However, the distances travelled were small, averaging less than 1,000 m, and the majority of movements were between neighbouring social groups. The annual proportion of badgers moving between social groups was significantly positively related to cub mortality in the same year, but was not related to population density, sex ratio or group size. There was also a significant statistical relationship between the annual proportion of badgers that moved and the incidence of disease in the population in the following year. Hence, years of high movement rates were followed by an increase in the number of new cases of disease detected in the population (Figure 13:2).

Culling operations themselves can leave social disruption in their wake (see also Cheeseman *et al.*, 1993), and if they thereby increase rates of movements between groups, as evidence from Tuyttens *et al.* (2000) suggests, then this control strategy may precipitate new infections in the remaining population.

Immune status

A fourth potential problem with culling relates to the badger's natural defences against the *M. bovis* pathogen, which are still poorly understood. An intensive diagnostic investigation was carried out from 1981 to 1995 on 44 badgers from a social group at Woodchester Park that was persistently associated with endemic infection (Newell *et al.*, 1997). About 10% of the badgers showed no evidence of infection despite exposure to endemic *M. bovis* within the group. Furthermore, transient seropositivity (determined by ELISA test) was detected in several cubs during the first six to eight months, most of which then remained culture-negative for up to five years.

The culling of an indiscriminate proportion of the local badger population may not serve long-term disease control if animals with a degree of natural protection from infection are amongst those killed. Newell *et al.* (1997) identified badgers in which transient antibody responses were associated with persistent subsequent freedom from detectable infection. One possible explanation for this is that these animals have some degree of immunological protection from disease, although further research is required to confirm or refute this contention. If a degree of immunological protection against *M. bovis* does exist in some badgers, then disease control in the population would be best served by increasing the proportion of animals exhibiting this trait.

Recolonisation following culling

The fifth line of evidence that suggests badger culling may not provide a sustained reduction in risks to cattle comes from studies of post-removal recovery. Cheeseman *et al.* (1993) observed the process of recolonisation that followed the complete clearance of badgers from two clusters of social groups peripheral to the core of the Woodchester Park study area. Social organisation was severely disrupted, with badgers travelling over greater distances and using more main setts than usual, for several years following the removal. It took approximately ten years for the population to recover to pre-removal density, and infected badgers were not detected until ten years after removal. However, immigrant badgers were identified in the removal areas within one year of the operation, including cubs in one instance (CSL unpublished data). Cheeseman *et al.* (1993) suggested that, following culls that did not remove all the badgers (such as routine MAFF badger removal operations), the rate of population recovery would be more rapid. Tuyttens *et al.* (2000) recorded almost complete recovery to pre-cull population levels within the two years following MAFF badger removal operations at North Nibley, Gloucestershire.

These examples illustrate the ability of badger populations to recover from culling operations, and suggest that they may only provide transient reductions in population density. Nevertheless, following the Woodchester Park removal, infection (identified by either ELISA test or culture) was not observed in the area for ten years (Cheeseman *et al.*, 1993). However, disease may persist in social groups where only partial removal of badgers is achieved. In badger social groups that had been subjected to culling at North Nibley, for example, ELISA-positive individuals were identified within one year, and culture positives within three years of the badger removal operation (D.W. Macdonald, unpublished data).

Behaviour of infected badgers

It is still unclear under what circumstances cattle are most at risk of infection from badgers, although it is likely that contamination of pasture with infectious excretory products is important. Nevertheless, behavioural observations of infected badgers suggest that contamination of farm buildings and stored feeds may be another potentially important route of infection.

Cheeseman & Mallinson (1981) described the detailed movements of three known infected badgers on a farm in the Woodchester Park study area. All three animals were mainly solitary, and appeared to partially lose their fear of humans, one being persistently associated with farm buildings. The authors also found that in the Gloucestershire and Avon area the prevalence of infection amongst badgers found dead or in extremis in farm buildings (64%) was significantly higher than that in road casualties (21%). Ongoing behavioural studies shows that some infected badgers are habitual visitors to farm buildings throughout the year, and have been observed to contaminate cattle feed with excretory products (Garnett, *et al.*, 2002).

For certain cattle herds on farms where badgers can gain frequent access to buildings and feed this has the potential to be the single most important source of infection, because of the potential for bacilli to accumulate in an area where cattle may be concentrated. Further work is currently in progress to investigate badger access to buildings and stored feed in the hope of identifying farm management practices to minimise risks of infection to cattle.

CONCLUSIONS

The evidence from studies at Woodchester Park and elsewhere suggests that piecemeal culling of badgers, in the absence of an accurate diagnostic test, is unlikely to provide a long-term solution to reducing the incidence of TB in cattle. The use of lethal control to reduce the prevalence of disease in badger populations is hampered by an inability to target infected individuals, the potentially disruptive consequences of culling operations on badger social organisation and potentially rapid population recovery. Furthermore, indiscriminate culling may not be the best option for controlling disease in a population that may contain an unknown proportion of individuals with some measure of protective immunity. On the other hand, extensive indiscriminate removal of badgers from large areas is unlikely to be politically or socially acceptable policy for the control of TB in cattle. Clearly, both the extent to which badgers may be responsible for TB in cattle and the effects of reactive localised badger culling, need to be quantified. These are both central objectives of the field experiment currently being implemented by DEFRA (Bourne *et al.*, 1999; Woodroffe *et al.*, chapter 11). If it transpires that badgers do represent a significant risk of infection to cattle then the remaining options include vaccination, limited population control and changes to farm management practices, or a combination of these measures.

ACKNOWLEDGEMENTS

Many thanks to Paul Spyvee, Dave Handoll, Pete Mallinson, John Howell, Anton de Leeuw and Matt Waldram for their expertise in fieldwork and sampling badgers. Thanks also to Chris Kilner (DEFRA) and Richard Clifton-Hadley (VLA) for providing cattle information, the staff of the Epidemiology and Bacteriology departments at VLA for their laboratory expertise and collaboration over the years, and to Fiona Stewart for helpful comments. We are also indebted to all the farmers and landowners in the Woodchester Park study area for their co-operation during fieldwork. The projects described here were funded by the Animal Health and Veterinary Group of MAFF (now DEFRA).

14

Rodenticide use against farm rat populations: Biological constraints on effectiveness and safety

D P Cowan & R J Quy

INTRODUCTION

The Norway rat *Rattus norvegicus* thrives on many British farms. It was present on 42% of agricultural properties surveyed in 1993 (Meyer *et al.*, 1995) and just over half of the farm grain stores surveyed in 1987 (Prickett, 1992). The numbers of rats living in and around a group of farm buildings range from a few, which may go unnoticed, to many hundreds, but rarely exceed 400 (Quy *et al.*, 1993). Occasionally, however, unusually large rat populations in excess of 1,000 individuals can build up (Quy *et al.*, 1995). Although rats can cause economic losses through damage to growing or stored crops and to structures, the main concerns are potential threats to human and animal health posed by zoonotic diseases such as leptospirosis (Webster & Macdonald, 1995), hantavirus (Pether & Lloyd, 1993), salmonellosis (Taylor, 1969) and cryptosporidiosis (Quy *et al.*, 1999). Rat numbers on farms thus need to be managed to minimise these health risks.

Traditionally, this has meant resorting to lethal control techniques. Fumigation is only rarely used given restrictions on its utility close to buildings, but it can be useful in dealing with populations living away from buildings. Trapping definitely has a role to play in some settings, for instance, where it is absolutely essential to remove the last rodent. It is, however, very labour intensive and requires considerable skill to set traps in a safe but effective manner. Hence, lethal control generally means a treatment carried out with baits containing rodenticide poisons. Indeed, rodenticides have almost become the first, last and only resort for many.

Here, we first describe what is known about the patterns of rodenticide use on farms. We then consider aspects of the biology of rats that constrain the effectiveness of rodenticide treatments and how these constraints also influence environmental safety in terms of possible risks to non-target wildlife. Finally, we discuss how rodenticide use might be optimised alongside alternative approaches to managing farm rat populations.

PATTERNS OF RODENTICIDE USE ON FARMS

Before considering the effectiveness and safety of rodenticide use it is necessary to consider the properties of the various rodenticides used to control farm rats. It is usual to distinguish fast-acting, 'acute' rodenticides, such as zinc phosphide, from slow-acting, so-called 'chronic' rodenticides. These slow-acting materials are the anti-coagulants, which disrupt blood-clotting activity.

The first generation of anticoagulants arrived in the 1950s, led by warfarin and followed by others, including chlorophacinone, coumatetralyl and diphacinone. These materials revolutionised rodenticide use, the key being the delay of a number of days between consumption of the poison and the onset of symptoms. There is thus little opportunity for individuals to learn to avoid baits after consuming a sub-lethal dose. This is in marked contrast to the fast-acting rodenticides, where learned aversion, commonly called bait shyness, is a very serious constraint on effectiveness (Chitty & Southern, 1954). The availability of anticoagulants led to enhanced effectiveness and efficiency, particularly as pre-baiting prior to the presentation of baits containing rodenticide was no longer necessary to overcome learned aversion (Buckle, 1994).

The non-anticoagulant ergocalciferol or cholecalciferol, also known as vitamin D2 or D3, is in an intermediate, 'sub-acute' category. Symptoms appear at around 24 hours but Prescott *et al.* (1992) showed that small doses could cause bait shyness, and Quy *et al.* (1995) presented strong evidence that survivors of a field trial had developed an aversion towards baits containing calciferol.

The more potent second-generation anticoagulants, difenacoum and bromadiolone, arrived in the 1970s with the promise of being able to kill animals that were resistant to first-generation materials. Subsequently, the even more potent brodifacoum and flocoumafen have become available in the UK. Fatal doses of these rodenticides for susceptible rats are contained in amounts of bait equivalent to less than 10% of an animal's daily food requirement, hence these two compounds are now generally known as 'single-feed' anticoagulants.

Table 14:1 shows that the use of rodenticides on farms is dominated by the anticoagulants, with over 99% of bait used containing these compounds. The amount of active substance used is important from the perspective of the quantity of potentially hazardous material placed in the environment. However, the amount of bait used is a more useful figure for comparative purposes, as this takes into account different rodenticide potencies, as reflected in varying concentrations of active substance in bait formulations. The amounts of non-anticoagulant active substance used represent 7.4% and 3.1% of total rodenticide use on arable, or grassland and fodder crop farms, respectively. However, because of the relatively high concentrations of non-anti-coagulants in baits, they represent only 0.3% of total bait use on both farm types. In terms of the amounts of bait used, the two most extensively used rodenticides are the second-generation anticoagulants difenacoum and bromadiolone. Single-feed anticoagulants represent only very small percentages of overall use.

The data shown in Table 14:2 summarise the results of successive usage surveys, and suggest that there has been an appreciable shift in recent years away from using first-generation anticoagulants. However, there have been no increases in the relative amounts of non-anticoagulants and single-feed rodenticides used. The switch to

Table 14:1 Amounts of rodenticide active substance (AS) and rodenticide bait used on farms in Great Britain growing arable crops in 1996 (De'Ath et al., 1999) and grassland and fodder crops in 1997 (Garthwaite et al., 2000). (Fumigants have been excluded from these tables as has the narcotic alphachloralose, which is only used against house mice).

| | | Farms Growing Arable Crops | | | | Farms Growing Grassland and Fodder Crops | | | |
| | | Amount of AS used | | Amount of bait used | | Amount of AS used | | Amount of bait used | |
Active substance		Kg	%	Kg	%	Kg	%	Kg	%
Anticoagulant:	Brodifacoum	0.13	0.1	5,163	0.4	0.05	0.2	1,986	0.6
	Bromadiolone	14.17	12.6	276,690	23.7	6.72	23.0	131,926	37.8
	Chlorophacinone	12.39	11.0	243,604	20.8	3.54	12.1	63,192	18.1
	Coumatetralyl	39.91	35.6	60,339	5.2	8.79	30.2	5,328	1.5
	Difenacoum	28.66	25.5	550,052	47.0	6.65	22.8	134,240	38.5
	Diphacinone	0.45	0.4	8,901	0.8	0.27	0.93	5,434	1.6
	Flocoumafen	0.01	<0.1	284	<0.1			91	<0.1
	Warfarin	8.23	7.3	20,544	1.8	2.22	7.6	5,964	1.7
	Total Anticoagulant	103.95	92.6	1,165,577	99.7	28.24	96.9	348,161	99.7
Non-anticoagulant:	Calciferol	0.03	<0.1	30	<0.1				
	Calciferol+difenacoum	3.33	3.0	3,253	0.3	0.86	3.0	840	0.2
	Cholecalciferol+difenacoum					0.05	0.2	64	<0.1
	Zinc phosphide	4.89	4.4	243	<0.1				
	Total non-anticoagulant	8.25	7.4	3,526	0.3	0.91	3.1	904	0.3
	Total rodenticide	112.20	100	1,169,103	100	29.15	100	349,065	100

Table 14:2 Average amounts of rodenticide bait used per farm and amounts of bait containing first-generation anticoagulants (chlorophacinone, coumatetralyl, diphacinone and warfarin), second-generation anticoagulants (bromadiolone and difenacoum), single-feed anticoagulants (brodifacoum and flocoumafen) and non-anticoagulants (calciferol, cholecalciferol and zinc phosphide) expressed as percentages of total bait used from each survey. (Fumigants have been excluded from these tables as has the narcotic alphachloralose which is only used against house mice).

Farm Type	Year	Kg of bait per farm	% First-generation anticoagulant	% Second-generation anticoagulant	% Single-feed anticoagulant	Non-anticoagulant
Arable crops in England[1]	1988	32.48	51.07	47.40	0.39	1.14
Grassland and fodder crops in England and Wales[2]	1989	27.49	42.49	55.91	0.84	0.76
Arable crops in England[3]	1990	31.11	52.04	46.40	0.68	0.88
Arable crops in GB[4]	1992	18.99	41.28	56.70	1.64	0.38
Grassland and fodder crops in GB[5]	1993	21.33	40.41	58.44	0.84	0.32
Arable crops in GB[6]	1994	22.52	54.23	43.33	1.54	0.9
Arable crops in GB[7]	1996	15.88	28.52	70.72	0.47	0.3
Grassland and fodder crops in GB[8]	1997	16.69	22.89	76.25	0.60	0.26

[1] Olney et al. (1991a)
[2] Olney et al. (1991b)
[3] Olney & Garthwaite (1992)
[4] Olney & Garthwaite (1994)
[5] Olney et al. (1994)
[6] Thomas & Wild (1996)
[7] De'Ath et al. (1999)
[8] Garthwaite et al. (2000)

Table 14:3 *Location of use of rodenticides expressed as a percentage of total use with respect to amount of active substance and amount of rodenticide bait used (Fumigants have been excluded from these tables as has the narcotic alphachloralose which is only used against house mice).*

Farm Type	Year	Inside Buildings		Around Buildings		Away From Buildings	
		% Active Substance Used	% Bait Used	% Active Substance Used	% Bait Used	% Active Substance Used	% Bait Used
Arable crops in England[1]	1988	42.0	-	35.3	-	22.6	-
Grassland and fodder Crops in England and Wales[2]	1989	46.6	-	40.8	-	12.7	-
Arable crops in England[3]	1990	38.2	-	48.3	-	13.6	-
Arable crops in GB[4]	1992	53.5	-	38.2	-	8.2	-
Grassland and fodder crops in GB[5]	1993	61.9	-	28.4	-	9.7	-
Arable crops in GB[6]	1994	39.4	38.6	35.6	30.2	25.0	31.2
Arable crops in GB[7]	1996	58.7	55.8	36.5	37.5	4.8	6.7
Grassland and fodder crops in GB[8]	1997	54.5	40.5	34.6	44.3	10.9	15.2

[1] Olney *et al.* (1991a)
[2] Olney *et al.* (1991b)
[3] Olney & Garthwaite (1992)
[4] Olney & Garthwaite (1994)
[5] Olney *et al.* (1994)
[6] Thomas & Wild (1996)
[7] De'Ath *et al.* (1999)
[8] Garthwaite *et al.* (2000)

second-generation anticoagulants has been accompanied by a reduction in the average amount of bait used per farm. This could reflect the greater potencies of the second-generation anticoagulants. Alternatively, it could represent increased pressure on farming costs, leading to reduced emphasis on rodent control.

Where rodenticides are used may influence the risks posed to non-target wildlife; for instance, the highly potent single-feed anticoagulants are restricted to indoor use only. Since the disappearance of corn ricks, significant populations of house mice *Mus domesticus* are rarely found away from buildings in agricultural habitats on mainland Britain (Richards, 1989; Berry, 1991). For instance, in a study of the house mouse on arable land by Rowe & Swinney (1977), 87% of house mice were caught inside farm buildings, while less than 5% were caught in hedgerows. Hence, it can be assumed that rodenticide use around and away from buildings is largely targeted at rats. No clear-cut trends in rodenticide placement patterns are apparent, although there is some suggestion that use around buildings has increased (Table 14:3). However, this is more at the expense of use away from buildings rather than that inside buildings. This suggests that rat control has become more focused on the populations living in and around farm buildings, instead of an increase in emphasis on rat rather than mouse control.

EFFECTIVENESS AND SAFETY

The usage surveys (Table 14:1 to Table 14:3) reveal extensive use of rodenticides on farms. How effective and safe is this use? The aim of rodenticide treatments must be to achieve the desired reduction in the target rat population as quickly as possible, while using the minimum amount of rodenticide. However, rodenticides are not species-specific but are inherently toxic to all vertebrates. Their use thus poses potential risks to non-target wildlife.

Wildlife might be put at risk through two routes. The first is the direct or 'primary' route, which is consumption of bait. Secondly, there is the indirect or 'secondary' route, where predators and scavengers are exposed to rodenticides by consuming prey containing rodenticide residues. These prey can be either target or non-target species. Use of rodenticides around farm buildings, and particularly away from farm buildings, may represent greater primary and secondary risks to non-target wildlife than use inside buildings, where other wildlife is rare (Taylor & Quy, 1978; Fenn et al., 1987; Quy et al., 1999). The actual risks to wildlife are difficult to quantify and will vary with a multitude of factors, including the methods and rodenticides used, and the behaviour, physiology and ecology of both target and non-target species.

The numbers of vertebrate incidents involving rodenticides recorded in the Wildlife Incident Investigation Scheme (WIIS) are modest (Table 14:4, Shore, Fletcher & Walker, Chapter 4). However, the majority of these incidents involve deliberate abuse or misuse, and many involve livestock or companion animals rather than wildlife. Some incidents are associated with urban or suburban settings. The WIIS does not, therefore, identify the extensive rodenticide use on farms as being a major threat to wildlife. Some concern has been expressed that the number of rodenticide incidents has increased, perhaps as a result of increased use of the more potent second-generation anticoagulants. However, no clear trend is apparent in Table 14:4 to support this view. Nevertheless, WIIS is reliant on incidents being recognised and reported, and there are incidents involving approved use (10.8% of all rodenticide incidents in Table 14:4, see also Luttik *et al.* 1999). In addition, there are consistent reports of rodenticide residues being found in both mammalian and avian predators and scavengers (Newton *et al.*, 1990; Shore *et al.*, 1996; Birks, 1998; Newton *et al.*, 1999; see also McDonald & Birks, Chapter 9). The significance of these residues is unclear in terms either of threats to the health of the individuals concerned, or for individuals where lethal exposure may have gone unnoticed. It is, therefore, reasonable to examine what factors are likely to lead to non-target exposures from approved rodenticide use, and consider whether these may be subject to change in the future.

Intuitively, the more rodenticide bait that is placed in the environment and the longer it is left there, the greater the potential risk to non-target wildlife. Effectiveness may also be reflected in these same parameters. Therefore, effectiveness and safety can be

Table 14:4 Number of wildlife incidents involving rodenticides recorded by the Wildlife Incident Investigation Scheme. (For 1998 and 2000 these are the numbers of incidents where death has been attributed to another pesticide but odenticide residues have also been found).

Year	Deliberate Abuse	Misuse	Approved Use	Un-specified Use	Other	Total	% Of all Vertebrate Incidents
1992[1]	7	5	3	8	0	23	15%
1993[2]	6	8	6	12	2	34	18%
1994[3]	6	11	3	15	0	35	18%
1995[4]	8	9	0	16	0	34	19%
1996[5]	12	12	6	13	0	43	22%
1997[6]	7	9	0	11	0	27	16%
1998[7]	7	20	2	9	0	38	22%
1999[8]	3	16	6	18	2	43	33%
2000[9]	10	10	8	8	8	36	24%

[1] Fletcher & Hunter (1992)
[2] Fletcher *et al.* (1994)
[3] Fletcher *et al.* (1995)
[4] Fletcher *et al.* (1996)
[5] Fletcher *et al.* (1997)
[6] Fletcher *et al.* (1998)
[7] Fletcher *et al.* (1999)
[8] Barnett *et al.* (2000)
[9] Barnett *et al.* (2002)

linked. In some cases, factors that constrain effectiveness will also constrain environmental safety. In other instances, there may be a trade-off between effectiveness and safety. The factors that are likely to have the most impact on treatment length and outcome (with origins in aspects of rat biology rather than baiting methodology) are feeding behaviour in relation to bait consumption, ranging behaviour in relation to recolonisation and physiological resistance to rodenticide.

Feeding behaviour

Neophobia is widely recognised as a characteristic of Norway rats, and is expressed as an initial wariness of both novel food, such as rodenticide bait, or a novel object, such as a bait container (Chitty & Southern, 1954; Inglis et al., 1996). The impact of such behaviour is initially to delay the onset of bait consumption and then to inhibit ingestion of a lethal dose of rodenticide, thereby increasing the length of the treatment.

In Hampshire, central-southern England, anticoagulants perform relatively poorly (Rennison & Dubock, 1978; Richards, 1981; Greaves et al., 1982). Quy et al. (1992a) suggested that this might have been due to lower bait acceptance. Furthermore, they suggested that this could have arisen from heightened neophobia that was heritable and had been selected for by intensive rodenticide use. Brunton et al. (1993) offered data in support of this behavioural resistance hypothesis based on observations of relatively poor performance of calciferol against rats from this area. However, there is now good evidence that calciferol can induce aversion towards baits (Prescott et al., 1992; Quy et al., 1995) and thus the poor efficacy demonstrated by Brunton et al. (1993) is explicable in terms of bait shyness, especially in the absence of pre-baiting, rather than genetic selection favouring behavioural resistance.

So, what was the underlying cause of poor anticoagulant performance in the county of Hampshire? The key factors concerned were revealed by a detailed comparison of treatments carried out on farms there and in the neighbouring county of Sussex. Full details of the methodologies and experimental design of this study can be found elsewhere (Quy et al., 1992b; Quy et al., 1993; Cowan et al., 1995). Briefly, a total of 32 standard rodenticide treatments were carried out in and around farm buildings. The treatments involved seven weeks baiting with a loose grain formulation containing either 0.005% difenacoum or 0.005% bromadiolone. Food sources for the rats on each farm were characterised, and censuses of the rat populations were carried out before, in the middle of and after each treatment. Daily estimates of the size of the population present on each farm were obtained by linear interpolation between each of the successive census estimates. Bait was laid throughout the infested areas (including along any infested hedgerows leading away from farm buildings) and the baits were placed in wooden bait containers. Chemical markers incorporated into the bait could be recovered from the bodies of rats that ingested them. Analyses of the bodies of survivors of the treatments revealed that, in general, they had eaten insufficient bait to ingest lethal doses of rodenticide (Quy et al., 1992b; Cowan et al., 1995). The study

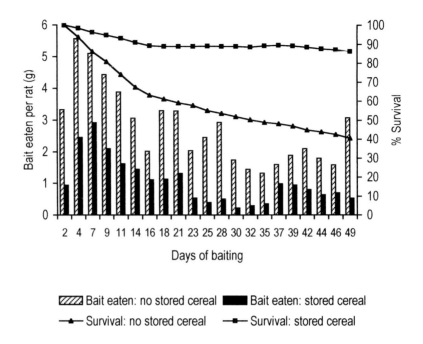

Figure 14:1 Estimated mean amounts of bait eaten per rat per day (bars) and mean percentages of populations surviving (lines) during 32 seven-week treatments with baits containing 0.005% w/w difenacoum (n=16) or 0.005% w/w bromadiolone (n=16) on farms in Hampshire and Sussex. The average consumption of bait per rat for each day for each treatment was calculated by dividing the total bait consumption recorded at each visit by the number of days between visits and the numbers of rats estimated to be alive at that time.

thus demonstrated that the key constraint on effectiveness was poor bait consumption. The main factors in generating poor bait consumption related to the predictability of food availability on the farm (Quy et al., 1992b; Quy et al., 1994). The particular importance of the availability of stored cereal in influencing effectiveness is illustrated in Figure 14:1. Treatments carried out in the absence of stored cereal generated greater bait consumption and thus had a greater impact on rat numbers than treatments carried out in the presence of stored cereal.

The rat thus responds differently towards bait in different habitats. Where a favoured food supply is consistently available, neophobia increases, while where it is restricted or frequently changing, it is reduced so that even unpalatable baits are widely accepted (Quy et al., 1996). This flexible behaviour is what might be expected from such an opportunistic species. It also occurs in captive colonies of wild rats where, against a background of constant change, rats readily enter novel bait containers that, under stable conditions, may be avoided for weeks (Watkins et al., 1999). On typical arable farms, which are particularly abundant in Hampshire, cereal is often stored and available to rats post-harvest for a number of months. Furthermore, there may be little subsequent disturbance in and around the farm buildings during autumn and winter.

Here, rats will have the opportunity to become familiar with a preferred, predictable and abundant food supply against which a rodenticide bait has little chance of competing successfully.

In terms of environmental safety, the length of treatments will be extended under ecological conditions that constrain bait consumption. In consequence, non-target wildlife has potentially greater access to bait. The main method of managing this primary risk is to restrict access. This can be achieved by placing the baits down rat burrows and lightly covering the entrance or using natural cover. Where such cover does not exist, bait containers can be used, although this can further constrain effectiveness and increase treatment length (Cowan *et al.*, 1994; Inglis *et al.*, 1996; Quy *et al.*, 1996). Such risk management may restrict access to larger birds, although small passerines have been known to be killed by calciferol baits placed inside bait containers (Quy *et al.*, 1995). The use of such practices is unlikely to restrict access to non-target small mammals, whose deaths are likely to go unrecorded. Such small mammals may in turn pose a secondary risk to predators or scavengers.

The consequences of constraints on bait consumption are different for secondary risks. If rats do not eat rodenticide then they clearly cannot transfer it to other wildlife. The actual secondary risk will depend primarily on three factors: firstly, the reliance of a particular predator or scavenger on the target rat population; secondly, the number of rats containing rodenticide residues; and thirdly, the distribution of those residues in the rat population. It is beyond the scope of this chapter to develop a full risk assessment, but we can begin to ask what combination of circumstances might generate significant risk. The unique data on daily bait consumption during rodenticide treatments shown in Figure 14:1 offer a starting point for this approach. These data can be used to estimate the average amounts of residue in the bodies of live rats during the course of the treatments (Figure 14:2). Here, the average anticoagulant residue per rat is derived, in terms of milligrams per kilogramme of body weight, by multiplying the amount of bait consumed per day (g as per Figure 14:1) by the amount of rodenticide in one gram of bait (0.05 mg) and then by four to scale for body weight (the average body weight of the rats is assumed to be 250 g). This figure is then added to the residue calculated to remain from that present on the previous day, scaled to allow for a plasma half-life of 15 hours for the cis-isomer of difenacoum (Atterby *et al.*, 2001), using the following equation:

$$\text{residue}_{(day\ x)} = (\text{bait eaten }_{(day\ x)}*0.2) + \text{residue}_{(day\ x-1)}(\text{EXP}(-((\text{LN}(2)*24)\ /15)))$$

The residues were consistently higher in the treatments carried out in the absence of stored cereal. These levels were marginally lower than those recorded by Poché (1988) for rats killed by bromadiolone (approximately 2 mg per kilogramme of body weight). They were also lower than the range of residues known to pose a threat to predatory mammals and birds that repeatedly consumed poisoned rodents under laboratory conditions (e.g. Luttik *et al.*, 1999; Newton *et al.*, 1999). However, these figures represent averages across a number of farms, and for some individual treatments the residues may have been considerably higher, at least for short periods, particularly during the

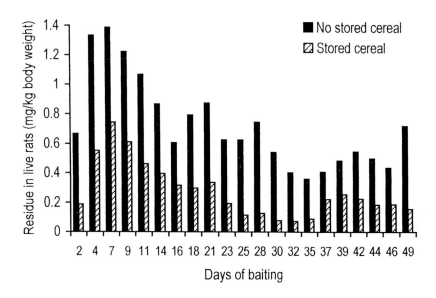

Figure 14:2 Estimated mean rodenticide residues in live rats (mg per kilogramme of body weight) during 32 seven-week treatments with baits containing 0.005% w/w difenacoum (n=16) or 0.005% w/w bromadiolone (n=16) on farms in Hampshire and Sussex (see text for details of method of residue estimation).

first week or so. Whether there is, in fact, a significant problem under some circumstances is a matter for a proper risk assessment, which would require measuring real residue levels during the course of treatments under different conditions (Luttik *et al.*, 1999).

Ranging behaviour

Rat movements, in relation to the recolonisation of cleared sites, may influence efficacy, not only in the short-term, but also in relation to the timing and frequency of re-treatments. This in turn may influence safety by determining the total amount of rodenticide used on a farm in a given period of time.

Quy *et al.* (1992a) showed that anticoagulant use against rat populations in and around farm buildings in Hampshire took longer to achieve success on farms with hedgerow rat populations than on farms without hedgerows. This probably reflected continual replacement of dead animals in and around farm buildings, by rats from reservoir populations in the hedgerows. One message from this is that control needs to be co-ordinated over a sufficiently wide area to minimise the recovery of rat populations through reinvasion. However, in the context of management of resistance to anticoagulants (see below), populations away from farm buildings may represent reservoirs of susceptible genes that can dilute the effects of selection favouring resistance arising from intensive treatments around farm buildings. Furthermore, rodenticide treatments against rat populations living in hedgerows, woodland and along watercourses may pose different risks to non-target wildlife than those in and around farm buildings.

Physiological resistance

MacNicoll *et al.* (1996) demonstrated that resistance to warfarin, and by inference other first-generation anticoagulants, is widespread in the UK. Warfarin-resistant rats can survive exposure to more than a hundred times the dose that is lethal to susceptible animals (Greaves & Cullen-Ayers, 1988), and so warfarin-resistant rats are unlikely to be able to ingest a lethal dose of first-generation anticoagulant under field conditions. Consequently, populations with a high prevalence of resistant animals are unlikely to be controlled by first-generation anticoagulants. Prolonged ineffective use in these circumstances might increase the primary risk to non-target wildlife, and there are also implications for secondary risks. MacVicker (1998) showed that residues of coumatetralyl found in rats from central-southern England were significantly higher than those from animals taken from an area of the east Midlands, where resistance has not been reported (see also Smith, 1999). Furthermore, some individuals contained body burdens of coumatetralyl at least fifty times more than the reported lethal doses. Under these circumstances, the secondary risk may be greater than for more potent but effective second-generation compounds (Cox & Smith, 1990; Smith, 1999).

The increased use of second-generation anticoagulants in recent years (Table 14:2) could reflect reduced efficacy of first-generation materials arising from widespread physiological resistance. However, the absence of any increased use of single-feed anticoagulants or non-anticoagulants suggests that this is not the case for second-generation compounds.

Concerns regarding resistance to second-generation anticoagulants were first raised for populations in central-southern England, particularly in the county of Hampshire (Redfern & Gill, 1978,1980). Initially, physiological resistance to anticoagulants was thought to underlie poor anticoagulant performance in this area (MacNicoll & Gill, 1987). However, Greaves & Cullen-Ayers (1988) pointed out that the lethal dose of second-generation anticoagulants to resistant rats from Hampshire was approximately only four times that for fully susceptible rats. They therefore argued that resistant rats could be expected to ingest a lethal dose under conditions favourable for bait uptake and, hence, physiological resistance was not, on its own, an adequate explanation for reduced anticoagulant effectiveness. As discussed above, the overriding factor causing ineffective treatments in this area was poor bait consumption, driven by ecological factors, and this low-degree of physiological resistance was not generating practical control failure (Cowan *et al.*, 1995). However, such resistance might increase treatment length. Furthermore, the average amounts of bait eaten per rat were higher on farms where physiological resistance was present (Cowan *et al.*, 1995) and thus the residues were also likely to have been higher. Hence, the presence of low-degree physiological resistance could have some impact on both primary and secondary risks to non-target wildlife.

The prevalence of low-degree physiological resistance was increased amongst survivors of treatments in Hampshire, thereby potentially sowing the seeds for selection of higher degrees of physiological resistance, which could lead to practical control problems (Cowan *et al.*, 1995). Populations of rats with high prevalences of higher degrees of resistance to difenacoum and bromadiolone have now been recognised in a limited area of north-west Berkshire, just to the north of the original focus of second-generation resistance in Hampshire (Quy *et al.*, 1995). These animals can survive very significant exposure, such that they are unlikely to succumb to practical treatments. For instance, one animal survived consumption of an estimated 450 g of bait over 23 days. A total of 248 kg of bromadiolone bait were eaten with no discernible effect on a population of in excess of 800 rats (Quy *et al.*, 1995). Animals consuming such large amounts of bait would carry very significant residues of rodenticide, which could be transferred to predators. The population was eventually controlled using calciferol baits (which involved some primary casualties in the form of small passerines) and intensive trapping.

Why this high prevalence of high-degree resistance has not spread further is unclear, but may be due to costs that the resistance trait represents for the fitness of resistant animals in the absence of anticoagulant use (Smith *et al.*, 1991). The degree of resistance to difenacoum fell over a 17-month period in which no anticoagulants were used against the Berkshire population (Quy *et al.*, 1995). However, for the fully susceptible trait to return, some susceptible genes need to be present, and it is possible that these no longer exist in the problem area. The longer high-degree resistance persists in this area the more likely it is to spread. Similarly, the longer it persists, the more likely it is that further selection will ameliorate the costs of resistance or even afford an advantage in the absence of anticoagulant use, as has been reported for some heterozygous warfarin resistant rats (Smith *et al.*, 1993; Smith *et al.*, 1994b). So at some stage a resistance management strategy will be required if the effectiveness of those second-generation anticoagulants currently approved for use outdoors is to be prolonged.

CONCLUSIONS

The use of rodenticides has tended to dominate the field of rodent control techniques, especially since the introduction of anticoagulants in the 1950s. However, the extensive rodenticide use on farms shown in Table 14:1 at best achieves only temporary reductions in rat numbers, unless steps are taken to reduce the carrying capacity of the habitat. A long-term solution needs to recognise that prevention is better than pesticide cure. The size of rat populations in and around farm buildings ultimately depends on the availability of food, but also on the amount of harbourage. Harbourage means safe nesting sites in which to breed successfully, and common recommendations made for eliminating rodents include proofing of buildings (Jenson, 1979; MAFF, 1996) and the removal of cover close to food sources (Wilkins, 1982; Meyer, 1994). This approach can be extended to environmental change designed to encourage natural enemies and competitors, so further limiting population growth (Green & Taylor, 1975; Taylor, 1975; Duckett, 1982; Engeman et al., 1997). This should not mean that one pest is replaced with another, or that there are adverse consequences in the structure of remaining populations (Lim et al., 1993); rather, that subtle changes in habitat over a closely defined area would expose rats to more predation from existing predators or more competition for resources. If successful, the effects would be seen over the long-term. These reductions in habitat quality could lead to less reliance on the use of pesticides. Ultimately the aim would be for rodenticide use to be seen as the last, rather than the first, resort in a coherent farm rat-management strategy. This would reflect the increasing emphasis on ecologically-based management of rodent pests that places less reliance on rodenticide use, and is built on a sound understanding of the biology, behaviour and habitat use of the species concerned (Singleton et al., 1999).

ACKNOWLEDGEMENTS

We gratefully acknowledge the Ministry of Agriculture, Fisheries and Food for funding this work. We would also like to thank Manuel Berdoy, Fran Tattersall and an anonymous referee for their helpful comments on this manuscript.

Deer damage to cereals: Economic significance and predisposing factors

R J Putman & P Kjellander

INTRODUCTION

Apparently in response to recent changes in agricultural practices and patterns of land use, populations of various species of deer are increasing both in number and in distribution within Great Britain and indeed throughout much of continental Europe (Gill, 1990; Harris *et al.*, 1995; Putman, 1995). In parallel with this, and perhaps also in response to increased public awareness, there has also been an increase in concern about damage to agriculture, forestry and conservation habitats (Putman, 1995, 1996; Key *et al.*, 1997; Putman & Moore, 1998; Doney & Packer, 1998).

In fact, there has, until comparatively recently, been very little objective data on the extent and significance of damage by deer to agriculture and horticulture. In this chapter, we review a number of more recent published studies to assess the actual economic significance and distribution of damage from grazing and trampling by deer, to cereals and other arable crops in the UK, and compare this to data available from other European countries.

Such data as are now available suggest that damage to agricultural crops tends to be of very local (farm-level) significance and shows high spatial and temporal variation. From our own field studies, we explore the various environmental factors that seem to be associated with damage level suffered in any instance, and briefly consider possible management options.

DAMAGE BY DEER TO ARABLE CROPS IN THE UK

The first attempt to assess, at a national level, the impact of deer (and other wildlife) on agricultural crops, was a database maintained by the Wildlife and Storage Biology section (WSB) of the Agricultural Development and Advisory Service (ADAS) between 1985 and 1989, on which were recorded all requests for assistance or advice received by the section. A smaller programme, precursor to the launch of the National database, was piloted in the West Country in 1984, providing some limited information for that year also.

Entries on this COSTER database (**C**omputerised **S**ummary of **T**echnical **R**eports) recorded date and reason for enquiry; in the case of reports of damage sustained, information was also recorded on type of damage, crop type and species believed to be responsible. We should note however that in each entry, the species recorded was that believed by the *enquirer* to be responsible for the damage; follow up visits by ADAS staff to confirm the record were only infrequently undertaken, nor was the actual severity of reported damage necessarily checked on the ground. The database thus records merely the number of cases of wildlife-induced damage reported to ADAS advisers; in addition, no distinction is made between reports of separate incidents and repeated complaints from the same individual about a continuing problem — which would be entered as if they represented independent reports. Finally, the database only logs cases of damage actually reported to WSB — and represents an (unknown) proportion of actual wildlife incidents. Despite such drawbacks, the COSTER database is one of the few sources available to us offering some relative index of the frequency with which serious deer damage is encountered — or at least reported — in different parts of England and Wales over the four-year period of its operation.

Requests for advice on deer damage account for only 1% of all wildlife 'incidents' received by WSB over the period of the survey (Trout *et al.* 1994). Over the initial period of the survey (1985 and 1986), numbers of requests received nationally for advice in relation to deer totalled 226. Regionally, these broke down as 35 in the Midlands and central regions, 51 in the east of England, 42 in the south-east, 85 in the south-west and 11 in Wales. The majority of reports referred to damage by red deer *Cervus elaphus*, fallow *Dama dama* and roe deer *Capreolus capreolus*. Overall, requests for advice related primarily to damage to grass, cereals and root crops (48%), damage to trees and forestry (10.5%), and damage to horticulture or nursery crops (10.5%). A further 65 requests over that period (31%) were for general advice on legislation, conservation and control, but did not specify the nature of damage prompting the request for information; we may presume that these, too, reflect some experience of damage, although the context is not specified.

The more detailed records for 1987 to 1989 are summarised in Table 15:1 (from Putman, 1995). Total numbers of requests for advice received during this period relating to red, roe and fallow deer was 220. Most (122) of these records related to fallow deer; 59 sought advice in relation to problems experienced with roe deer, and 39 for red. There was again considerable regional variation in reports received, reflecting differences in land use in different parts of the country, and the major crop types produced in different areas. Half of all reports concerning damage to standing cereals came from East Anglia and east England, while grassland damage was reported primarily from northern and south-western regions. Overall, damage to grass and cereals accounted for the majority of all enquiries (12 and 102 reports respectively, a total of 52% of all incidents reported).

Within this overall pattern, there were clear associations apparent between different types of damage and the deer species implicated. Thus, most reports of damage to oilseed rape or to nursery crops, garden shrubs and top fruit involved roe deer, which were rarely implicated in reports of damage to grass or other cereal crops. Reports of damage by red deer were largely in connection with pasture, silage crops or field

Table 15:1 Damage to agricultural crops in the UK and their association with the different species of deer. Data presented list the number of reports of damage to particular crop types attributed to red, fallow and roe deer in reports received by COSTER between January 1987 and March 1989.

Crop-type	Red	Roe	Fallow	Total
Forestry	4	4	12	20
Grass, cereals	11	20	65	96
Horticulture	0	9	3	12
Others	24	20	38	82
Total	39	53	118	210

cereals, while 76% of all complaints concerning damage to field cereals cited fallow deer as responsible. In general, there appears at present to be little damage to cereal agriculture in the UK from sika deer *Cervus nippon*, muntjac *Muntiacus reevesi* or Chinese water deer *Hydropotes inermis* (Putman, 1995; Putman & Moore, 1998).

These COSTER data provide a valuable preliminary insight into the geographical distribution of deer damage and the types of damage/problem species involved in each area; results of this more formal survey accord well with impressions gained from other, anecdotal sources (e.g. Mayle, 1994; Putman, 1995). However, the database records only reports received; as noted, these represent an unknown percentage of total damage caused by wildlife in any area. In consequence, the COSTER records may suffer some regional bias in recording; further, since the records do not represent 'blanket' coverage of any one area, returns may not even be representative of damage levels within each region. Finally, all incidents are recorded as reported, without further validation.

Economic significance

In a subsequent study (Doney & Packer, 1998), four specific areas of the country were selected for more detailed survey, and information on presence of deer and extent of damage to agricultural interests was sought in questionnaires distributed to approximately 2,590 farms in Gloucestershire/Somerset; Essex/Suffolk; lowland Yorkshire; and Northamptonshire. In this study, follow-up visits were also made to a number of the sites canvassed, in order to undertake independent validation of levels of damage reported.

In total, 1,192 returns were received to the questionnaire, with 69% of respondents reporting deer present on their holdings. Cereals were perceived by respondents to be the crop-type most likely to suffer damage from deer (362 respondents of the 822 reporting deer on their land), with damage also reported to grass crops (50/822 respondents), farm woodlands (241) and a lower incidence of damage reported to root crops, top fruit, vegetables and oilseed rape (Doney & Packer, 1998). In these geographic regions, deer species most commonly implicated in damage were roe and fallow.

Only about one-third of respondents considered losses reported as significant; for those farms concerned mostly with growing cereals, 17% claimed damage by deer was negligible in economic cost, while 85% overall assessed the significance of loss at £500 per annum or less for the whole farm. Farmers' assessments of damage were, however, not particularly accurate (although they were as likely to underestimate as to exaggerate economic loss: Doney & Packer, 1998). Subsequent visits to a sample of respondent farms to assess actual economic loss due to grazing of winter wheat (Doney, 1999), showed losses of up to 0.57 tons per hectare on farms that were visited on a daily or weekly basis by roe or fallow deer, but negligible economic loss, or an actual gain in yield, with lower intensity of grazing (Doney, 1999).

These results are similar to those reported earlier by Putman (1986) in an assessment of damage caused by roe deer to crops of winter- and spring-sown barley and wheat in southern England. Again, although quite high levels of grazing were recorded to vegetative parts of the crop early in the season, with up to 30% of the total crop area affected, this proved of no economic significance by harvest. In both Putman's and Doney & Packer's analyses, damaged areas of the crop showed evidence of a compensatory increase in rate of growth to catch up with ungrazed treatments by the time of harvest. Early grazing at low intensity also encouraged tillering within the crop, with an actual increase in the number of grain-bearing stems by harvest (Putman, 1986, 1989). Ears reached maturity at the same height and by the same date as those of undamaged plants; individual ears were somewhat smaller in size, but overall grain yields per square metre were not significantly reduced, and were in some cases increased (Putman, 1986; Doney, 1999).

Timing of damage in relation to the growth stage and growth characteristics of the crop will clearly affect the economic significance of any damage caused, since it will markedly influence the degree of crop recovery possible after grazing ceases. In a subsequent series of experimental trials, winter and spring-sown plots of wheat and barley were machine-mowed on a regular basis to simulate grazing. Significant reductions in grain yield at harvest were recorded where mowing continued until the end of May, or into June, for both cereal types (Putman, 1989) and *some* losses were sustained by wheat crops under any mowing regime. Under natural conditions, roe deer in southern England do not continue grazing in open fields much beyond mid-May, and thus cessation of grazing neatly coincides with the critical period beyond which it would impose lasting damage. In other areas, however, where grazing may continue for longer, or where deer of any species return to the crop to feed on the ripening grain, damage caused is absolute. Damage caused by trampling or rolling of the larger species (fallow or red deer) visiting cereal fields late in the season, when the opportunity for compensatory growth is past, may also be of real significance (Putman, 1989; Doney, 1999).

AGRICULTURAL DAMAGE BY DEER ELSEWHERE IN EUROPE

These conclusions of a generally rather low economic cost of damage by roe and fallow deer in cereal crops within the UK are not, in fact, unusual or atypical. Elsewhere in Europe it has been shown that loss of yield in maize or sugar beet crops due to grazing of vegetative parts of the plants by roe deer was also likely to be insignificant (Czech Republic: Obrtel & Holisova, 1983; Obrtel *et al.*, 1984). Thus, loss of vegetative parts of maize crops during the summer months resulted in a decrease in fresh weight of ears at harvest of only 2.6%. Since, in addition, grazing by deer affected less than 0.7% of the crop, the effective loss of yield for the crop as a whole was less than that, at 0.15% (Obrtel & Holisova, 1983). Kaluzinski (1982b) calculated that despite high densities of roe deer in agricultural areas in western Poland, consumption of vegetative parts of cereal crops outside the growing season was less than 1%, and would not significantly influence yield. Although later damage, involving direct removal of ripening ears, caused measurable and irrecoverable loss, it was nonetheless an insignificant proportion of the crop as a whole.

Recent data recorded for damage to cereals in different counties in Sweden confirm that here, too, the overall area of crops reported as suffering damage (as a proportion of the area grown in any region) never exceeds 5% and is usually lower than 1%. Over the eight-year period 1980 to 1987, the total compensation awarded for damage from red deer, fallow deer and moose *Alces alces*, averaged 8,700,000 Skr (equivalent to £670,000). Again, however, damage at a local or farm level can be significant, with up to a 26% loss of yield in unprotected oat crops against fenced controls (P Kjellander, unpublished data). We also note that while within the UK and through most of southern and central Europe, we are dealing with damage largely due to red, fallow and roe deer, in Sweden, as elsewhere in Scandinavia, much of the damage recorded is due to the significantly larger moose. Of damage reported here for Sweden, over 98% is attributable to moose — and as elsewhere in Europe, recorded damage from other species is low.

FACTORS AFFECTING DAMAGE LEVELS SUSTAINED

Although it would seem clear that damage to growing cereal crops from deer is not of economic significance on a national scale, on certain farms and in certain years, losses may be significant. Indeed one of the characteristic features emerging from all studies to date is that damage levels are extremely variable, both spatially and temporally. The degree of damage suffered, even in the same crop-type, may vary considerably between farms as well as between fields within a farm (Doney & Packer, 1998); it is equally apparent that levels of damage at any given site vary considerably from year to year.

It has become increasingly apparent in relation to the impact of browsing upon woodlands and commercial forestry, that damage levels are not related in any simple way to density (Putman, 1994; Reimoser & Gossow, 1996). Thus, at least for most species, damage levels tend to remain low and relatively constant, until the population density passes a certain breakpoint, when impact suddenly and dramatically increases (Gill; 1992, Putman, 1996). However, even at these higher densities, damage levels caused by deer show very substantial variation depending on a number of environmental and cultural factors. These include crop type, distance of the sensitive crop from cover, size of the planted area, availability of alternative preferred forages and habitat structure.

It would appear that the severity of damage caused to agricultural crops is likewise related to animal density as only one of a suite of determining factors. While damage levels recorded by Doney & Packer (1998) in the UK show a general increase with increasing deer presence, there were no significant correlations between any measure of damage caused and an index of more local deer population density based on census of pellet groups and visible trackways (Mayle et al., 2000). Equivalent studies in Sweden of the severity of damage caused to cereals by red deer, roe deer and moose have also showed no clear relationship between damage and local population density (P Kjellander, unpublished data).

As with forestry, in an agricultural context actual levels of damage sustained in any location and in any season, relate to several factors. These include overall habitat structure, proximity of the 'target' crop to cover, availability of alternative forage and the spring and summer climatic conditions in any year, which will affect not only the availability of alternative feeds overall, but also the timing of use of arable crops and thus their growth stage and the potential for recovery from early damage.

We report here an analysis of the relative importance of these various environmental factors and deer density, on damage levels sustained by a number of different cereal crops in Sweden.

The dataset

Data were available on damage to oats, barley, rye, winter- and spring-sown wheat (as well as oilseed rape) in four separate counties in Sweden: Uppsala county (60° 10'N: data from 1983 to 1985 and 1987 to 1992); Ostergotlands county (58°30'N: 1983 to 1992); Orebro county (59°30'N: 1987 to 1992); and Malmohus county (55° 40'N: 1982 to 1988). Data are based on complaints received under a compensation scheme operated in each of these counties. Since each county operates completely independent schemes of compensation, the data sets assembled here for each of the four counties are completely independent, so may be regarded as true replicates. The counties are spread from the extreme south to the north of Sweden.

Since they represent actual claims for compensation, each claim was ground-truthed before payment, so that levels of damage recorded have, to some extent, been validated. County officers visited all farms reporting claims; damaged areas were visited as soon as possible after damage and always before subsequent harvest. The recording officer reported the crop type and area damaged and also returned an estimate of the likely loss of yield within the crop by comparison to undamaged fields in the same area for that particular year. Given the somewhat subjective nature of such estimates however, our analyses here are based simply on the size of damaged areas (in hectares). In the same way, our analyses here consider damage caused from all species of deer combined, although surveying officers do record the species believed to be implicated on each occasion. Most damage, however, was caused by roe deer or moose, since red and fallow deer are very locally distributed and at low density.

No direct estimates of deer densities were available for the areas concerned; however, records are maintained in the same areas of car accidents involving deer, which, within counties, offers some index of relative abundance across years. All incidents that lead to any damage to people or vehicles must be reported by law. We do accept that such an index is imperfect: the precise relationship between deer density and accident frequency is unknown and is not necessarily linear. Direct data on deer densities in different areas were, however, unavailable.

Relationship between crop damage, deer abundance and spring temperature

With the exception of rye, damage levels recorded in all crops showed a strong and highly significant positive correlation across years. Correlations were separately calculated in each county, since climatic conditions and sampling years varied. For example, in the data for Malmohus County, annual levels of damage (in hectares) were significantly correlated between oats and barley (r=0.73, p=0.06), and between wheat and barley (r=0.91, p=0.002). Levels of damage were not significantly correlated between wheat and barley in Malmohus (r=0.50, p=0.26), but were significant in Ostergotlands county (r=0.70, p=0.02). This pattern of correlation was apparent in all four independent data sets and suggests that a high damage year is a high damage year for all crops, and *vice-versa*. The only exception to this pattern is, as noted, rye, which consistently showed a negative correlation in deer damage with that recorded in other crop-types. Rye is, however, a minor crop in Sweden, which is not grown in every year, and then only in small quantities; the apparent anomaly in patterns of damage may simply be due to small sample size.

Given the close correlation in levels of damage recorded in all other crop types, further analyses are focused primarily on oats, the most economically important and most widespread crop in Sweden. Damage levels sustained by oat crops showed no relationship with deer abundance, as reflected in the number of reported road traffic accidents (Table 15:2). As above, we acknowledge that such analysis does not definitively preclude the possibility of a relationship between damage and deer abundance, since the measure we have used here as an index of deer abundance (frequency of road traffic accidents) may not be a direct indicator of deer densities. Indeed, in other contexts, a significant relationship has been recorded between damage levels and frequency of road traffic accidents. An analysis, similar to our own here, of national statistics on farmland damage by sika deer in Japan (damage claims made over an eight year period to the Ministry of Agriculture, Forestry and Fisheries) in relation to the frequency of traffic accidents (records from the Japan Highways Public Corporation) shows a significant positive correlation (R^2=0.712; p=0.045: M Saeki, personal communication)

While the actual role of deer densities in determining levels of damage thus remains unresolved, it is clear that it is not simply density alone that affects the extent and severity of damage. In our Swedish datasets, there was a very significant relationship between the amount of damage suffered by oat crops in any one year, and spring temperature. Average temperature in March (Malmohus, the most southerly county) or March/April

Table 15:2 Relationship between the amount of damage caused to oat crops (% total crop area) in four counties in Sweden, an index of deer abundance and spring temperatures. Damage by moose and other deer species is pooled; the index of abundance used is recorded frequency of road traffic accidents (RTAs).

County	R²	p	Number of years' data	Frequency range of recorded RTAs
Uppsala	0.15	0.30	9	112-253
Ostergotlands	0.25	0.14	10	98-168
Orebro	0.16	0.44	6	163-207
Malmohus	0.32	0.11	7	2- 59

a) Regression of percentage area of oat crop damaged and road traffic accidents (RTAs).

County	R²	p	Number of years' data	Range of mean daily temperature between years
Uppsala	0.64	<0.009	9	-3.4 to +8.9°C
Ostergotlands	0.63	<0.006	10	+0.9 to +14.3°C
Orebro	0.89	<0.005	6	-4.0 to +8.3°C
Malmohus	0.86	<0.003	7	-2.6 to +3.2°C

b) Regression of percentage area of oat crop damaged and mean daily temperature in March (Malmohus) or March/April (other counties).

(other counties) accounted for between 63% and 89% of year-to-year variation in damage levels recorded, with damage reduced in years with higher spring temperature (Table 15:2; Figure 15:1). Comparable data for wheat and barley are also summarised in Table 15:3, but note that sample sizes for these more minor crops are reduced.

There are two possible explanations for such a close relationship between damage and spring temperatures. The lower damage recorded in milder years may relate to different cultural practices in such years. Seed of spring-sown cereals is generally sown earlier during milder seasons, and the crop thus starts to grow earlier; it also grows faster in milder conditions. In consequence, it matures earlier and can be harvested quickly, before, or only just within, the period when moose and other deer species would start to exploit the crop. In colder springs, the crop is sown later and also develops more slowly. As a result, it not only enters the period of maturation significantly later, but the period of maturation itself is more protracted. In such years, harvests are thus substantially delayed, and the crop remains in the fields for an extended period during the time when moose and other deer species actively graze cereals, and thus suffers greater damage before it is harvested.

An alternative explanation for the observed relationship between damage and spring temperatures might be that a warmer spring promotes the growth, availability and quality of alternative natural forages through the summer and into late summer/

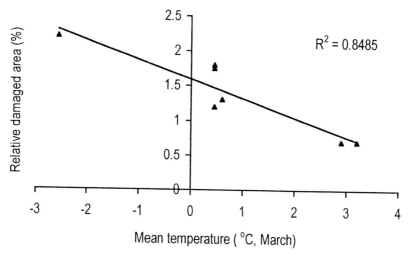

Figure 15:1 Relationship between the level of damage suffered by oat crops in any year and early spring temperature. By way of illustration, data are shown here for Malmohus County with the proportion of total crop area damaged in any year plotted against the mean daily temperature in March.

Table 15:3 Relationship between spring temperatures and the amount of damage caused to wheat and barley crops (% total crop area) in Sweden. Damage by moose and other deer species is pooled. Note that the total number of fields growing these alternative cereals is far lower than that for oats, the primary crop, and significance levels of correlation are thus lower because of reduced sampling efficiency.

County	R^2	p	Number of years' data	Range of mean daily temperature between years
Uppsala	0.60	<0.01	9	-3.4 to +8.9°C
Ostergotlands	0.52	<0.02	10	+0.9 to +14.3°C
Malmohus	0.47	<0.09	7	-2.6 to +3.2°C

a. Regression of percentage area of barley crop damaged and mean daily temperature in March (Malmohus) or March/April (other counties).

County	R^2	p	Number of years' data	Range of mean daily temperature between years
Uppsala	0.06	0.06	9	-3.4 to +8.9°C
Ostergotlands	0.37	<0.06	10	+0.9 to +14.3°C
Malmohus	0.82	<0.02	7	-2.6 to +3.2°C

b. Regression of percentage area of wheat crop damaged and mean daily temperature in March (Malmohus) or March/April (other counties).

autumn, increasing both the overall supply and the time for which alternative natural forages are available, and thus diverting feeding from agricultural crops.

Clearly, these two explanations are not mutually exclusive. We suggest that levels of damage relate both to the length of time the maturing crop is standing available in the field before it is ripe enough for harvest, and the availability throughout the season of alternative natural forages, with less damage in those years when natural foods as well as cereals, show better growth.

MANAGEMENT OPTIONS

It would appear that, in the UK and elsewhere in Europe, costs of damage by deer to cereals are generally low, and that damage of economic significance is very locally distributed and highly variable in both space and time. Management strategies for reducing impact need to be flexible and inexpensive if they are to be cost-effective. Given that there appears to be no simple relationship between damage levels sustained in any crop and deer density in the local area, management based solely on reduction of local deer populations is unlikely to be rewarded with an equivalent reduction in damage, unless deer populations are reduced to a minimum presence. Permanent fencing (deer-fencing) of individual fields is uneconomic and likely to be unproductive, since the location of damage commonly varies from year to year even on farms where a given crop may be grown in the same field in successive years.

Deer of most species implicated in damage, however, are largely dependent on woodland cover, and forage out into agricultural crops from the harbourage of nearby areas of woodland. Thus, while all species show some measure of flexibility in habitat occupance, all include at least some area of woodland within their range where possible (Putman, 1988). Even species such as roe, which may become established in agricultural areas in largely 'open-field' situations (e.g. Bresinski, 1982; Kalusinski, 1982a) are not entirely independent of some form of cover, even if only as small copses or overgrown hedgerows. Deer-fencing of the perimeter of woodlands in arable areas, at least along that edge adjacent to vulnerable crops, may help to reduce damage; Pepper (1992) provides specifications of suitable fences.

Permanent fencing is inevitably costly; some consideration may be given to protection of crops during the sensitive period of growth by electric or other temporary fencing. In general, trials of electric fences in the UK have not been successful. Fences are prone to breakdown or to shorting out on ground vegetation, and any interruption of supply will allow deer to enter the protected area. Further, such fencing does not reliably exclude deer, even when operating effectively; within the constraints of animal welfare and safety legislation, current equipment cannot produce a sufficient shock to deter red, fallow, sika or roe deer (Pepper et al., 1992). By contrast, in Sweden, fences with a measured voltage of 4.5-6.5 kV, with wires at 0.8 m and 1.7 m, and a high-visibility electric rope at 1.3 m, produced a consistent reduction in the number of sightings of moose, roe deer and fallow deer in electrically-fenced fields of oats relative to matched controls. Fenced areas showed a 26% increase in yield over unfenced fields, averaging 3,051 kg per hectare, compared with 2,264 kg per hectare in unfenced controls (Kjellander & Ahlquist, 1996).

Conflicting results have also been obtained in the use of chemical repellents or anti-feedants. P Kjellander (unpublished data) found a reduction in the use of experimental one-hectare plots in oat fields where some formulations of repellent were painted on perimeter posts, and also onto posts positioned at intervals through the crop itself, although trials have not been extended to test whether or not the repellent may effectively exclude animals from a whole field. Successful compounds included perfumed soap flakes (manufacturer's waste) and Cone oil (a waste product of fish processing, distributed by Skaraborgs skogsagare, Skovde, Sweden). Visits by moose to areas treated with soap flakes were reduced by up to 70% for the two-week period of application. In separate trials of Cone oil, visits by moose, roe and fallow deer to treated areas were reduced by 50% for up to four weeks following treatment, again with a demonstrable increase in yield (5,349 kg per hectare compared with 2,199 kg per hectare; P Kjellander, unpublished data). However, other workers have found no effect of treatment, even using the same repellents. Only one barrier repellent (Renardine) is, we understand, currently approved for use under the Control of Pesticides Regulations (1986); it appears to be ineffective against deer (Pepper, 1999).

One further management option, which may be considered to try to reduce the impact of deer grazing on crops is habitat manipulations designed to make the target crop less attractive, or to enhance the supply of alternative, preferred natural, forages in the immediate vicinity (Putman, 1998). While the use of such techniques has so far been explored largely within the context of reduction of damage within woodlands, changes in cultivation methods within the crop itself, or manipulation of the productivity of adjacent habitats, may reduce impact on agricultural crops as well. Putman (1986) noted that the timing of use of cereal crops by roe deer in the south of England corresponded very closely to a time when they were the only vegetational communities showing new growth and thus offering maximum concentration of digestible nutrients to this obligate concentrate selector. As soon as the grasses of woodland glades or rides began to flush, or buds burst on woody browse species, the deer switched their attention to these preferred foodstuffs and left the crops.

In a number of pilot trials conducted by R J Putman (unpublished data) on private estates in Hampshire, existing woodland clearings were enlarged, reseeded with early maturing varieties of ryegrass and fertilised, to provide an early flush of growth within the preferred woodland cover; use of arable fields by roe declined significantly after such treatment. B Mayle (personal communication) also reports areas where mowing woodland rides in midsummer to promote regrowth has been successful in reducing damage by fallow deer to growing and maturing cereal crops in adjacent farmland. Such techniques are still very much at the experimental phase, but may offer real potential in reduction of grazing impact on economic crops (Putman, 1998).

Overall, no single approach is likely to be effective, and in most situations managers should move towards adoption of some system of integrated management, involving both direct control of the pest population itself, physical or chemical protection of vulnerable crops and imaginative cultural changes aimed at 'diversion' of grazing or browsing pressure (Putman, 1989, 1998).

CONCLUSIONS

All available evidence suggests that damage by deer to cereals within the UK is rarely of economic significance, although deer grazing may cause some local problems in certain areas and in some years. Damage to cereal crops elsewhere in Europe is similarly rarely significant, where the main deer species present are red, fallow, roe or sika. (In those countries where moose are resident, damage may be more substantial but still tends to be only of local significance).

Despite the fact that as much as 30% of a crop area (Putman, 1986; Doney & Packer, 1998) may be affected by grazing during the vegetative phase, there is substantial compensatory growth, and actual losses at harvest are commonly greatly reduced or completely redressed. Because of this potential for recovery, it is important not to equate apparent damage early in the season to actual loss at harvest, although later damage to ripening grain stems through feeding or trampling may have more serious implications. Where significant damage does occur, it tends to be extremely local, and occurrence is highly variable in space and time. Factors affecting the severity of damage within a crop appear to include proximity of the 'target' crop to cover, availability of alternative forage and the spring and summer climatic conditions in any year; damage to crops appears to be only weakly related to deer density.

We suggest that effective management might include both physical and chemical protection of the crop (or at least the fencing, or application of chemical repellents along woodland boundaries), as well as habitat manipulations designed to increase the supply of alternative natural forages and local population reduction. Any investment in management measures designed to reduce damage should be carefully costed against the true economic loss at harvest; otherwise, costs of prevention may exceed actual costs of damage in many instances.

ACKNOWLEDGEMENTS

We would like to thank Will Manly and Fran Tattersall for inviting us to present this paper within the present volume, and would thank Jochen Langbein and an anonymous referee for comments on an earlier version of the manuscript. We thank Inga Ahlquist for all her efforts in production of the database used for our analyses of damage to cereal crops in Sweden, in entering and checking data from over 6,000 compensation claim forms; we would also acknowledge the Swedish Environmental Protection Agency for financial support for this part of the research. Finally, we would like to thank Midori Saeki for making available to us the results of her analyses of relationships between farmland damage by sika and frequency of road traffic accidents involving deer, in Japan.

16

Rabbits in the farmland ecosystem

R Trout

INTRODUCTION

The rabbit *Oryctolagus cuniculus* is an introduced species that has adapted well to farmland ecosystems. Farming operations cover 75% of the 24.1 million hectares of the UK (Nix & Hill, 2000), including 4.7 million hectares of arable, 6.6 million hectares of grassland, 5.8 million hectares of rough grazing, 0.6 million hectares of set-aside, 0.2 million hectares of horticulture and 0.5 million hectares of farm woodlands. An additional 2 million hectares in the UK is under forestry. Farmland management to produce crops for human consumption, for feeding stock or for timber, is often compromised by being very palatable to rabbits, and their damaging activities can necessitate complete replacement of crops in the worst case or render a crop uneconomic.

Rabbits became widespread in the wild in Britain only in the eighteenth century, long after their first introduction into the UK by the Normans in the twelfth century. They were originally husbanded as a highly regarded food resource by careful management in enclosed warrens. In medieval times larger areas of poor ground were managed for a useful income and rabbits were encouraged to spread and breed over large areas of poor ground, such as Breckland sites of 1,000 acres or more (Sheail, 1971). From these locations, where local damage by rabbits was severe, there was an inevitable spread to many agricultural areas following the enclosure awards. The final fragmentation of the major forested areas that probably had provided, together with major rivers, a reasonable barrier to rabbits, was also completed. Before myxomatosis became established in 1953, rabbit numbers were estimated at around 100 million (Thompson & Worden, 1956) and the damage caused estimated at £50 million annually. Snaring and gin trapping were popular methods of cropping populations, and before the Second World War, millions of rabbits were taken annually — over 4,000 tons of rabbit meat was shipped by rail to London from Wales alone.

Myxomatosis appeared in 1953 and reduced the rabbit population by over 99%, despite initial attempts by MAFF to eradicate the disease and legislation making its spread illegal. In practice, the numbers of rabbits, the market for rabbit meat and the felt hat trade all disappeared. For almost a generation, farmers and foresters were able to operate without the implications of rabbit damage and many have not been prepared for, or passed on the skills necessary, to counter the current resurgence in numbers.

Conservation & Conflict
ISBN 1 84103 001 5

The reasons for the long delay in recovery are uncertain, but are likely to include factors such as heavy predation pressure, as rabbits started to spread and excavate new warrens (since most of the old warrens had disintegrated in the intervening 25 years); repeated outbreaks of myxomatosis (though gradual replacement of virulent strains by less virulent ones and genetic resistance in rabbits were taking place); changes in the agricultural landscape (harbourage and hedgerow removal) and agricultural intensification. By the mid 1970s, rabbit populations had again started to become a problem at a local level.

Surveys by MAFF and game bag data from the Game Conservancy Trust indicated a resurgence of rabbits from the mid 1970s, and an estimated increase at a national level of about 2% per year from the 1980s (Trout *et al.*, 1986). By the mid 1980s, estimates put the damage caused by rabbits at up to £120 million per year (Mills, 1986), of which about £40 million was direct loss of cereals. Rabbit numbers are not equally abundant over Britain and some work has attempted to map this, and to evaluate those factors responsible (Trout *et al.*, 2000).

This chapter reviews the information available relating to the role of rabbits in the various agricultural ecosystems in the UK (see also Dendy *et al.*, Chapter 17 for an experimental assessment of rabbit damage). It necessarily concentrates on the direct and insidious damage caused by rabbits, but includes other impacts on the farmland ecosystem as well, including areas of conservation where their activities may be crucial in maintaining rare species and habitats.

THE IMPACTS OF RABBITS

Arable crops

It was commonplace until the 1950s to see fields of cereal with bare headlands and grazing across substantial parts of fields, especially adjacent to woodland. The greatest reported losses involved cereals. With the move towards winter wheat in the 1960s, rather than spring sowing as the main arable crop, the temporal availability of food for rabbits in the countryside has been greatly increased; previously, the food supply was low during the late autumn and winter months because of the large proportion of fallow or tilled ground.

Various estimates of the damage caused by rabbits to individual crops have been published. Gough & Dunnet (1950) suggested, on the basis of fenced plots in East Anglia, that major losses were attributable to rabbits, as up to an 80% reduction in wheat plant density was recorded. Church *et al.* (1953) analysed results from a survey of paired fenced/unfenced plots and obtained an average loss of grain of 204 kg per hectare (a 6.5% loss of yield). A repeat of the survey (Church *et al.*, 1956) indicated an 8% figure. For spring wheat, a figure of 17% loss was reported for fields sampled in Kent (Thompson & Worden, 1956). Corner (1960) measured grain and straw losses of 22% and 40% respectively due to rabbits grazing winter wheat in 1954. The spread of myxomatosis in 1953 and 1954 was followed by reported increases of up to 30% in crop yield, and an estimate by MAFF of an extra value for the national cereal crop in

1955 of £15 million, principally attributed to the elimination of rabbits. The virtual elimination of damage for 20 years reduced the interest in rabbits as a pest.

During the 1970s, basic research recommenced as reports of damage increased. The average loss in the 1976/1977 season for a sample of 17 sites with fenced exclosures in winter wheat was 12% (J Vaughan, personal communication). Crawley (1989) used one field with a wild rabbit population fenced away from growing plots at different times of year. The results showed a greater decline in yield following early grazing post-emergence (0-60 days) than was evident at other times. Unfortunately, the density of rabbits was not measured. He also described a similar pattern in a small enclosure trial with 22 rabbits per hectare. There was a sudden cut-off towards zero yield after 200 days of continuous grazing. The author was able to demonstrate significant declines in individual grain weights resulting from early damage by rabbits. The conclusion was that a newly emerged crop, after early grazing for about 60 days, was unable to compensate in yield, even though vegetative regrowth was often strong after rabbit exclusion occurred and the crop looked acceptable at harvest. Early grazing also caused significantly increased weediness, which has cost implications for arable maintenance. Bell et al. (1998) measured losses from high densities of rabbits in winter wheat varieties subjected to ungrazed controls, autumn, spring or constant grazing. Early grazing caused a reduced number of grains in this study also, and losses were 26-35% according to variety. (Dendy et al., Chapter 17 record the yield losses from different rabbit densities to growing winter wheat, spring barley and silage in fenced enclosures. However, rabbits were not introduced until two months after sowing, so plants had the capacity to build up shoots and roots before being grazed, thus missing the important impact of early grazing on wheat as shown by both Crawley, 1989 and Bell et al., 1998).

Bell & Watson (1993) measured the impact on spring barley of 20 rabbits in a 0.3 ha plot, and noted that some varieties were preferentially grazed over others, though they did not present the actual yield losses. Once plants were reduced in height, repeated grazing occurred to maintain a low height. Bell et al. (1998) compared losses from an enclosure with controlled numbers of rabbits with a regime of ungrazed, autumn, spring or constant grazing plots. Winter grazing significantly lowered the number of heads, and yield loss was most affected in early grazed plots.

Experimental work on rabbit damage to other crops has not been published, but significant losses have been reported to the author for sugar beet, soft fruit, field vegetables — especially crucifers — and maize (for stock or human consumption). A single bite on a vegetable destined for human consumption severely reduces the value; the same action in hardy nursery stock is even more costly because the stock cannot be sold for another year. In some instances, rabbit damage necessitates complete re-drilling or replanting of fields, which is an extremely costly addition to the planned husbandry (e.g. 'greens' may cost £750-1,000 per hectare to establish; Nix & Hill, 2000) .

Crops reportedly not susceptible to significant rabbit damage include potatoes, rye, triticale and there are anecdotal reports of some varieties of oilseed rape and linseed being only lightly damaged. Experimental evidence of damage to novel crops, such as lupins, remains to be evaluated.

Grassland

Most authorities agree that the losses to rabbits feeding on grassland (at around 120 g dry weight per day per adult rabbit) are underestimated by farmers. Analysis of feeding rates from Australasia and Europe indicate as a rough rule of thumb that the offtake by 6-14 rabbits is equivalent to one sheep (Thompson & King, 1994). Precise levels will be influenced by sheep breed, age and breeding status of both species, vegetation nutritive values and so on. Phillips (1953) reported that twice the density of lambs could be kept on a re-seeded pasture, and their weight gain was almost exactly twice that of lambs grazed on the adjacent rabbit-infested sward. A similar study in Kent in 1950/1951 showed a 20% and 64% lessening of sheep live weight gain in successive years (Thompson & King, 1994). Additionally, sheep could not be put onto the rabbit-grazed plots until June, whereas two 14-day grazing periods were possible on the protected plots. Fencing one rabbit infestation where the author had counted 196 rabbits resulted in calves being able to graze three months earlier than in previous years (S Jamerson, personal communication). This loss of the 'early bite' and the consequent need for extra feed is a typical scenario currently reported by stock farmers with a chronic rabbit infestation.

Bell et al. (1999) reported yield loss experiments and sward degradation caused by rabbits grazing a ryegrass re-seed. The experimental regime included replicates of winter grazing only, spring grazing only, constant grazing and ungrazed controls. Grazing by rabbits commenced in November 1998, two months after sowing, and their density was manipulated from 16-55 per hectare over the year, to mirror the naturally occurring annual fluctuations in population density. All-year grazing showed the greatest reduction in silage yields. Grazing on emergence significantly reduced the ryegrass component (by 13% by February 1999), compensated for by the ingress of other grass and weed species. Yields in the ungrazed control plots were about 15% higher than in the constant-grazing plots in the first and second years after sowing (10.9:9 tons dry matter per hectare and 11.5:10 tons dry matter per hectare respectively). Plots grazed only in winter or in spring had intermediate yields, being, respectively, 0.5-1 ton per hectare higher in the first year, and 0.25 ton per hectare each higher in the second year than the permanently grazed plots (but were still significantly lower than the ungrazed plots). In the final year, subplots of all the treatments were protected, and the yields of these were similar at 10.5 tons dry matter per hectare, whilst the grazed plots again yielded significantly less (only 61-67% as much). When dry weather conditions caused crop stress, the grass yield in the rabbit-grazed plots was only half. An important practical finding was that protection of previously grazed grassland resulted in a complete recovery of yield for the following growing season. However, rabbit grazing during ryegrass establishment had a permanent deleterious impact on the sward composition.

Trees, woodlands and plantations

The farmland ecosystem includes woods and plantations. Rabbits frequently live along the edges of these and move into adjacent fields to cause damage to agricultural and other crops. In addition, rabbits have historically been (and have again become) a major problem in damaging not only unprotected newly planted broadleaves and conifers but also affecting older and even mature trees (Gill *et al.*, 1995). Such damage has a long-term economic cost because any replanting delays the initial anticipated time to timber recovery, and a sudden loss of timber quality, through bacterial and fungal action, may render the crop completely uneconomic many years after it was planted and the establishment costs incurred. In the early 1950s, the Forestry Commission routinely rabbit-fenced all plantings and employed over 250 full and part time warreners, costing about £500,000 annually (Thompson & Worden, 1956). In the late 1990s, fencing, individual tree protectors and other management to protect new plantings is again becoming common (Pepper, 1998).

Under a Government policy initiative, grants to encourage planting small areas of trees under the Farm Woodland Premium Scheme have aimed to increase the number of small woods on farmland. Approximately a third of the cost of the setting up of the planting is recognised as protecting against mammal damage — principally rabbits (Nix & Hill, 2000). Work in fenced enclosures (Gill *et al.*, 1997) showed an increasing proportion of transplants with severe damage (necessitating replacing trees) with higher rabbit density. Thin-barked trees were especially vulnerable, such as beech *Fagus sylvatica*, ash *Fraxinus excelsior* and cherry *Prunus avium* but almost all species may be attacked when small. Work with clones of poplar *Poplus nigra* and willow *Salix* sp. showed variation in susceptibility to bark stripping by rabbits (Trout & MacVicker, 1994). Tests with wooden pegs showed that even the timber from some species was gnawed more heavily than others during penned trials (Hamilton & Hartley, 1997). Barkstripping of saplings or larger trees, in severe winters for example, can also result in the invasion of fungi and bacteria causing canker, rotting or staining of the wood as well as stem breakage in later years or even directly to death. The loss of yield or subsequent major loss of timber quality can severely affect the economics of the crop.

In some large areas of commercial forestry plantings within more extensive farmland ecosystems, the repeated replacement of eaten trees in replanted coups has been necessary over five successive years because the rabbit problem was not adequately managed (personal observations). The practice of destumping conifer plantations following clear-felling, and creating windrows of them, creates an ideal dry rabbit refuge that causes serious difficulties for rabbit management in future years. Other common forestry practices such as leaving brashings or lop/top may make future rabbit management difficult (Pepper, 1998). Similarly, suddenly fencing rabbits inside a young wood, to protect arable crops, can result in severe damage to trees.

Orchard trees can be damaged or even killed by rabbits and individual tree protection or fencing of orchards is again commonplace. Occasionally rabbits may even climb fruit trees in winter; the author has seen them remove the bark on the upper side of the branches up to 2 m above ground.

Upland rough grazings

Farm holdings with extensive grazing in the upland Less Favoured Areas of the UK frequently have rabbit populations, either living in open warrens or harboured under walls, bracken banks, woodland areas and rocky cliffs. Obvious runways show that these rabbits may well travel over a kilometre to feed on better quality forage in the lower inbye land as well as grazing the local grasses and heather. The impact of rabbits is often under-appreciated and consists mainly of the loss of grazing — especially the 'early bite' when stock are first let out over the hill in spring to eat the first flush of growth. Rabbit offtake may be equivalent to the sheep offtake (Iason & Hester, 1999). In an area of the Yorkshire Dales, over 5,000 rabbits were shot one summer — and an unknown percentage was not shot (personal observation). This equates to 500-800 sheep grazing continuously in the area. The sheep stocking ceiling for the same area was 400 ewes, so the offtake by rabbits was greater than that of the sheep. Rabbits in the same valley were seen to leave the hill daily during winter and feed on the food placed in the inbye land for the sheep. Some even congregated in the small hay barns there.

The second impact is the effect of rabbit grazing on the botanical composition of the area. When sheep and rabbits graze on the same area, the heather/grassland interface in particular may result in fragmentation, the separation of tightly grazed heather tumps, and eventual and elimination of heather plants (Hester & Iason, personal communication). Both preferential grazing of heather, and interspecific competition between grasses and heather are likely. Rabbit grazing on rough grazing land at lower altitude, such as chalk downlands, heathlands and sand dunes, are dealt with under the Nature Conservation heading below.

Set-aside

Set-aside (arable land temporarily removed from food production) can be utilised to assist in temporarily reducing losses from rabbits. Putting a field beside a rabbit-infested wood into set-aside for a year can result in subsidy income, minor inputs and no crop loss. The use of set-aside strips along field headlands has also been used to reduce the grazing losses in the remainder of a field, though rabbits frequently cross the strip and continue to feed farther out in the crop. A survey of 119 pairs of arable fields located near fields in set-aside for four or five years had twice the mean indices of rabbits in the set-aside perimeter than around the arable fields sampled (Trout *et al.*, 1994). In a sub-sample of the set-aside sites, rabbit activity was also measured in immediately adjacent fields. The indices of activity (abundance of droppings, number of plants grazed and the severity of grazing) in winter wheat were each significantly higher up to 30 m from the headland beside the set-aside fields than the indices in other fields sampled adjacent to other arable crops (Figure 16:1). This strongly suggests that whilst

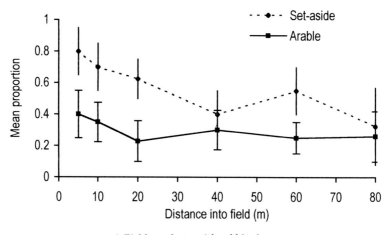

a) Field quadrats with rabbit signs

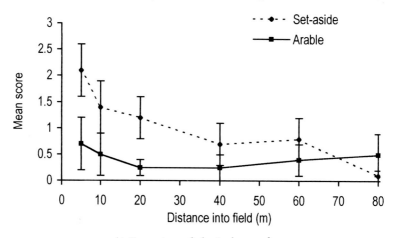

b) Percentage of plants damaged

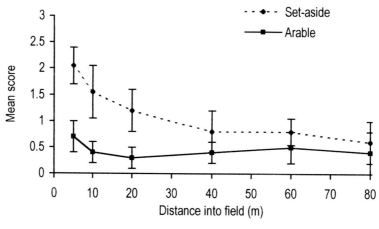

c) Degree of damage to plants

Figure 16:1 Increased rabbit sign and winter wheat damage recorded adjacent to long-term set-aside fields compared to samples from fields adjacent to other arable fields.

set-aside may have benefits in avoiding rabbit damage to crops in the short term, there is a longer-term cost unless rabbit management is continued. The gradual build-up of a rabbit population will include an increase in burrows in the harbourage (as well as perhaps within the field) and entail extra management resources to prevent serious crop damage in the future.

Infrastructure

The farmland habitat includes an infrastructure not directly related to cropping and this often contains a major resource of harbourage where rabbits can hide, have burrows and rear their young. Hedgerows, rough patches of scrub, ditches, stone walls, earthfilled walls, buildings, banks, embankments, quarries and rough grazings may all provide an opportunity for rabbits to burrow. In many instances where soil conditions cause difficulties in constructing or maintaining burrows or where much of the ground becomes waterlogged these may be only the safe habitat. Rabbits can cause chronic damage to earth or stone structures such that restoration is difficult and costly. For example the restoration of even a small length of collapsed wall is time consuming and costly. On Ramsey Island, of the Welsh coast, a length of 4 km of earth-core double-faced stone wall is being completely reconstructed following rabbit digging, at a cost of £20 per metre (I Bullock, personal communication). Rabbits burrowing into embankments can seriously weaken the structure. Chronic damage to young or semi-mature hedges can reduce biodiversity, and weaken or kill small sections that then require physical patching to retain stock. The presence of rabbits seriously increases the costs of establishment of new hedge plantings, and the use of tree guards reduces the attractiveness. Finally, infestations within unrelated infrastructures beside farmland belonging to a different owner, such as rail and road edges, frequently cause friction with neighbours over nuisance, crop damage, and occasionally result in litigation and pressure under the Pests (1954) Act.

Areas of nature conservation

Many sites of nature conservation importance are located within the farmland ecosystem, often on poorer land where intermittent stock grazing is the only agricultural activity. The rabbit is by no means important to them all but in many calcareous grassland, heathland and dune systems, rabbit grazing is crucial to maintaining some of the important habitats and species contained in the UK Biodiversity Action Plans (Department of the Environment, 1995). Sumption & Flowerdew (1985) and Flowerdew *et al.* (1992) highlighted many examples when reviewing the population crash in rabbits from myxomatosis. Ironically, this seriously reduced the biodiversity of many sites of conservation value and led to a decline of some valued species such as the large blue butterfly *Maculinea arion*, buzzard *Buteo buteo*, stone curlew *Burhinus oedicnemus* and polecat *Mustela putoris* (see MacDonald & Birks, Chapter 9). It also resulted, in association with changes in agricultural (including stock welfare) practices, in gradual, long term fragmentation and serious loss of habitat by scrub and woodland encroachment. As an indication of the consequences, the man-days devoted to clearing areas of degraded calcareous grassland and heath and managing the re-establishment of these habitats may well exceed half a million (M Oates, personal communication). Fifty years ago, special maintenance of chalk downland habitat was hardly necessary, as the stock and rabbit grazing between them maintained large areas of this classic habitat. In the Breckland farms of East Anglia, Dolman & Sutherland (1992) showed that the biodiversity of some heathlands was greatest where rabbits and stock both graze and this was significantly higher than where stock only are present.

In some sand dune areas, rabbits may be the major grazers, being present all year round, whilst farm stock are used intermittently, and often only on fenced sections. The appearance of Rabbit Haemorrhagic Disease (RHD) in some conservation sites has again shown how important (the lack of) rabbit grazing can be (Trout, 1998, 1999). At Dawlish Warren, Devon, which has had outbreaks of the disease each year for five years, staff now regularly have to cut vegetation in and around the dune slacks to prevent them becoming completely overgrown with rank vegetation. This has important implications for early flowering rarities and sward composition. In a sand dune site in Cumbria the estimated consequence of a 95% drop in rabbit activity for producing fresh sand outside burrows is apparently impacting on the maintenance of important ruderal plant colonisers and even the provision of shelter for newly emerged natterjack toads *Bufo calamita* (P. Burdon, personal communication).

Rabbits are not normally able to remove established scrub and woodland that has invaded previously open conservation areas; this needs to be approached by a combination of man management and stock grazing, currently including the experimental use of specialised breeds. Rabbits may then contribute to creation of, and then maintain, the required specialist sward characteristics. Meanwhile, the farmers on the other side of the boundary will become more concerned at the potential for agricultural crop damage, and so a managed balance between objectives on and off farmland is required.

MANAGEMENT OF RABBIT POPULATIONS

Methods

An important aspect of rabbits in the farming scene is the need for and methods of active management to reduce crop losses. There are three major strategies of rabbit management:

- Keeping them away from valued plants or crops, for example by fencing, tree guards or repellents.

- Making the locality less attractive for rabbits to live and feed there, for example by changes in cropping, the deliberate use of set-aside or removal of (potential) harbourage.

- Removing rabbits by fumigation, shooting, trapping and so on.

Which strategy or which combinations should be used depends on the objective, local circumstances and constraints. There are over 30 methods for managing rabbit damage, each of which may play its role within a strategy designed to reduce damage towards acceptable levels, but may be inappropriate, unwanted or too inefficient in another particular plan. For example, methods that target rabbits underground (fumigation, ferrets, killing traps in burrows) tend to select more females, whereas shooting selects mainly males (Smith *et al.*, 1994a). A survey of methods used by 1,000 landowners in a stratified random sample in England and Wales in the mid 1980s (Figure 16:2) revealed that three of the first four methods widely used were generally the most enjoyable (sporting) but least effective. The effort expended for successful management will vary markedly depending upon the potential for losses anticipated, and the size of the area. A horticultural enterprise with a few hectares of commercial roses, worth many pounds each, or a vegetable grower with a field of cauliflower for human consumption, will generally be prepared to consider much higher inputs per unit area to manage a small number of rabbits than a farmer with hundreds of hectares of moorland. Working at a larger scale than one farm is frequently a better option than uncoordinated action, provided that planning and adequate resources are available.

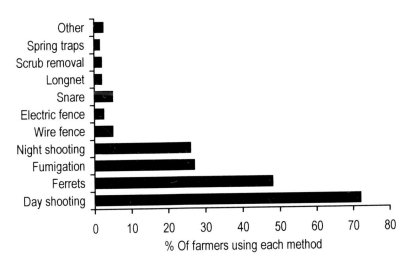

Figure 16:2 Methods of rabbit management used by c.1,000 farmers in 1986.

Sport as rabbit management

The rabbit has traditionally been a less important sporting quarry on farmland in the UK than game birds or hares (see, for example, Oliver-Bellasis & Sotherton, Chapter 3), though the reverse is the case for some countries in Europe, such as France and Spain. Rabbits were an important part of the local rural economy during the early part of the twentieth century in the UK and the sporting aspect included the poaching of rabbits from the local estates. Data gathered by the Game Conservancy Trust for a large number of game estates indicate a slow recovery of the numbers taken per annum since myxomatosis (Tapper, 1999). Whilst the differentials across the country may be real, interpretation of the actual numbers killed are well recognised as only partly a reflection of the local density of rabbits.

Statistics gathered by the British Association for Shooting and Conservation indicate that 60% of a sample of their members hunted rabbits (Reynolds & Batley, 1999). Members favoured day and night shooting — a popular choice of farmers also — over other forms of management. For daytime shooting, half of those used dogs: labradors, spaniels, terriers and lurchers for flushing and retrieving or else catching rabbits. Most rabbit carcasses were put to good use, either eaten by humans or by pet dogs or ferrets. Relatively few were sold or left, suggesting a 'sustainable' cropping regime. The scale of this removal of rabbits is, however, likely to be an order of magnitude less than in the 1940s, when 'free' meat was an important component of the diet of countryfolk, and perhaps 10 million rabbits were cropped annually (Thompson & King, 1994).

Rabbit removal for sport has the intrinsic implication that enough should be left for similar sporting activities the next year. Since the evidence suggests that a 90% kill rate is required to maintain a low level of rabbits for longer than a year (Smith & Trout, 1994), sporting actions alone against rabbits rarely yield adequate long-term results for farmers.

TIMING OF MANAGEMENT

It has traditionally been suggested that rabbit management is most cost-effective during the winter months, when myxomatosis has reduced numbers, so as to remove rabbits that would form the breeding stock for the spring. This remains the preference where practicable, but is not always followed nor always best practice. Game estate staff in particular frequently insist on rabbit management being delayed until the end of the pheasant shooting season (end of January) to avoid the possible disturbance to birds. Unfortunately, by this time the earliest litters of rabbits are almost independent, and it is these that have been shown to have the highest survival rate (Smith & Trout, 1994; Figure 16:3), so delaying rabbit control leads to more rabbits. There are, however, clear circumstances when winter control may not be best. These include early autumn-sown cereals and grass where the early damage causes the greatest subsequent loss (as reported above). The other instances relate to either high value crops, such as horticulture, field vegetables or damage to farm infrastructure where damage is immediately unacceptable. Many farmers start to shoot rabbits in mid-summer, when the numbers seen are greatest and crop damage becomes very obvious; had they started a campaign against adults before they began producing the 15-22 young per female per year (Trout & Smith, 1995) the numbers in summer may have been considerably lower. The costs of emergency action, including the lost or damaged crop, shows that prior planning of a rabbit management campaign the previous year would have probably been a wiser choice.

Rabbit population parameters and estimates of mortality from different control methods are being integrated at the more theoretical level to potentially provide a management decision system for the future (Smith & Trout, 1994; Smith, 1997). The combined impact of RHD and myxomatosis is unmeasured but evidence suggests that some populations remain reduced by over 75% for at least five years. Research at the international level continues towards developing new control methods such as fumigants (Ross *et al.*, 1998), reproductive inhibitors and repellents, whilst the potential for RHD to provide additional 'natural' free control is also being evaluated (Trout, 1999).

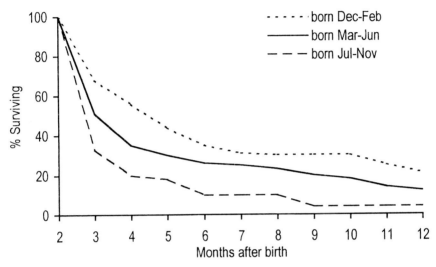

Figure 16:3 Survival rates for rabbits born in the early, middle and late parts of the breeding season.

CONCLUSIONS

Rabbits are currently an integral part of most of the whole-farm ecosystems over much of the UK. Damage levels to many crops are unacceptably high, and probably result in many fields being uneconomic (other than from EU subsidy payments). The levels of rabbit damage experienced justify higher, but better targeted, inputs than are currently accepted by many farming enterprises. This is one major challenge for advisors. Nevertheless, although most farmers would prefer to have many fewer rabbits, few would opt for no rabbits at all, recognising the important interlinking of rabbits with sport, food for predators and wider conservation. Myxomatosis, once imagined as the 'magic bullet' ceased to be the complete natural controlling agent after 40 years — though it is still an important mortality factor. The current evidence for RHD indicates very reduced populations in some areas in the short and medium term, but as yet unexplained variability both in immediate impact and time before rabbit populations recover. Careful planning of integrated rabbit management strategies, using more than one method and deliberately adding control operations to the impacts of natural diseases, remains the best management solution. Some management tools are under threat, such as snares, or are rarely used because few people have the skills. Rabbit management involving killing techniques over large parts of some farms will also become very much more difficult to implement if legislation banning hunting with dogs is enacted. Advisors and practitioners will need to be focused and creative in the future or the rabbit will continue to adversely affect a variety of ecosystems in and around farmland.

ACKNOWLEDGEMENTS

I am grateful to the organisers of the conference for commissioning this paper as part of the conference proceedings. Some of the work reported here was originally funded by several MAFF Policy Divisions and would not have been possible without the co-operation of colleagues in the former ADAS, within the former CSL Worplesdon Laboratory and many farmers.

17

Quantifying the costs of crop damage by rabbits

J Dendy, I G McKillop, S M Fox & G J Western

INTRODUCTION

Rabbits have now re-established themselves as a major vertebrate pest in the UK (see Trout, Chapter 16 for an overview). In the 1950s, the disease myxomatosis wiped out 99% of the UK rabbit population (Lloyd, 1970) and resulted in the rabbit no longer being considered a pest of agricultural crops. Since that time, the effects of the disease have waned and it is now thought to kill only 20% of the population each year (Ross *et al.*, 1989). In fact, rabbit numbers have recovered to such an extent that the rabbit population is now thought to have returned to at least one third of pre-myxomatosis levels and is increasing at 2% a year (Trout *et al.*, 1986; Harris *et al.*, 1995). As myxomatosis is having less of an effect controlling rabbits, numbers will continue to increase, and could well double in the future. Obviously, as rabbit numbers rise, the threat of extensive damage to arable and other crops is considerable. Farmers must be made aware of the costs of the potential problem, as a good understanding of the economics of rabbit damage is essential to make informed decisions regarding rabbit management strategies. In this chapter, we briefly describe a series of experiments aimed at accurately assessing the costs of rabbit damage.

THE EXPERIMENTS

Yield loss due to rabbit grazing can be calculated by protecting areas of crop from grazing, and comparing the yields of protected and unprotected areas within fields (see examples in Trout, Chapter 16). Until recently, it has been difficult to conduct such field research, as it has been hard to determine accurately the numbers of rabbits grazing a field. However, the development by the Central Science Laboratory (CSL) of a research facility in the south of England has overcome this problem. At the CSL Field Station, a number of rabbit-proof enclosures were created, in which cereals or other crops could be grown and known numbers of rabbits introduced, thus enabling accurate yield loss models to be developed. The aim is to provide farmers with simple models to determine the size of the losses they may face and to help them design appropriate control programmes for each situation.

Between 1989 and 1999, experimental trials were carried out to develop yield loss models for winter wheat, spring barley and grass silage. The effects on crop yields of grazing by different densities of rabbits were determined over three-year periods. In all the trials, there were replicated, experimental, enclosures containing rabbits at various densities, and control enclosures without rabbits.

All the experiments were conducted in 100 m by 85 m (0.85 ha) rabbit-proof enclosures. Rabbit-proof fencing, designed to prevent movement of rabbits between treatments, separated enclosures from each other. The fences were regularly checked throughout the course of the experiment and any damage repaired when necessary. Rabbit boxes were put into each of the enclosures to provide shelter. Crops were grown in some enclosures (see details for each trial, below), and ADAS farm staff or contractors conducted crop husbandry. All enclosures received identical agro-chemical treatments and standard rates of inorganic nitrogen when necessary.

In all of the trials, rabbits were put into experimental enclosures when the crops had reached growth stage 1 (GS1), about two months after sowing, when the height of the crop was between 80-100 mm. Single-sex populations were established, using adult rabbits, to ensure that there was no breeding during the course of the experiment. Every rabbit was fitted with a coloured, numbered tag in each ear, and enclosures were checked at about weekly intervals, when any dead rabbits were replaced. Rabbits were removed from the cereal trials after harvest. In the grass trial however, cage trapping to monitor numbers took place over a two-week period following the last cut each year. Tag numbers were noted and rabbits released back into their enclosures. Any rabbits that had not been captured by the end of this period were assumed to be missing and were replaced, ensuring rabbit numbers remained the same.

In the winter wheat and spring barley trials, different densities of rabbits were used each year (see details below). The densities used were considered to be representative of the over-winter densities present in the wild in the major cereal growing regions of England. One treatment density was retained each year to provide a standard measure of the effects of grazing between years. Densities were allocated to enclosures in a random manner at the start of each year. In the silage trial, rabbit densities were higher than in the cereal trials, and remained the same each year.

The effect of grazing on individual plants was determined by erecting 4m by 4m rabbit-proof pens within the enclosures at varying times during the growing season. Pen position within the enclosures was allocated using a randomisation plan. Rabbits only rarely managed to enter the fenced pens but data from those pens where this did occur were excluded from the analyses. At various times of the year, crop samples were taken in order to provide an estimate of damage and to calculate crop dry matter. Crop samples were also taken from the enclosures just prior to harvest to provide estimates of the level of damage to the plants and for quality analysis. In each trial, the crop was harvested from each enclosure and weighed separately. It was later adjusted to a standard moisture content of 15%, and yields could then be compared to provide an estimate of loss. Regression analyses were carried out on the reduction in yield with rabbit density. In the graphs regression lines were constrained to 100% where y=100% and x=0.

IMPACTS OF RABBIT GRAZING ON WINTER WHEAT

Winter wheat is the main winter cereal crop grown in the UK, and appears to suffer badly from rabbit grazing. In our first trial, winter wheat (Mercia) was sown in 12 enclosures. There were three replicates at each of three different densities of rabbits, and three control enclosures without rabbits. In December of each of the three years, populations were established using female rabbits, one replicate of which always had a density of 20 per hectare, to provide a standard measure of the effects of grazing between years; the others ranged from 5-40 rabbits per hectare. In the first year of the trial, three fenced pens were erected at random locations before rabbits were placed in the enclosures, and therefore contained plants which were always protected from grazing ('control' pens). Three 4 m by 4 m areas were also marked out but not fenced ('always-grazed' pens) at random locations in each enclosure and these therefore contained plants which were never protected from grazing. In the second and third years of the trial, the number of control and always-grazed pens was doubled. In each enclosure, data from plants in the always-grazed pens were compared with data from plants in the control pens. At harvest, two 0.5 m² quadrat samples of plants were taken from within each pen and the following parameters were measured: grain weight; plant number; number of ears; number of grains per ear; grain-specific weight; grain size (1,000-grain weight); plant height; grain quality (Hagberg value); grain nitrogen content; and the nitrogen content of the straw. Severe weather damage to the crop at the end of the third year led to the exclusion of these data from the analysis.

The amount of grain harvested from the enclosures and from the always-grazed pens decreased as rabbit density increased, and in each case the regression equation indicated about a 1% yield loss per rabbit per hectare (enclosures: 0.92%; pens: 0.81%) (Figure 17:1). At the higher densities, this relationship broke down as grazing resulted

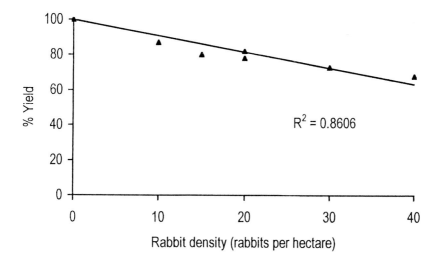

Figure 17:1 Reduction in percentage yield of winter wheat with rabbit density.

in death of all plants in an enclosure, and hence zero yield, rather than a continuing decrease in yield. This occurred in 1990/1991 at 40 rabbits per hectare, when all plants were killed in two of the three enclosures, whereas in the third, yield was reduced by about 35%. This also occurred in 1991/1992 but at only 25 rabbits per hectare, when plants were killed in all three enclosures.

Increasing the number of rabbits grazing the crop did not appear significantly to reduce plant numbers until the threshold was reached where all plants were killed at the higher densities (Table 17:1). It did, however, significantly reduce the number of ears produced per square metre and the yield per ear (Table 17:1). Increasing the number of rabbits grazing the crop did not significantly affect the specific weight of the grain or 1,000-grain weight (Table 17:1).

Table 17:1 Percentage decrease in harvest variables per rabbit per hectare, calculated from regression equations of each variable.

Harvest variable	% Decrease per rabbit per hectare	R^2	F	df	p
Enclosure yield	0.92	0.96	164.3	1,4	<0.0001
Pen yield	0.81	0.88	59.3	1,6	<0.0001
Plant numbers	0.10	0.04	0.3	1,6	ns
Ears/m²	0.36	0.67	17.3	1,6	<0.01
Yield/ear	0.46	0.70	19.3	1,6	<0.01
Specific weight	0.08	0.21	3.1	1,6	ns
1,000-grain weight	0.01	0.14	0.1	1,6	ns
Plant height	0.20	0.66	16.6	1,6	<0.01

Plant height at harvest decreased as grazing intensity increased (Table 17:1). However, although this result was statistically significant, the extent of the reduction was very small. The mean (±SE) height of ungrazed plants was 766 mm (±34) and therefore, using the above equation, plant height would be reduced by only 1.5 mm per rabbit per hectare. (There were, however, considerably greater height differences between grazed and ungrazed plants during the earlier part of the growing season, for example, about 50% in April at 20 per hectare, but these had disappeared by harvest after stem extension was complete). In the enclosure where plants survived at 40 rabbits per hectare, plant height at harvest averaged 782 mm whereas that in enclosures with no rabbits averaged 811 mm.

The yield of grain from the control enclosures was about 6.5 tons per hectare, and using the model developed it can be predicted that at this site each rabbit caused a loss of 65 kg of grain, equivalent to a loss of about £6.50 per rabbit at 1998 prices (£100 per ton).

IMPACTS OF RABBIT GRAZING ON SPRING BARLEY

In earlier research, (J A Dendy, I G McKillop and S M Fox, unpublished data) spring barley was identified as a crop whose yield was relatively unaffected by grazing when rabbits could choose to graze from seven different cereal crops and oilseed rape. Our second trial was therefore conducted to develop a model to predict the effects of rabbit grazing on the yield of spring barley when grown alone.

In February of each year, a spring barley crop (Alexis) was established in six enclosures. Two treatment densities ('high' and 'low') were used each year, with two enclosures being used as controls. Rabbit populations were established using adult male rabbits. Treatment densities ranged from 11 to 35 rabbits per hectare, with 18 rabbits per hectare being used as the standard in each year. Every year, pens were erected in each enclosure to prevent rabbit grazing. Each pen was erected in one of three rows, using a randomisation plan to determine its position. 'Always-grazed' and 'control' pens were established as before, and in addition 'partly-grazed' pens were erected six weeks after the introduction of rabbits, mid-way to harvest; except for the timing of their erection, these were the same as control pens. At the time of pen erection, crop samples were taken from each enclosure and were used for damage assessment and to determine dry weight. At harvest, a Sampo small-plot combine was used to cut a 2 m strip of crop within each pen. Grain samples were taken to determine moisture content, specific weight, grain quality and 1,000-grain weight.

There were no differences in the numbers of plants growing in the different enclosures, indicating that increasing rabbit numbers did not reduce plant numbers. However, significant differences were found in the plant heights: plants in the high-density treatments were significantly smaller than those in the controls (Table 17:2). Results from this trial showed there was a reduction in yield with increasing rabbit numbers (Figure 17:2). Yield loss due to grazing was calculated to be 0.33% per rabbit, less than that of winter wheat. The mean weight of grain from the control enclosure was about 5 tons per hectare. Therefore, the losses caused by each rabbit were about 17 kg of grain, which at 1998 prices (£80 per ton) was approximately £1.40 per rabbit.

Table 17:2 Percentage change in harvest variables per rabbit per hectare, calculated from ANOVA data. High densities were 18, 28 or 35 per hectare; low densities were 11 or 18 per hectare.

Harvest variable	% Change per rabbit per hectare		p
	High density	Low density	
Plant numbers	+0.04	+0.06	ns
Plant height	-0.37	-0.52	>0.0001
Ears/m^2	+0.03	+0.06	ns
Yield/ear	+0.03	-0.19	>0.01
1,000-grain weight	-0.04	-0.38	>0.0001

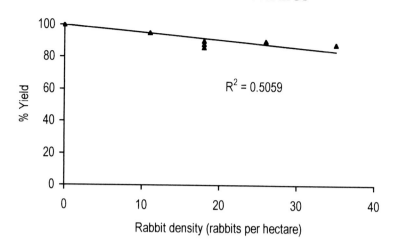

Figure 17:2 Reduction in percentage yield of spring barley with rabbit density.

IMPACTS OF RABBIT GRAZING ON GRASS

Grass for use as silage is widely grown in the UK, but little is known of the effects of rabbit grazing on yields. A perennial rye grass crop was sown in six enclosures; two were used as controls, two had rabbit densities of 50 per hectare, and two had densities of 25 per hectare. The densities remained the same throughout the three-year trial and were established using female rabbits. Pens were erected in each enclosure at various times during the growing period in randomly selected positions. Three silage cuts were taken each year in May, July and September. In the first year, control, partly-grazed and always-grazed pens were erected, and control pens were replaced after the May cut. However, as the time interval between the second and third cuts was shorter, partly-grazed pens were not erected. As rabbits were not removed from their enclosures following the final cut in September (except for the short period of cage trapping) the period of grazing was longer in years two and three, allowing for two sets of fenced pens to be set up at six week intervals. (These additional pens were called 'December' pens). Prior to the erection of the pens, grass samples were taken from each enclosure and used to determine the dry weights of the crop and for nutritive analysis. At silaging, all the grass was cut within each enclosure to give a total yield value for each enclosure. A Haldrup sampler was used to cut the grass within the pens and weighed to give a yield value. Samples were again taken to determine a dry weight value and for nutritional analysis for the material within the pens.

Results from the trial showed that there was a significant reduction in yield with increasing population density (Figure 17:3). Although this grazing effect was shown to be significant each year, its effect was most marked in May. The total annual yield loss at the end of the trial was about 1% per rabbit per hectare (Figure 17:3). At 1998 prices (approximately £17 per ton), the loss can be calculated as approximately £3.40 per rabbit. Overall, there were no differences between the yields taken from the small

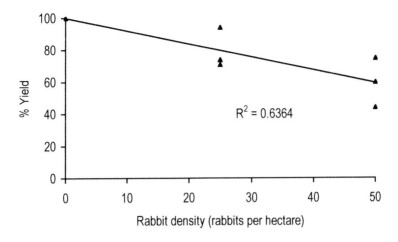

Figure 17:3 Reduction in percentage yield of grass with rabbit density.

fenced pens, and chemical analysis showed there to be no differences in their nutritional values (Table 17:3).

CONCLUSIONS

Conducting research to determine and predict the level and costs of damage is extremely important to farmers because it enables them to determine the cost-effectiveness of different control strategies, and to make more efficient use of their land. The development of CSL's unique field research site has enabled us to conduct these trials, and has for the first time led to the development of a series of models relating yield loss of crops and the costs of these losses to rabbit numbers.

In the wild, however, rabbit numbers will increase during the crop growing season as a result of breeding, rather than remain constant as in our trials. In the south of England, each female rabbit can produce between 14 and 22 young each year (Trout & Smith, 1995) and over-winter numbers can increase five-fold by the summer (Trout & Tittensor, 1989). In addition, we did not put rabbits into our enclosures until about two months after the crop had been sown, whereas in the wild, rabbits would have access to the crop from emergence. On the other hand, our animals did not have access to any other feeding area, which would not always be the case in the wild. However, rabbits are highly territorial. Most do not move more than about 50 m from their burrows and are rarely recorded at distances greater than 150 m from their burrows (e.g. Southern, 1940; Cowan *et al.*, 1987) and so they are likely to have access to only a few fields on which to feed. Therefore, this final difference between the conditions in our trial and the wild may be less important than the two previously discussed. More trials are currently underway to produce estimates of loss taking into account each of these factors. The results from these trials will then be incorporated into the current model to provide more accurate estimates of yield loss figures.

Table 17.3 *Nutritional assessment (means over all three years) of samples taken from small fenced pens with high or low rabbit density, and from control pens with no rabbits. C = control area (unfenced); AG = always grazed; C = grazed; D = December.*

Analysis	Control				High Density				Low Density			
	C	AG	G	D	C	AG	G	D	C	AG	G	D
Dry matter (%)	25	25.6	20.9	22.2	24.5	24.5	20.3	22.3	23.7	23.6	24	23.5
Soluble carbohydrate	14.6	15.3	20	17.5	14.3	14.5	20	18.8	15.7	16.9	22.4	12.5
Protein content	15.2	15.8	14.5	11.6	15.8	17.0	14.7	12.8	15.5	20.1	11	15.7
Modified Acid Detergent	26.6	27.2	26.2	65.7	26.2	26	26.6	28	26.8	26	27.4	28
Digestibility (%)	67	67	68.6	66	67.3	68.1	68	66.2	67	68.1	67.5	65.5
Metabolisable energy (MJ/kg)	10.7	10.6	10.9	10.5	10.7	10.9	10.9	10.5	10.6	10.8	10.7	10.4

Plant height through the growing season was affected in each of the cereal trials, and although shown to be statistically significant, the extent of the reduction overall was very small (1.5-2 mm per rabbit). Thus, it is important to recognise that plant height at harvest was relatively unaffected by grazing. The main component of plant height at harvest is the height of stems rather than leaves. Any damage to the stem growing point would kill the stem but, if it were undamaged, grazing of leaves would not be expected to have a marked affect on stem extension or height. Comparing the losses found for each of the cereal crops in economic terms, it would appear more profitable for farmers to grow winter wheat in areas with a low rabbit infestation. Growing spring barley on land with higher rabbit populations would result in greater profits, as they would incur less yield loss.

Many farmers carry out rabbit control only as a result of seeing grazed plants following emergence of the crop, and judge their success by subsequent changes in plant height. The results of these trials indicate that many farmers are therefore probably overestimating the success of their control operations and underestimating the amount of damage rabbits have caused to their crops. The effectiveness of control operations should instead be judged by, for example, counting the numbers of rabbits grazing on the field at night before and after control. This would provide the farmer with a more accurate estimate of the size of the rabbit population grazing his crop, and using our models, would enable him to determine the level of loss with which he may be faced.

ACKNOWLEDGEMENTS

This research was funded by the Rural Division of MAFF. We thank the ADAS farm staff and employees of Crop Fosters for their technical assistance.

The bio-economics of fox control

D W Macdonald, J C Reynolds, C Carbone, F Mathews & P J Johnson

INTRODUCTION

Is predator control cost-effective? This is an applied question, but tackling it requires answers to fundamental questions about the relationship between predators and prey. Consider the sequence linking humans, red foxes *Vulpes vulpes* and their prey, some of which are perceived as pests and others as assets. While predation by foxes of pests like rabbits *Oryctolagus cuniculus* might be welcome, predation of game species or rarities might be unwelcome (e.g. Reynolds & Tapper, 1996; Macdonald *et al.*, 1999). Because of the expectation of unwelcome impacts (among other reasons), people sometimes seek to limit numbers of foxes by killing them. The sequence people-foxes-rabbits illustrates clearly that questions of wildlife management are not answered by biology alone. For example, economics is relevant insofar as (i) predation by foxes is claimed to cause financial loss and killing foxes aims to reduce such loss by limiting fox numbers; and (ii) crop damage by mammalian herbivores is claimed to cause financial loss and predation by foxes is claimed to mitigate that loss by limiting pest numbers. Indeed, it has been suggested that killing foxes is financially counter-productive to farmers who suffer rabbit damage (e.g. McDonald *et al.*, 1997; see also Newsome *et al.*, 1989; Trout & Tittensor, 1989; Macdonald *et al.*, 2000c). Rabbits rank very high on the farmer's list of mammalian pests (Atkinson *et al.*, 1994; Baker & Macdonald, 2000; see also Macdonald & Johnson, Chapter 1 and Trout, Chapter 16).

We may expect the various components of this mix — the biology of predator and prey, economics and human sociology — to interact in complex ways and to vary with place and time. A full account of the economics of attempts to control foxes would require us to assign economic value to all the various attributes and effects of the hunt (in the wide usage of that word). Economic theory is concerned with the optimization of 'satisfaction' or 'utility'. It is the pursuit of these — not profit alone — that drives the economy and results in a particular distribution of resource use (J Marsh, personal communication). Terms like 'financially counter-productive', in the sense used above, therefore do not provide a full account of economic behaviour. Because people hunt for reasons other than to kill foxes, a full account of its economics would include assessing the value of hunting as a leisure activity. One approach would involve establishing how much a particular group of hunters would be prepared to pay in order to retain their style of hunting. This has the advantage of recognizing that the activity has non-exchange costs as well as non-exchange values (that is, no open market transactions occur). However, it also involves hypothetical questions where no real costs or benefits are at stake. Furthermore, it relates only to *perceptions* of costs and benefits; these may differ from the actual costs and benefits.

The foregoing difficulties would beset a full analysis; however, we do not aim to present an exhaustive analysis of the economic implications of all aspects of fox control here. It is indeed debatable if the methods are yet available to achieve such an analysis. Instead, we examine the costs and benefits of fox control in terms of impacts on the accounting profits of particular businesses. We attempt an exploratory analysis of how some relevant factors might be measured and reconciled, to create a balance sheet that informs judgment of whether attempted fox control is cost-effective. We draw on case studies (e.g. Baines *et al.*, 1995; Heydon & Reynolds, 2000ab; Heydon *et al.*, 2000) and our earlier syntheses (Macdonald *et al.*, 2000c, Reynolds 2000), together with some previously unpublished analyses, to illustrate such an approach. At the outset, we stress that in the absence of much more empirical work, the results we derive are certainly imprecise; we present them partly to illustrate that gathering the necessary additional data would be extremely worthwhile. This, then, is a specific case to illustrate a general approach that we advocate, and so we will begin with some generalities about attempted predator control.

EFFECTIVENESS AND EFFICACY

'Effectiveness' is an important term in any debate about pest control, and is often used synonymously with 'efficiency'. For clarity, however, and in line with the shade of meaning recognized by Fowler's *Modern English Usage* (1981), we distinguish between efficiency in the performance of control methods relative to that of other methods, and their effectiveness in achieving a given aim. 'Efficiency' therefore expresses performance of a management technique (e.g. in terms of animals killed) relative to cost (e.g. in time, effort or money). 'Effectiveness', on the other hand, applies to the aims of management rather than the action. Thus, if the aim is to achieve population control, then culling is effective if it results in the required population density. If the aim is to control damage, however, successful population control may or may not be effective, depending on precisely how damage is related to population density. Returning to efficiency, one method may be more efficient than another if it achieves the same aim in a shorter time, or more cheaply.

Concepts of 'strategy' and 'targeting' arise from considerations of both efficiency and effectiveness. Consciously or not, the landowner or his representative adopts some strategy to manage wildlife, embracing the choice of methods, and the amount and timing of effort. In principle, accurately targeted management strategies are likely to be more effective and cost-efficient than less focussed strategies, if targeting is not itself costly. For instance, if lamb killing is a characteristic of individual foxes, then a strategy of selectively removing those individuals — if feasible — might be more likely to reduce lamb losses cheaply than would a strategy to reduce fox numbers overall. On the other hand, if it is not possible to identify lamb-killing foxes until damage has already been done, a more costly strategy of controlling the fox population in its entirety might be the more effective option. Lacking detailed knowledge, the best choice is not obvious.

Problems in assessing the efficiency of fox control methods

Comparing fox culling methods, as components of a control strategy, is fraught with difficulties. Part of the problem is measuring efficiency. Often, the comparison is of foxes killed in a given time-span. This 'catch-per-unit-time' is not generally a reliable measure of a method's efficiency because it does not take into account the number of animals available to be culled and their accessibility. For any method, a high catch-per-unit-time is possible only where there are many foxes to cull. Fox density typically varies three-fold within a year, and at least six-fold between regions of Britain (Lloyd, 1980), clearly distorting comparisons of catch-per-unit-time between seasons and regions even for the same method. It will particularly distort comparisons of methods that differ in their seasonal or regional use. For example, winter months suit 'lamping' (with spotlight and rifle) better than snares because the lack of vegetative cover makes concealment difficult for fox and snare alike. Conversely, spring and early summer, when cover is high, favour snares over lamping, but the fox population is at its minimum and capture rates are inevitably lower. Thus snaring will *de facto* prove less efficient than lamping when the two methods are compared. Nevertheless, fox control at this time using snares might contribute substantially towards a culling strategy that is both targeted (on adult territory holders) and effective (there may be seasonal variability in the value of culling). Another problem is that time may not be the best measure of cost. Snares have a low daily capture rate, yet function largely in the absence of the operator.

Perhaps the most serious problem in comparing efficiencies is that, because culling methods are rarely used exclusively, any given fox is at risk from several methods. A fox killed by one method is no longer available to another, so there is interdependence between the culls obtained by each method. If one method is reduced or abandoned, then, all else being equal, culls taken by other methods inevitably increase, without any increase in effort, simply because more foxes are available to be killed. Because of this interdependence, comparisons between methods may be confounded by the mix of methods used, and the relative effort applied to each.

Problems in assessing the effectiveness of fox control methods

Assessing the effectiveness of culling strategies is no easier. The major problem is the difficulty in accurately assessing the number of foxes present, and the significance of the damage they cause. The best indication most operators have of the size of the fox population is the number of encounters with foxes, either in everyday life, or through damage caused, or while culling. However, levels of damage may not relate well to abundance, and numbers of foxes culled is a poor indicator of effectiveness. Even for professional wildlife biologists, estimating the number of foxes present is a very difficult task (Lloyd, 1980).

Effectiveness is also scale-dependent. The wider the scale over which population control is considered, the less reliable any conclusions will be for any specific context. The aims of control are usually defined for rather small areas: for any land-holder suffering damage, the question of whether they are in England, Wales or Scotland is less important than whether they farm in uplands or lowlands. Nevertheless, management patterns at a large scale may be an important input to economic decisions at a smaller scale. The Game Conservancy Trust's National Game Bag Census shows significant regional variation in the number of foxes culled on shooting estates (Tapper, 1992). Data from gamekeepers on foxes sighted during culling operations, and direct field surveys of fox abundance, confirm that this variation reflects regional differences in fox abundance (Reynolds & Tapper, 1995ab; Heydon et al., 2000). In regions with low fox abundance, it is clearly easier to attain both local and regional control over fox numbers (Heydon & Reynolds, 2000b; Reynolds, 2000).

To illustrate these points, we consider, below, the costs to some human enterprises that result from fox predation, and the efficiency of two different fox control scenarios in relation to these. We use case studies from two contrasting areas of Britain, where very different forms of hunting with dogs operate. In the final section of this chapter, we present two independent statistical models for estimating the benefit of fox predation on rabbits.

FINANCIAL EFFICIENCY OF HUNTING FOXES WITH DOGS FOR THE SHEEP FARMING COMMUNITY IN MID-WALES

In the 1,366 km² region of mid-Wales studied by Heydon & Reynolds (2000ab; Heydon et al., 2000), at least 968 foxes were culled in the survey year, and 73% of these were killed by methods involving dogs (mounted hunts, terrier groups, fox destruction clubs or gun-packs). These figures were derived from a combination of questionnaire surveys of farmers and bag records of hunts. Extrapolation to account for non-respondents suggests that up to 2,602 foxes were killed in total. To calculate the efficiency of these methods for the sheep farmer, we assumed that the entire cull was taken in these ways; this avoids the unknown cost per fox for other methods, and the unknown dependability for the sheep farmer of other fox control efforts, most of which are inspired by game management.

The cost of culling

The Afonwy Hunt, for which accounts were available, operated partially within the mid-Wales region studied by Heydon & Reynolds (2000ab). This hunt had an expenditure of £12,411 (gross) for the season 1998/1999, in which they accounted for 259 foxes (Federation of Welsh Packs, 2000), implying a cost of £47.90 per fox. Apportioning the incomes listed by the hunt, at most, 80% of its costs were born by the area's sheep farmers (a further 6% was paid as a good neighbour policy by forestry bodies; Federation of

Welsh Packs 2000). Taking these costs as typical, if the entire cull for mid-Wales had been taken by methods involving dogs, the likely operational cost would have been between £46,382 and £124,675, and the maximum likely cost to the sheep farming community in the region would have been £99,258. (The Afonwy accounts included costs for only one part-time working participant plus hounds and terriers; by implication additional time and labour costs are found within the community.)

Of 830 farmers in mid-Wales, 94% had sheep, averaging 558 ewes, implying a total flock of 435,352 ewes. Ewes averaged 1.24 lambs, of which 40.5% were born indoors. Total reported pre-weaning losses among all 522,422 lambs (indoor and outdoor) were 6%, and farmers attributed 0.6% (3,134 lambs) to foxes. (If fox losses were confined to outdoor lambs, this loss amounts to 1% of outdoor lambs.) These losses occurred despite the fox culling regime prevailing at the time. At a typical market price in 1996 of £31.50 per lamb (Nix, 1996; subsequently, the market value of lambs has fallen drastically), the total regional loss of income would have been £98,738, a figure comparable with the cost to the farmer of fox culling (80% of the cost of culling is £99,740). To set these costs into perspective, the annual veterinary costs associated with each ewe and her lambs were £2.80 (Nix, 1996). The regional fox cull, expressed per ewe, costs a maximum of £124,675 for 435,352 ewes, equivalent to 28p per ewe, or one-tenth of the veterinary costs. These expenditures do not take account of the ultimate benefit to the farmer in terms of saved productivity.

Clearly, these are simple extrapolations, and ignore such possibilities as that those lambs lost may have been less viable, or that their early loss might involve some savings in food or labour. Also, a proportion of the outdoor lambs taken by foxes (and we should emphasise that evidence that they were taken by foxes is not conclusive) would have died from other causes — that is, removing foxes entirely would not reduce lamb losses by 0.6%.

For comparison, an experimental study in New South Wales, Australia (Saunders et al., 1997) sought to control fox numbers at a farm scale (not a regional scale as discussed here) by means of '1080' poison baits. Despite high fox densities, lamb mortality was reduced by poisoning foxes at the start of lambing. However, the estimated cost-effectiveness of poisoning ranged from clear loss (Australian $1566) to clear return (ca. $2365) for a typical 2,000 ha farm, depending on the starting density of foxes, ewe condition and productivity, lamb health, and lamb sale prices.

The cost of not culling

The fox population found in mid-Wales by Heydon & Reynolds (2000ab) was below capacity, probably because of the heavy culling pressure. To calculate the lamb losses prevented by culling, it is first necessary to postulate what might happen to fox density if culling ceased. Because we cannot predict what fox density might become if culling ceased, we consider what might happen if density doubled. We assumed that twice the density of foxes causes twice the level of losses (1.2% of lambs or 6,268), totalling £197,476. Balanced against a control cost of £99,740, the net benefit of control (or loss in the absence of control) is £97,736. In other words, although lamb losses represent a small percentage of gross income, fox culling is financially efficient under these assumptions. These figures are, of course, sensitive to variation in lamb prices, and assume that the costs of culling remain fixed.

This is a simplification: other possibilities are that the vulnerable portion of the sheep population is already exploited and additional foxes could be sustained only by exploiting a different diet; given the timing of the hunting season with respect to lambing, and the low proportion of lambs taken, this is probably not the case. Alternatively, competition among foxes might leave the rate of lamb losses static. This is also unlikely as foxes are currently well below carrying capacity. Our best estimate is, then, that lamb losses would be in proportion with the overall density of foxes.

Further complications follow. Suppose that after a suspension of culling, fox density has increased, and culling is perceived to have been effective after all, and is reinstated. (This actually happened during the Second World War in a nearby region of mid-Wales hunted by the Plas Machynlleth Foxhounds: The Federation of Welsh Packs, 2000). If the fox population doubled — as we posited above — a zero population growth rate would require annually the deaths of at least twice as many foxes as previously. Although we cannot predict to what extent other causes of mortality would also increase with fox density, it seems plausible to suppose that twice as many foxes might have to be culled annually to begin to reduce the population. We do not know to what extent the hunt's costs are fixed (one possibility is that while the costs of encountering and hunting foxes may be fixed, the costs of digging them out may be variable). If costs of killing increased *pro rata* with fox density then the killing of twice as many foxes annually would cost an extra £99,258 per annum in addition to current expenses (while it might prove much easier to kill foxes at higher density, we note that 45-90% of foxes taken in this region by packs are dug out). In this scenario, culling would no longer be cost-effective on an annual basis, but instead would become a burden for the sheep farming community. Nevertheless, if the community risks not culling, fox density and lamb losses might increase still further.

COST EFFICIENCY OF HUNTING WITH HOUNDS IN THE ENGLISH MIDLANDS

As a first attempt at considering the cost-effectiveness of a typical Midlands mounted hunt, we adopt the premise that the hunt exists solely for the purpose of culling foxes. Again, there are two parts to this exercise. First, the cost of killing each fox by hunting. Second, and much more problematically, estimating the economic impact that these foxes might have had on farming enterprises had they not been removed from the population.

The cost of culling

For the first part of this problem, what are the elements of the hunt and how much do they cost? The principal sources of expense are plainly the horses, the hounds and their associated costs. The following account is based on costs and Figures for 2000.

We estimated costs for a hunt with two full-time employees, costing £20,000-£25,000 annually, including whatever accommodation is maintained by the hunt for their staff (but not the cost of maintaining any property owned by the hunt, nor kennels). PACEC (2000) found that the average annual wage for a full time employee (or Full Time Equivalent, FTE) was £8,000, implying a total cost to their employer of about £10,000 including National Insurance. For comparison, the MFHA (2000) quote a range of one to five full-time staff, and a recent questionnaire survey of hunt Masters suggests an average of 2.6 (PSL, 2000).

From hunt Masters' responses, Macdonald & Johnson (1996) reported a median field of 50 mounted followers. However, a high proportion of these may be considered superfluous to the task of fox culling (they might be considered as a 'moving grand-stand'). The MFHA cite a range of 4-18 'hunt horses'. We used a figure of ten horses as our baseline. Livery for a hunting hack costs £90-£140 per week (IAMBH, 2000), which we used to approximate the annual cost per horse. Additional costs include veterinary attention, riding equipment, transport to meets, tack repair and farriery. There is also the initial cost of buying a horse (which, given an initial purchase price of £5,000 and the prospect of perhaps ten seasons hunting for an 8-year-old suggests about £500 per year; a retired hunting horse might be worth £500 as a 'hack'). For the total of these additional costs, we used the IAMBH figure of 'not less than' £600 per year per horse. Informal inquiries among a sample of equestrian hobbyists suggest this to be realistic. In running a hunting stable rather than commercial livery, the hunt may save a proportion of this, and we incorporated a notional 20% discount, giving £48,000 per year for a stable of ten horses. This results in an annual cost of approximately £68,000 (our range is £68,000-£73,000). The total figure for all hunts of circa £14,000,000 cited by Produce Studies Limited (2000) implies a mean per hunt of £50,000, which they described as 'remarkably consistent'.

A typical Midlands hunt kills in the order of 100 foxes in a season (averaging 71), so we estimate a cost to the hunt of £680 per fox. Burns *et al.* (2000) cite a national average cost of £930 from hunt funds per fox killed by mounted foxhunts (range: £380 in Wales to £1,500 in the south of England), excluding the expenses of followers. Seven hunts incurred costs of over £3,000 per fox.

The cost of not culling

How to set a hunting cost of around £700 per fox against the likely cost of not culling by hunting, is a complex question. We know that other methods of culling are wide-spread in the Midlands. The survival prospects of foxes surviving foxhunting are unknown. We assumed that, without hunting, the foxes would remain in the population and on average, be equally likely to cause offence to farmers as unhunted foxes. How might this be translated into a financial impact on farmers?

The hunt cull amounts to 10% of the pre-breeding fox population (Macdonald & Johnson, 1996). If there were a linear relationship between fox density and damage (and this is merely a convenient assumption), ceasing hunting would result in 10%

more damage. This would translate (according to the damage estimates by the farmers) to an increase in cost to each of the farmers suffering fox damage of the order of £20 (at 1991 rates).

In a typical Midlands hunt country there are about 600 farmers (Macdonald & Johnson, 1996). About a quarter of farmers in the Midlands report damage due to foxes (Heydon & Reynolds, 2000b). If 25% were susceptible to fox damage, the extra total cost to farmers in one hunt country would be approximately £3,000. This calculation, based on many assumptions and unverified estimates of fox damage, serves to reveal that even taking the largest available estimate of fox damage, the entire hunt-scale impact would be of the order of £3,000 against an expenditure of at least ten times greater.

The foregoing discussion, as we have said, rests on the assumption that hunting exists solely to control foxes. If, instead, we assume mounted hunting is purely recreational, then plainly it makes no sense to include its costs as any part of a farming enterprise (irrespective of whether hunting has any impact on fox numbers). Further-more, there is clear evidence (Heydon & Reynolds, 2000b; Reynolds, 2000; Macdonald *et al.*, 2000c) that some (but not all) hunts exert influence to restrain the culls taken by all interest groups. Thus, it becomes relevant to ask by how much hunts could increase their own effort if desired, and how much the culls taken by other interest groups might increase. These questions are beyond the scope of this chapter but are discussed further in Macdonald *et al.* (2000c).

RABBIT CONTROL BY FOXES, AND THE COST OF KILLING FOXES

The nuisance value of foxes will differ markedly between arable farmers and those whose principal interests are livestock or game. Indeed, because they kill pests, foxes might even be an asset to some farmers (Trout & Tittensor, 1989). Here, and in the absence of definitive field data, we attempt to synthesise the best available fragments of data to answer the question: how much do foxes save farmers by killing rabbits? We investigate this question first by asking whether foxes can limit rabbit populations, and second by asking whether such an effect might financially benefit arable farmers. We do this using two separate modelling exercises. Our first approach uses a stochastic simulation model of fox population dynamics (see below), incorporating a simple model of rabbit population growth (Smith & Trout, 1994) to predict the effect of fox predation on rabbit populations in the medium term (ten-year periods were modelled). An important outcome of this first model was that above a certain rabbit density, foxes are unable to regulate rabbit numbers. A second model used simple growth rate calculations to convert rabbit numbers into a financial impact in terms of crop damage within a range of rabbit densities suggested by the first model where foxes could have an impact. This second model was also manipulated to assess the impact of fox control, and the effect of crop type, on the financial implications. Our preliminary findings suggest that arable farmers might benefit economically from fostering foxes on their holdings. At the least, they should keenly support fieldwork to explore whether our calculations accurately reflect reality.

Can foxes limit rabbit populations?

To model fox predation on rabbits, we first modelled fox population dynamics in a notional 500 km². An important feature of our simulation model was that breeding was limited to 200 territories, and only individuals that had resided in a territory for four weeks were able to breed. Litter size was set to an average of four, across all territory-holding females. Annual mortality rates were 40% for adults and 65% for juveniles (Pech *et al.*, 1997); these annual rates were converted to constant weekly rates over the year. The initial number of foxes was 550 adults and 250 juveniles. Given the mortality rates used, this resulted in a seasonal pattern ranging from a low of about 600 in mid-spring, to a high of approximately 1,200 individuals after breeding (mid-summer) (Figure 18:1). We also compared the effects of two methods of fox control (Figure 18:1): fox hunting and fertility control (based, respectively, on the accounts of Macdonald & Johnson, 1996; and Pech *et al.*, 1997). During simulations in which foxhunting was added to the baseline mortality rates, approximately 50 and 75 foxes were taken annually from the population to represent moderate and heavy hunting, respectively. Since this model was stochastic, the exact number of individual foxes killed varied from run to run. In simulations with fertility control, the percent of infertile females (those with breeding territories) was increased from a baseline of 10% to 50%.

We then modelled the number of rabbits killed by foxes in each four-week period, and subtracted these from the modelled rabbit population. Rabbits were assumed to comprise half the foxes' diet (Kolb & Hewson, 1979; Reynolds & Tapper, 1995b; see also Baker & Harris, Chapter 10). In the model, rabbits were classed as juveniles (less than 20 weeks) and adults, and 77% of animals taken were juveniles (after J Calzada, unpublished data). Each fox randomly selected prey from the prey distribution for that season until its weekly food requirements were met. Adults required 3 kg per week; juvenile intake was increased linearly each week so that at 12 weeks intake equalled that of adults. Cub intake was calculated from birth because it was assumed that the adults would increase their intake to compensate for feeding the cubs.

The impact of foxes on rabbits was estimated in terms of monthly patterns in numbers killed, seasonal (three-monthly) patterns in numbers of foxes and numbers of rabbits killed, and the cumulative numbers of rabbit-weeks lost due to fox predation. The model calculations were simulated for one year starting on 1 April. Means were estimated from 100 runs of the model. In the simulations, maximum numbers of rabbits were killed in the autumn and spring, due partly to seasonal variation in diet, and partly due to variation in fox numbers and control (Figure 18:1). A seasonal index of the impact of fox predation can be expressed in terms of 'rabbit-weeks'. By this measure, maximum impact occurred in autumn.

Finally, we modelled rabbit population dynamics, based on stage-specific estimates of fecundity and mortality (by four-week intervals) provided by Smith & Trout (1996). We also set mortality rates to reflect poor (40%), average (48%) and good (52%) survivorship conditions for rabbits. In years of poor survivorship, the rabbits went extinct even without fox predation; this might, for example, be the case during an

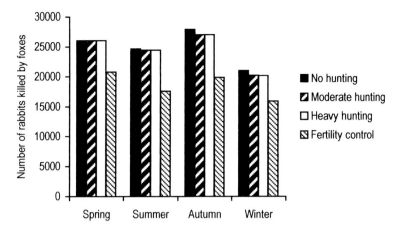

Figure 18:1 The predicted number of rabbits eaten by foxes in each season under four fox control scenarios: no control, moderate hunting, heavy hunting and fertility control. Values are based on the mean of 100 runs of the model.

outbreak of myxomatosis or during periods with several years of drought. When survivorship was good, rabbit numbers would have increased 22-fold over nine years in the absence of fox predation. Under these conditions, where initial rabbit densities were low (5 rabbits per hectare), their population growth was substantially reduced by fox predation (Figure 18:2a), but when initial densities were high (20 rabbits per hectare) rabbit population growth was so rapid that the effect of predation was marginal (Figure 18:2b). Under average survivorship and lower rabbit densities (starting with 5 or 10 rabbits per hectare), however, simulated fox predation did limit and stabilize the growth of rabbit populations (Figure 18:2c).

Figure 18:3 illustrates how the time it took for fox predation to drive the rabbit population to extinction in the model varied with initial rabbit population size and fox control. (Although extinction times are a convenient measure of fox impact in the model, in reality we do not expect foxes to drive rabbits extinct insofar as in reality the foxes themselves would probably change their diet as rabbits diminished). In the model, if foxes remained at a constant density, and 50% of their diets always consisted of rabbit, rabbits became extinct within nine years when rabbit survival was average and their initial density was less than seven per hectare. Time to extinction was longer when rabbit survivorship was improved, or fox reproductive rates reduced (in this case through simulated fertility control). Foxes are therefore most valuable if farmers find some other means to reduce rabbits to a density at which foxes can then regulate them further.

Figure 18:4 shows the overall ratio of expansion of rabbit numbers over nine years (the ratio of expansion is the number at the end of the first year divided by the number at the end of the tenth year). The maximum rates of expansion (i.e. in the absence of fox predation) for average and good years are indicated by the dashed horizontal line.

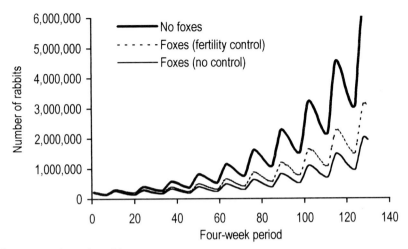

a. Five rabbits per hectare, good survivorship

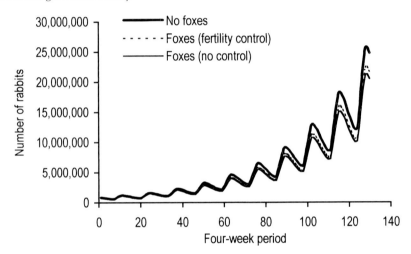

b. Twenty rabbits per hectare, good survivorship

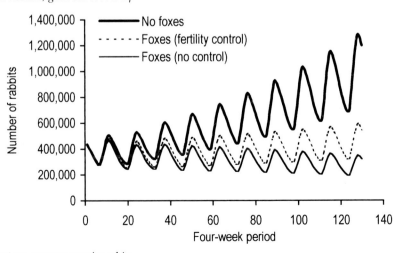

c. Ten rabbits per hectare, average survivorship

The curves indicate the reduction in rabbit population growth resulting from the erosion in rabbit survival due to fox predation (but ignoring any density dependent reductions in rabbit fecundity or survival).

What are the financial implications of fox predation on rabbits?

What are the financial implications of these levels of fox-induced rabbit mortality? Our answer is based on simple growth rate calculations. In short, using a spreadsheet to simulate the known population dynamics of foxes and rabbits, we calculated the number of rabbits that would be alive in a population under conditions of favourable versus average population growth (λ=1.61 and 1.24 respectively, see Smith & Trout, 1994), in the absence of fox control. We then calculated the likely impact on rabbit population growth of fox control at levels documented as realistic for farmland in the English lowlands. This enabled us to calculate, for both wheat and barley, the quantity of crops that would have been eaten by rabbits, had the rabbits not been eaten by the foxes. We also contrast scenarios under which the rabbits graze exclusively on arable fields or distribute their grazing evenly throughout the landscape.

Numbers of foxes alive each year were used to estimate annual predation rates on rabbits, the deaths of which provided a value, per head, of crop damage averted (Table 18:1). These estimated benefits were projected over three years to account for the corresponding reduction in future rabbit population growth. Assumptions concerning the predation of rabbits by foxes were as for the previous model, described above.

Table 18:1 Value (pounds sterling per hectare per year) of savings due to diminished crop damage following fox predation on rabbits. The results explore scenarios of average and high rabbit population growth, duration in years, crop type (wheat or barley) and the presence or absence of fox control.

		Crops Only				Even Distribution			
Growth		No Control		Shooting		No Control		Shooting	
(lambda) Year		Wheat	Barley	Wheat	Barley	Wheat	Barley	Wheat	Barley
1.24	1	22.1	13.6	13.8	8.5	4.4	2.7	2.8	1.7
1.24	4	42.0	25.9	26.4	16.2	8.4	5.2	5.3	3.2
1.61	1	24.8	15.3	15.6	9.6	4.9	3.0	3.1	1.9
1.61	4	103.5	63.7	64.9	40.0	20.6	12.7	12.9	8.0

Figure 18:2 (left) The predicted number of rabbits in a hunt country against time (four week period). These results are based on the Leslie matrix model presented in Trout & Smith 1994). Similar trends are shown with the added effect of fox predation imposed on Trout & Smith's model, one with foxes uncontrolled, and the other with foxes exposed to fertility control. Three scenarios are shown, a) a starting number of rabbits of 5 per hectare with growth under good survivorship conditions; b) a starting number of rabbits of 20 per hectare with good survivorship conditions; and c) a starting number of 10 per hectare with average conditions.

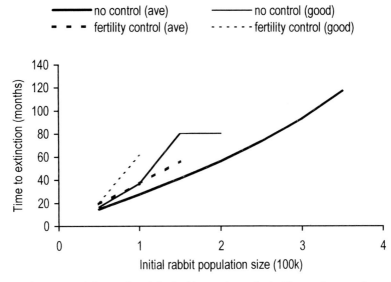

Figure 18:3 *Summary of the results of the Leslie matrix analysis. Figure shows estimated time to 'extinction' against the initial rabbit population size.*

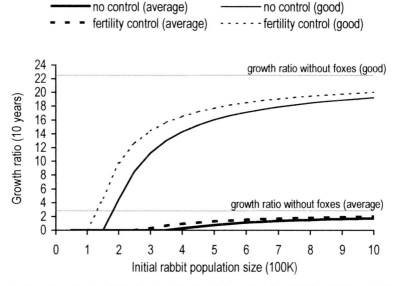

Figure 18:4 *Growth ratio of rabbits (see text for details) against the initial rabbit populations size.*

We thus calculated the damage to wheat and barley per fox (£6.5 and £4 per head). The overall costs averted per hectare were calculated assuming that agricultural land comprised 12% wheat and 8% barley. These calculations were repeated (i) in the absence of fox control, (ii) under foxhunting (50 foxes were killed across the year, 43% during cub hunting, assumed to be juveniles between August-October; the remainder of the foxes killed were evenly distributed between mid-November-end March), and (iii) shooting, with 45% of mortality in April, and the rest spread evenly throughout the year.

If foxes were not controlled, and with average rabbit population growth, the rabbits that would have been eaten by foxes consumed £22.05 per hectare of wheat annually, at 1997 prices. The saving due to foxes would be much more dramatic if circumstances were such that rabbit population growth was fast ($\lambda = 1.61$), and they fed entirely on wheat or barley (see Table 18:1). Similarly, if foxes continued eating rabbits for, say, four years; in this extreme, but not implausible, case the mitigation of crop loss could escalate to a maximum saving of £103.53 per hectare per annum.

In the case of foxhunting, the financial impact of lost fox predation on rabbits is minimal, because at the intensity modelled here, the impact of foxhunting on fox numbers is negligible: we calculate that without foxhunting the mean number of foxes alive each month in 500 km² (the size of an average hunt country or territory) is 852 which falls only to 837 following simulated hunting of 50 foxes (a cull of 0.1 per square kilometre). Under this level of foxhunting, the 150,581 rabbits that foxes kill would be reduced by only 3,000 or so. However, were the farmers to shoot foxes at a rate of about 0.05 per square kilometre per annum, the vulpine tally of rabbits would be reduced by 56,120. Depending on which scenario of rabbit survivorship and dispersion we model, the result is that, in the first year, wheat worth between £1.63-9.25 per hectare would be consumed by rabbits that would otherwise have been eaten by foxes. In a more extreme scenario, where fox predation runs over three additional years, during a fourth year the savings due to fox predation would diminish by £3.12-38.58 per hectare. In the fourth year, the farmer not only saves on the rabbits that the foxes eat, but also on those that would have been alive had the fox not eaten their ancestors in the first three years. The farmer, about to kill a young fox just recruited into the rabbit-eating population, might pause to consider whether doing so is worth the £156-886 of saved rabbit grazing that he thereby may forfeit. (At 2.8 foxes per square kilometre, each fox accounts for rabbits on the equivalent of 35.7 ha, and the value of the fox is calculated as the damage the rabbits killed by the fox during its foraging life would have done had it not killed them).

To cull or not to cull?

In a survey of rabbit densities in Britain in relation to levels of predation, Trout & Tittensor (1989) found that where predator pressure was lowered by human perse-cution, rabbit densities were on average 1.5 times as large as where people exerted less predator control. Our simulations (Figure 18:2) show that, given average survivor-ship and moderate to low rabbit densities, foxes can prevent the population growth of rabbits, but have a marginal effect at high rabbit densities. Trout & Tittensor (1989) also found no evidence for a regulatory effect of predator numbers in areas with the highest rabbit densities. However, rabbits display unstable population fluctuations and may be limited by foxes during low density periods following a population crash. While foxes and people always 'work together' to retard rabbit population growth, our models suggest that the foxes' contribution is more likely to lead to lower average rabbit numbers where people have already limited the rabbit population. The crucial question is whether this positive effect of foxes is trivial or significant compared with the damaging effects of foxes (and of rabbits).

Our simple calculations to estimate the immediate annual benefits of fox predation on cropped areas suggest these to be marginal, but the benefits of fox predation in limiting rabbit growth over the longer term could be far greater in the absence of culling (the differences between the columns for savings with and without fox control are approximately £10-53 per hectare for wheat and £6-£28 per hectare for barley; Table 18:1). Similar calculations, with different values, could be made for pasture, where rabbit grazing competes with livestock.

Clearly, in deciding whether to treat foxes as pests or assets, farmers must offset these benefits of fox predation against losses, and both merit careful quantification. The almost overwhelming difficulty of arriving at generalizations in this context arises from the reality that farms, farming communities and landscapes are all exceptionally complicated. The tidy notion of a farmer whose sole interest is in cereals, and who thus perceives foxes only as a vehicle for limiting rabbit damage, is not often mirrored in reality. From the national perspective, the figures for 1997 (Digest of Agricultural Census Statistics, HMSO, 1997) suggest that 2.04 million hectares of the UK are under wheat, with 1.36 million hectares of barley. If we apply the calculations presented above and ignore other inputs, total national annual savings of approximately £30-£150 million worth of crops thanks to foxes killing rabbits would be suggested.

Few individual farmers may, however, be presented with such a simplistic scenario. To address just how frequently this is true, we re-analysed questionnaire data collected from a national sample of farmers in the early 1980s (Macdonald & Johnson, 2000; Chapter 1). To identify those farmers for whom arable farming is the single enterprise, we used answers to questions concerning farm composition (farmers were asked to state the areas of their farms used for different enterprise types), and the question: "Are pheasants or other game birds shot on your land?" Farmers recording zeroes in the 'pasture' and 'grass leys' boxes, and 'no' to the question about game were considered not to be vulnerable to any costs due to foxes. (This is an imperfect measure, as it takes no account of farmers with poultry or free-range pig interests; nor does it take account of farmers who allow some shooting on their land, but who have no financial interest in it). Of the 792 farmers who answered all the necessary questions, only 1.8% could be identified as not being susceptible to some damage by foxes by this index. However, if susceptibility was defined as having any sheep or answering 'yes' to the game shooting question (or both), then the number of farmers considered not to be susceptible rises to 7.2%. With the yet more restrictive condition that farmers had to report the existence of a paying game shooting syndicate on their land, and or to keep sheep, the proportion of farmers 'immune' to fox damage rises to 27.1%.

CONCLUSIONS

The analyses presented above illustrate that whether culling is in the interests of a particular farmer depends on a complex balance of issues, which include population dynamics, practical management techniques, and financial balances. A further complication is that population control is not simply a choice for the individual farmer. Local (farm-scale) attempts to control fox numbers actually engage the fox population of a larger area. Widespread local control efforts can amount to regional control of fox numbers (Heydon & Reynolds, 2000b). The density of foxes will therefore reflect the prevailing aims of the land-holding community, as will their impacts on livestock, game and rabbits, and the effort required to control their numbers. This is the context in which individual land-holders must make decisions about fox management. Although beyond the scope of this chapter, these will include judgements on ethics and the non-exchange value of wildlife (the latter, thus far intractable, topic is introduced in Macdonald, 2001).

We do not pretend in this chapter to have brought our analysis to a point where the balance of pros and cons is clear for any one farmer or for any one aim. Far from it: there are many elements that we have not addressed here, and most aspects remain inadequately explored in terms of scientific study. All aspects of the question are susceptible to scientific research, but few have been carefully studied. A key question is the relationship between perceived and actual damage levels: very few studies have directly measured impacts. The impact of foxes on wild game birds is relatively well studied, but there remain many gaps in our knowledge of it. However, we do claim, through these analyses, to have revealed a level of complexity and inescapable inter-disciplinarity, which is often not adequately considered in discussions of wildlife management in general, or fox control in particular. In doing so we hope to contribute to a model for rational decision-making.

The investigator in this arena must wrestle with some difficult problems. One is that the system under study has already been influenced by prevailing land-uses for hundreds of years, not in a static way, but under a constantly changing set of environmental, economic and technological conditions. Thus it has a temporal dynamic that is often overlooked, although this should colour comparisons between scientific studies that are sometimes decades apart. The intensity of fox culling during recent history, and its regionally variable character, have also resulted in a regional pattern of fox abundance that complicates attempts to generalise about the impact of foxes on human aims, or about their control. Yet it is important to strive for generalisation, for without it investigation becomes complicated out of all proportion to the importance of the issue.

ACKNOWLEDGEMENTS

Much of our thinking for this chapter was consolidated while we worked within a larger team on a research report (Macdonald *et al.*, 2000c) for Lord Burns' Inquiry into Hunting with Dogs (see Burns, 2000). We warmly thank our colleagues on that team: Fran Tattersall, Steve Rushton and Mark Shirley together with Rob Atkinson from whose advice our team benefited. The groundwork on which we have drawn was undertaken partly with sponsorship from The Countryside Alliance and the International Fund for Animal Welfare. In this respect, we are indebted to collaborators Matthew Heydon and Mike Short, and thank Stephen Tapper, Richard Ryder and Georgina Mace for their advice and encouragement. We particularly thank those colleagues who made helpful criticisms of earlier drafts of this paper: Sandra Baker, Christl Donnelly, Dieter Helm, Ian Hodge, John Marsh, Steve Rushton, Roger Trout and Michael Winter. We also thank Will Manley for his advice.

REFERENCES

Adrian MI, Wilden W, Delibes M. 1985. Otter distribution and agriculture in southwestern Spain. In: *International Congress of Game Biology*. Brussels: International Congress of Game Biologists. 519-26.

Aebischer NJ. 1991. Twenty years of monitoring invertebrates and weeds in cereal fields in Sussex. In: Firbank LG, Carter N, Darbyshire JF, Potts GR, eds. *The Ecology of Temperate Cereal Fields*. Oxford: Blackwell. 305-331.

Aebischer NJ, Robertson PA. 1992. Practical aspects of compositional analysis as applied to pheasant habitat utilisation. In: Priede IG, Swift SW, eds. *Wildlife telemetry, remote monitoring and tracking of animal*. Chichester: Ellis Horwood. 285-296.

Aebischer NJ, Robertson PA, Kenward RE. 1993b. Compositional analysis of habitat use from animal radio-tracking data. *Ecology* **74**: 1313-1325.

Aebischer NJ, Marcström V, Kenward RE, Karlbom M. 1993a. Survival and habitat utilisation: A case for compositional analysis. In: Lebreton JD, North PM, eds. *Marked Individuals in the Study of Bird Populations*. Switzerland: Birkhäuser Verlag. 343-353.

Aldridge HDJN. 1985. On the relationships between flight performance, morphology and ecology in British bats. [PhD thesis]. UK: University of Bristol.

Aldridge HDJN. 1987. Turning flight of bats. *Journal of Experimental Biology* **128**: 419-425.

Altman DG. 1994. *Practical statistics for medical research*. 4th. Ed. London: Chapman & Hall.

Alterio N. 1996. Secondary poisoning of stoats (*Mustela erminea*), feral ferrets (*Mustela furo*), and feral house cats (*Felis catus*) by the anticoagulant poison, brodifacoum. *New Zealand Journal of Zoology* **23**: 331-338.

Alterio N, Brown K, Moller H. 1997. Secondary poisoning of mustelids in a New Zealand *Nothofagus* forest. *Journal of Zoology, London* **243**: 863-869.

Andrews J. 1992. Some practical problems in set-aside management for wildlife. *British Wildlife* **3**: 329-336.

Andrews J, Rebane M. 1994. *Farming & Wildlife. A practical management handbook*. Sandy, Bedfordshire: RSPB.

Andrews E, Howell P, Johnson K. 1993. *Otter survey of Wales 1991*. London: The Vincent Wildlife Trust.

Animal Health Board 1997. *Annual report for the year ending 1997*. New Zealand: Ministry of Agriculture and Forestry.

Anon. 1990. *Handbook for Phase 1 habitat survey – a technique for environmental audit*. Peterborough: English Field Unit, Nature Conservancy Council.

Anon. 1994. *Biodiversity: The UK Action Plan*. London: HMSO.

Anon. 1998. *Managing landscapes for the greater horseshoe bat*. Peterborough: Information and Marketing Team, English Nature.

Anon. 1999. *Otters and river habitat management*. Bristol: Environment Agency

Arnold HR. 1978. *Provisional atlas of the mammals of the British Isles*. Monks Wood: Institute of Terrestrial Ecology.

Arnold HR. 1993. *Atlas of mammals in Britain*. London: HMSO.

Atkinson RPD, Macdonald DW, Johnson PJ. 1994. The status of the European mole *Talpa europaea* L. as a pest and its management. *Mammal Review* **24**:73-90.

Atterby H, Kelly MJ, MacNicoll AD. 2001. Difenacoum resistance in rats is not a consequence of increased metabolism and excretion. In: DP Cowan, CJ Feare, J Pelz, eds. *Advances in Vertebrate Pest Management II*: 193-201.

Attuqayefio DK, Gorman ML, Wolton RJ. 1986. Home range sizes in the Wood mouse, *Apodemus sylvaticus*: habitat, sex and seasonal differences. *Journal of Zoology, London.* **210**: 45-53.

Baines R, Baker S, Hallett J, Macdonald, DW. 1995. *The Impact of Foxes and Foxhunting on the Management of Wiltshire County Farms Estate*. A report to Wiltshire County Council.

Baker SE, Macdonald DW. 2000. Foxes and foxhunting on farms in Wiltshire: a case study. *Journal of Rural Studies* **16**: 185-201.

Balfour E. 1957. Observations on the breeding biology of the hen harrier in Orkney. *Bird Notes* **27**: 6 and 7, 177-183 and 216-224.

Balharry D. 1993. Factors affecting the distribution and population density of pine martens (*Martes marles*) in Scotland. [PhD thesis]. UK: University of Aberdeen.

Balharry E. 1998. How to exclude pine martens from game and poultry pens. London: Vincent Wildlife Trust.

Balharry EA, Macdonald DW. 1999. Cost-effective electric fencing for protecting gamebirds against pine marten *Martes martes* predation. *Mammal Review* **29**: 67-72.

Balharry EA, McGowan GM, Kruuk H, Halliwell E. 1996. *Distribution of pine martens in Scotland as determined by field survey and questionnaire.* Scottish Natural Heritage Research, Survey and Monitoring Report No. 48. Perth: Scottish Natural Heritage

Banks PB. 2000. Can foxes regulate rabbit populations? *Journal of Wildlife Management* **64**: 401-406.

Banks PB, Dickman CR, Newsome AE. 1998. Ecological costs of feral predator control: foxes and rabbits. *Journal of Wildlife Management* **62**: 766-772.

Barataud M. 1993. L'activité crépusculaire et nocturne de 18 espèces de chiroptères, révélée par marquage luminescent et suivi acoustique. *Le Rhinolophe* **9**: 23-57.

Barlow ND. 1991a. A spatially aggregated disease/host model for bovine TB in New Zealand possum populations. *Journal of Applied Ecology* **28**: 777-793.

Barlow ND. 1991b. Control of endemic bovine TB in New Zealand possum populations: results from a simple model. *Journal of Applied Ecology* **28**: 792-809.

Barnes RFW, Tapper S, Williams J. 1983. Use of pastures by brown hares. *Journal of Applied Ecology* **20**:179-185.

Barnett EA, Hunter K, Fletcher MR, Sharp EA. 2000. Pesticide poisoning of animals 1999: investigation of suspected incidents in the United Kingdom. London: Ministry of Agriculture Fisheries and Food.

Barnett EA, Fletcher MR, Hunter K, Sharp EA. 2002. Pesticide poisoning of animials 2000: investigation of suspected incidents in the United Kingdom. London: DEFRA.

Barr CJ, Bunce RGH, Clarke RT, Fuller RM, Furse MT, Gillespie MK, Groom GB, Hallam CJ, Hornung M, Howard DC, Ness MJ. 1993. *Countryside Survey 1990 Main Report.* London: Department of the Environment.

Barreto GR, Macdonald DW. 2000. The decline and local extinction of a population of water voles, *Arvicola terrestris* in southern England. *Zeitschrift für Saugetierkunde* **65**: 110-120.

Barreto GR, Macdonald DW, Strachan R. 1998. The tightrope hypothesis: an explanation for plummeting water vole numbers in the Thames catchment. In: Bailey RG, José PV, Sherwood BR, eds. *United Kingdom floodplains.* Yorkshire: Westbury Academic and Scientific Publishing. 311-327.

Bateson P, Bradshaw EL. 1997. Physiological effects of hunting red deer (*Cervus elaphus*). *Proceeding of the Royal Society of London (B)*, **264**: 1707-1714.

Beardall C. 1996. Sensitive ditch work. *Farming & Conservation* **April 1996**: 18-21.

Beck A, Bontadina F, Gloor S, Hotz T, Lutz M, Mühlethaler E. 1994. *Jagdhabitatwahl und nächtliche Aufenthaltsgebiete der Grossen Hufeisennase* (Rhinolophus ferrumequinum) *in Raum Castrisch/GR.* Switzerland: Arbeitsgruppe zum Schutz der Hufeisennasen Graubündens ASHG, Encarden 51, 7152 Sagogn.

Bell AC, Watson S. 1993. Preferential grazing of five varieties of spring barley by wild rabbits. *Annals of Applied Biology.* **122**: 637-641.

Bell AC, Byrne PM, Watson S. 1998. The effect of rabbit (*Oryctolagus cuniculus*) grazing damage on the growth and yield of winter cereals. *Annals of Applied Biology.* **133**: 431-442.

Bell AC, Byrne PM, Watson S. 1999. The effect of rabbit grazing damage on the growth, botanical composition and yield of a ryegrass re-seed. *Annals of Applied Biology.* **135**: 417-424.

Bence SL, Stander K, Griffiths M. 1999. Nest site selection by the harvest mouse (*Micromys minutus*) on arable farmland. *Aspects of Applied Biology* **54**: 197-202.

Berny PJ, Buronfosse T, Buronfosse F, Lamarque F, Lorgue G. 1997. Field evidence of secondary poisoning of foxes (*Vulpes vulpes*) and buzzards (*Buteo buteo*) by bromadiolone, a 4-year survey. *Chemosphere* **35**:1817-1829.

Berry RJ. 1991. The house mouse. In: Corbett GB, Harris S, eds. *The Handbook of British Mammals.* Oxford: Blackwell. 239-247.

Birks JDS. 1998. Secondary rodenticide poisoning risk arising from winter farmyard use by the European polecat *Mustela putorius. Biological Conservation* **85:** 233-240.

Birks JDS. 2000. The recovery of the polecat, *Mustela putorius,* in Britain. In: Griffiths HI, ed. *Mustelids in a Modern World.* Leiden: Backhuys Publishers. 141-152.

Birks JDS, Kitchener AC, eds. 1999. *The distribution and status of the polecat* Mustela putorius *in Britain in the 1990s.* London: The Vincent Wildlife Trust.

Blackmore DK. 1963. The toxicity of some chlorinated hydrocarbon insecticides to British wild foxes (*Vulpes vulpes*). *Journal of Comparative Pathology and Therapeutics* **73:** 391-403.

Blandford PRS. 1987. Biology of the polecat *Mustela putorius:* a literature review. *Mammal Review* **17:** 155-198.

Blunden J, Curry N. 1988. *A future for our countryside.* Oxford: Blackwell.

Boatman ND. 2000. The costs of conservation on arable land. *Annual Review of The Game Conservancy Trust of 1999,* 103-107.

Boatman N, Stoate C. 1999. Arable farming and wildlife – can they co-exist? *British Wildlife* **10 (4):** 260-267.

Boatman ND, Wilson PJ. 1988. Field edge management for game and wildlife conservation. *Aspects of Applied Biology* **16:** 53-61.

Bontadina F, Beck A, Gloor S, Hotz T, Lutz M, Mühlethaler E. 1995. Jagt die Grosse Hufeisennase *Rhinolophus ferrumequinum* im Wald? – Grundlagen zum Schutz von Jagdgebieten der letzten grösseren Kolonie in der Schweiz. *Der Ornithologische Beobachter* **92:** 325-327.

Bontadina F, Hotz T, Gloor S, Beck A, Lutz M, Mühlethaler E. 1997. Protection of feeding areas of *Rhinolophus ferrumequinum.* An action plan based on the results of a radio-tracking study in an alpine valley of Switzerland. In: Ohlendorf B, ed. *Zur situation der hufeisennasen in Europa.* Germany: IFA-Verlag GmbH. 33-39. (In German with English and French summaries.)

Bourne J, Donnelly C, Cox D, Gettinby G, Morrison I, McInerney JP, Woodroffe R. 1998. *Towards a sustainable policy to control TB in cattle – A scientific initiative.* First Report of the Independent Scientific Group on Cattle TB. London: Ministry of Agriculture, Fisheries & Food.

Bourne J, Donnelly C, Cox D, Gettinby G, Morrison I, McInerney JP, Woodroffe R. 1999. *An epidemiological investigation into bovine tuberculosis - towards a sustainable policy to control TB in cattle.* Second Report of the Independent Scientific Group on Cattle TB. London: Ministry of Agriculture, Fisheries & Food.

Boyce C. 1991. Water vole. In: Corbett GB, Harris S, eds. *The handbook of British Mammals,* Oxford: Blackwell.

Brosset A, Barbe L, Beaucournu JC, Faugier C, Salvayre H, Tupinier Y. 1988. La raréfaction du rhinolophe euryale (*Rhinolophus euryale* Blasius) en France. Recherche d'une explication. *Mammalia* **52:** 101-122.

Brazil MA, Shawyer CR. 1989. *The Barn Owl, The Farmer's Friend Needs a Helping Hand..* London: The Hawk Trust.

Bresinski W. 1982. Grouping tendencies in roe deer under agrocenosis conditions. *Acta theriologica* **27:** 427-447.

Bright PW. 1993. Habitat fragmentation: problems and predictions for British mammals. *Mammal Review* **23:** 101-110.

Bright PW. 2000. Lessons from lean beasts: Conservation biology of the mustelids. *Mammal Review* **30:** 217-226.

Bright PW, Harris S. 1994. *Reintroduction of the pine marten: feasibility study.* English Nature Research Report No. 84. Peterborough: English Nature.

Bright P, Morris P. 1989. A practical guide to dormouse conservation. *Occasional publication of the Mammal Society.* No. 11. London: The Mammal Society.

Bright PW, Morris PA. 1996. Why are dormice rare? A case study in conservation biology. *Mammal Review* **26 (4):** 157-187.

Bright PW, Mitchell P, Morris PA. 1994. Dormouse distribution: Survey techniques, insular ecology and selection of sites for conservation. *Journal of Applied Ecology* **31:** 329-339.

Brown JH, Lasiewski RC. 1972. The metabolism of weasels: The cost of being long and thin. *Ecology* **53:** 939-943.

Brown KP, Alterio N, Moller H. 1998. Secondary poisoning of stoats (*Mustela erminea*) at low mouse (*Mus musculus*) abundance in a New Zealand *Nothofagus* forest. *Wildlife Research* **29:** 419-426.

Brown JA, Harris S, White PCL. 1994b. Persistence of *Mycobacterium bovis* in cattle. *Trends in Microbiology* **2:** 43-46.

Brown ED, Macdonald DW, Tew TE, Todd IA. 1994a. *Apodemus sylvaticus* infected with *Heligmosomoides polygyrus* (Nematoda) in an arable ecosystem: epidemiology and effects of infection on the movements of male mice. *Journal of Zoology, London* **234:** 623-640.

Brunton CFA, Macdonald DW, Buckle AP. 1993. Behavioural resistance towards poison baits in brown rats, *Rattus norvegicus. Applied Animal Behaviour Science* **38:** 159-174.

Bryder L. 1988. *Below the Magic Mountain: a Social History of Tuberculosis in Twentieth-Century Britain.* Oxford: Clarendon Press.

Buckle AP. 1994. Rodent control methods: chemical. In: Buckle AP, Smith RH, eds. *Rodent pests and their control.* Wallingford: CAB International. 127-160.

Bunce RGH, Barr CJ, Clarke RT, Howard DC, Lane AMJ. 1996. Land classification for strategic ecological survey. *Journal of Environmental Management* **47:** 37-60.

Burns T, Edwards V, Marsh J, Soulsby L, Winter M. 2000. *Report of the Committee of Inquiry into Hunting with Dogs in England and Wales.* Norwich: The Stationery Office.

Burrows R. 1968. *Wild fox: a complete study of the red fox.* Newton Abbot, Devon: David & Charles.

Butet A. 1985. Régime alimentaire d'une population de mulots sylvestres (*Apodemus sylvaticus* L. , *1758*), dans une lande xéro-mésophile en cours de recolonisation végétale. *Bulletin Ecologie* **17:** 21-37.

Butler DJ. 1980. Feeding ecology and management of foxes, (*Vulpes vulpes*) in coastal Aberdeenshire. [Unpublished PhD thesis.] UK: University of Aberdeen.

Calzada J, Palomares F. 1996. Frequency of occurrence of different rabbit remains in lynx and fox faeces. *Doñana Acta Vertebrata* **23:** 243-252.

Campbell LH, Avery MI, Donald PF, Evans AD, Green RE, Wilson JD. 1997. *A review of the indirect effects of pesticides on birds.* JNCC Report No 227. Peterborough: JNCC.

Canova L, Maistrello L, Emiliani D. 1994. Comparative ecology of the wood mouse *Apodemus sylvaticus* in two differing habitats. *Z. Säugetierkunde,* **59:** 193-198.

Carss DN, Nelson KC, Bacon PJ, Kruuk H. 1998. Otter (*Lutra lutra*) prey selection in relation to fish abundance and community structure in two different freshwater habitats. In: Dunstone N, Gorman M, eds. *Behaviour and Ecology of Riparian Mammals,* Cambridge: Cambridege University Press.

Caughley G, Sinclair ARE. 1994. *Wildlife Ecology and Management.* Oxford: Blackwell Scientific Publishing.

Cavillini P, Volpi T. 1995. Biases in the analysis of the diet of the red fox *Vulpes vulpes. Wildlife Biology* **1:** 243-248.

Chamberlain DC, Crick HQP. 1999. Population declines and reproductive performance of skylarks *Alauda arvensis* in different regions and habitats of the United Kingdom. *Ibis* **141:** 38-51.

Chamberlain DE, Wilson JD, Fuller RJ. 1999. A comparison of bird populations on organic and conventional farm systems in southern Britain. *Biological Conservation* **88:** 307-320.

Chanin PRF, Jefferies DJ. 1978. The decline of the Otter (*Lutra lutra*) in Britain: an analysis of hunting records and discussion of causes. *Biological Journal of the Linnean Society* **10(3):** 305-328.

Chapman PJ, Duvergé PL, Morris CJ. 1997. Research and conservation work on the greater horseshoe bat. In: *The Vincent Wildlife Trust Review of 1996.* London: The Vincent Wildlife Trust. 47-57.

Cheeseman CL, Harris S. 1982. Methods of marking badgers (*Meles meles*). *Journal of Zoology, London* **197:** 289-292.

Cheeseman CL, Mallinson PJ. 1981. Behaviour of badgers (*Meles meles*) infected with bovine tuberculosis. *Journal of Zoology, London* **194:** 284-289.

Cheeseman CL, Jones GW, Gallagher J, Mallinson PJ. 1981. The population structure, density and prevalence of tuberculosis (*Mycobacterium bovis*) in badgers (*Meles meles*) from four areas in south-west England. *Journal of Applied Ecology* **18:** 795-804.

Cheeseman CL, Little TWA, Mallinson PJ, Page RJC, Wilesmith JW, Pritchard DG. 1985. Population ecology and the prevalence of tuberculosis in badgers in an area of Staffordshire. *Mammal Review* **15:** 125-135.

Cheeseman CL, Mallinson PJ, Ryan J, Wilesmith JW. 1993. Recolonisation by badgers in Gloucestershire. In: Hayden TJ, ed. *The Badger.* Dublin: Royal Irish Academy. 78-93.

Cheeseman CL, Wilesmith JW, Stuart FA. 1989. Tuberculosis: the disease and its epidemiology in the badger, a review. *Epidemiology and Infection,* **103:** 113-125.

Chesson J. 1983. The estimation and analysis of preference and its relationship to foraging models. *Ecology* **64**: 1297-1304.

Chinery M. 1973. A Field Guide to the Insects of Britain and Northern Europe. London: Collins.

Chitty D, Southern HN. 1954. *Control of rats and mice, 3 Volumes.* Oxford: Clarendon Press.

Church BM, Jacob FH, Thompson HV. 1953. Surveys of rabbit damage to wheat in England and Wales 1950-52. *Plant Pathology* **2**: 107-112.

Church BM, Westmacott MH, Jacob FH. 1956. Surveys of rabbit damage to winter cereals in England and Wales 1953-54. *Plant Pathology* **5**: 66-69.

Clarke J, ed. 1992.*Set-aside.* British Crop Protection Council Monograph. **No. 50.** Farnham: British Crop Protection Council.

Clifton-Hadley R, Wilesmith, J. 1995. An epidemiological outlook on bovine tuberculosis in the developed world. In: Griffin F, de Lisle G, eds. *Tuberculosis in Wildlife and Domestic Animals*, Otago Conference Series No. 3. Dunedin: University of Otago Press. 178-182.

Clifton-Hadley RS, Sayers AR, Stock MP. 1995a. Evaluation of an ELISA for *Mycobacterium bovis* infection in badgers (*Meles meles*). *Veterinary Record* **137**: 555-558.

Clifton-Hadley RS, Wilesmith JW, Stuart FA. 1993. *Mycobacterium bovis* in the European badger (*Meles meles*): Epidemiological findings in tuberculous badgers from a naturally infected population. *Epidemiology and Infection* **111**: 9-19.

Clifton-Hadley RS, Wilesmith JW, Richards MS, Upton P, Johnston S. 1995b. The occurrence of *Mycobacterium bovis* infection in and around an area subject to extensive badger (*Meles meles*) control. *Epidemiology and Infection* **114**: 179-193.

Clover C. 1999. When 'Democracy' becomes mob rule. *The Field,* October 1999.

Clutton-Brock TH, Rose KE, Guiness FE. 1997. Density related changes in sexual selection in red deer. *Proceedings of the Royal Society London B.* **264(1387):** 1509-1516.

Clutton-Brock TH, Stevenson IR, Marrow P, MacColl AD, Houston AI, McNamara JM. 1996. Population fluctuations, reproductive costs and life history tactics in female Soay sheep. *Journal of Animal Ecology* **65**: 675-689.

Corbet GB. 1971. Provisional distribution maps of British mammals. *Mammal Review* **1**: 95-142.

Corner HH. 1960. Rabbit damage to winter wheat. *Scottish Agriculture* **39**: 170-181.

Corp N, Gorman ML, Speakman JR. 1997. Ranging behaviour and time budgets of male wood mice *Apodemus sylvaticus* in different habitats and seasons. *Oecologia* **109**: 242-250.

Côté IM, Sutherland WJ. 1997. The effectiveness of removing predators to protect bird populations. *Conservation Biology* **11**: 395-405.

Cowan DP. 1991. Rabbit *Oryctolagus cuniclus.* In: Corbet GB, Harris S, eds. *The handbook of British mammals,* 3rd Edition. Oxford: Blackwell Publications. 146-154.

Cowan DP, Bull DS, Inglis IR, Quy RJ, Smith P. 1994. Enhancing rodenticide performance by understanding rodent behaviour. In: *Brighton Crop Protection Conference – Pests and Diseases.* Farnham: The British Crop Protection Council. 1039-1046.

Cowan DP, Dunsford G, Gill JE, Jones A, Kerins GM, MacNicoll AD, Quy RJ. 1995. The impact of resistance on the use of second-generation anticoagulants against rats on farms in southern England. *Pesticide Science* **43**: 83-93.

Cowan DP, Vaughan JA, Christer WG. 1987. Bait consumption by the European wild rabbit in southern England. *Journal of Wildlife Management.* **51**: 386-392.

Cox PR, Smith RH. 1990. Rodenticide ecotoxicology: assessing non-target population effects. *Functional Ecology* **4**: 315-320.

Cox M, White PCL, Barbier EB. 1999. Ecological economics of bovine tuberculosis in badgers and cattle. In: Cowan DP, Feare CJ, eds. *Advances in Vertebrate Pest Management.* Fürth, Germany: Filander Verlag. 131-145.

Crawford A, Evans D, Jones A, McNulty J. 1979. *Otter Survey of Wales 1977-78.* Nettleham: Society for the Promotion of Nature Conservation.

Crawley MJ. 1989. Rabbits as pests of winter wheat. In: Putman RJ, ed. *Mammals as Pests.* London: Chapman and Hall. 168-177.

Cresswell P, Harris S, Jefferies DJ. 1990. *The history, distribution, status and habitat requirements of the badger in Britain.* Peterborough: Nature Conservancy Council.

Cresswell WJ, Harris S, Cheeseman CL, Mallinson PJ. 1992. To breed or not to breed: an analysis of the social and density-dependent constraints on the fecundity of female badgers (*Meles meles*). *Philosophical Transactions of the Royal Society of London B.* **338:** 393-407.

Critchley CNR, Hodkinson DJ, McKenzie SE. 1999. Potential benefits to water voles (*Arvicola terrestris*) of waterside buffer strips in an agri-environment scheme. *Aspects of Applied Biology* **54:** 1-6.

Critchley CNR, McKenzie SE, Hodkinson DJ. 1998. Water voles. In: McLaren RP, ed. *Monitoring of the Water Fringe Area Option of the Habitat Scheme 1994-97.* Report to Ministry Of Agriculture, Fisheries And Food. Cambridge: ADAS. 23-47 (Unpublished)

Dalyell T. 1999. Westminster Diary. *New Scientist* 11 September 1999. 55.

Davison A, Birks JDS, Griffiths HI, Kitchener AC, Biggins D, Butlin RK. 1999. Hybridization and the phylogenetic relationship between polecats and domestic ferrets in Britain. *Biological Conservation* **87:** 155-161.

Day MG. 1968. Food habits of British stoats (*Mustela erminea*) and weasels (*Mustela nivalis*). *Journal of Zoology, London* **155:** 485-497.

De'Ath A, Garthwaite DG, Thomas MR. 1999. *Rodenticide Usage on Farms in Great Britain Growing Arable Crops 1996.* Pesticide Usage Survey Report 144. London: Ministry of Agriculture, Fisheries & Food Publications.

Delahay RJ, Brown J, Mallinson PJ, Spyvee PD, Handoll D, Rogers LM, Cheeseman CL. 2000a. The use of marked bait in studies of the territorial organisation of the European badger (*Meles meles*). *Mammal Review* **30:** 73-87.

Delahay RJ, Langton S, Smith GC, Clifton-Hadley RS, Cheeseman CL. 2000b. The spatio-temporal distribution of *Mycobacterium bovis* (Bovine Tuberculosis) infection in a high density badger (*Meles meles*) population. *Journal of Animal Ecology* **69:** 428-441.

Delibes M, Macdonald SM, Mason CF. 1991. Seasonal marking, habitat and organochlorine contamination in otters (*Lutra lutra*): A comparison between catchments in Andalucia and Wales. *Mammalia* **55(4):** 567-578.

Dell'Omo G, Shore RF. 1996a. Behavioural effects of sublethal exposure to dimethoate on wood mice, *Apodemus sylvaticus*: I- Laboratory studies. *Archives of Environmental Contamination and Toxicology* **31:** 91-97.

Dell'Omo G, Shore RF. 1996b. Behavioural effects of acute sublethal exposure to dimethoate on wood mice, *Apodemus sylvaticus*: II-Field studies on radio-tagged mice in a cereal ecosystem. *Archives of Environmental Contamination and Toxicology* **31:** 538-542.

Denholm-Young PA. 1978. Studies of decomposing cattle dung and its associated fauna. [DPhil thesis.] UK: University of Oxford.

Department of the Environment. 1995. Biodiversity: the UK Steering Group Report. Volume 2: Action Plans. London: HMSO.

Department of the Environment. 1997. *The Hedgerow Regulations.* London: DOE publication.

Dolan L. 1993. Badgers and bovine tuberculosis in Ireland: a review. In; Hayden TJ, ed. *The Badger.* Dublin: Royal Irish Academy. 108-116.

Doll R, Hill AB. 1950. Smoking and carcinoma of the lung. Preliminary Report. *British Medical Journal* **3:** 739-748.

Dolman PM, Sutherland WJ. 1992. Ecological changes in Breckland grass heaths and their management for conservation. *Journal of Applied Ecology* **29:** 402-412.

Donald PF. 1998. Changes in the abundance of invertebrates and plants on British Farmland. *British Wildlife* **9(5):** 279-289.

Doney J. 1999. Validation of responses from the agricultural sector targeted by the lowland deer questionnaire. Appendix 4. In: Packer JJ, Doney J, Mayle B, Palmer SCF, Cope M, eds. *Field and Desk Studies to Assess Tolerable Damage Levels for Different Habitats and Species of Deer.* London: Report to Ministry of Agriculture, Fisheries & Foods.

Doney J, Packer J. 1998. An assessment of the impact of deer on agriculture. In: Goldspink CR, King S, Putman RJ. eds. *Population Ecology, Management and Welfare of Deer.* Manchester: Manchester Metropolitan University, Universities' Federation for Animal Welfare. 38-43.

Douglas MJW. 1965. Notes on the red fox *Vulpes vulpes* near Braemar, Scotland. *Journal of Zoology* **147:** 228-233.

Dover J, Sotherton NW, Gobbett K. 1990. Reduced pesticide inputs on cereal field margins: the effects on butterfly abundance. *Ecological Entomology* **15:** 17-24.

Driver A. 1997. River and wetland rehabilitation in the Thames Catchment. *British Wildlife* 8 (6) 362-372

Duckett JE. 1982. Barn owls (*Tyto alba*) - a proven natural predator of rats in oil palm. In: Pushparajah E, Chew, PS eds. *The Oil Palm in Agriculture in the Eighties*. Kuala Lumpur: Incorporated Society of Planters. 461-472.

Dunnet GM, Jones DM, McInerney JP. 1986. *Badgers and Bovine Tuberculosis*. London: HMSO.

Duvergé PL. 1996. Foraging activity, habitat use, development of juveniles, and diet of the greater horseshoe bat (*Rhinolophus ferrumequinum*-Schreber 1774) in south-west England. [PhD thesis.] UK: University of Bristol.

Duvergé PL. 1997. *Foraging activity and habitat use of greater horseshoe bats* Rhinolophus ferrumequinum *in Wales; preliminary results from Slebech Hall*. London: The Vincent Wildlife Trust.

Duvergé PL, Jones G. 1994. Greater horseshoe bats- activity, foraging behaviour and habitat use. *British Wildlife* 6: 69-77.

Edwards R. 1997. Furry subversives undermine Britain's railways. *New Scientist* 153(2072): 6

Engeman RM, Barnes VG, Anthony RM, Krupa HW. 1997. Effect of vegetation management for reducing damage to lodgepole pine seedlings from northern pocket gophers. *Crop Protection* 16: 407-410.

Erlinge S, Frylestam B, Göransson G, Högstedt G, Liberg O, Loman J, Nilsson IN, von Schantz T, Sylvén M. 1984. Predation on brown hare and ring-necked pheasant populations in southern Sweden. *Holarctic Ecology* 7: 300-304.

Evans A, Wilson J. 2001. The implications of ten years of research on lowland from birds: An RSPB perspective. *RSPB. Conservation Review* 13: 7-17.

Feber RE, Johnson PJ, Smith H, Baines M, Macdonald DW. 1995. The effects of arable field margin management on the abundance of beneficial arthropods. *British Crop Protection Council Symposium Proceedings*, 63: 163-170.

Feber R, Smith H, Macdonald DW. 1996. The effects on butterfly abundance of the management of uncropped edges of arable fields. *Journal of Applied Ecology*, 33: 1191-1205.

Fedriani JM, Travaini A. 2000. Predator trophic guild assignment: the importance of the method of diet quantification. *Revue d'Ecologie (Terre Vie)* 55: 129-139.

Fenn MGP, Tew TE, Macdonald DW. 1987. Rat movements and control on an Oxfordshire farm. *Journal of Zoology, London* 21: 745-749.

Fernandez FAS, Evans PR, Dunstone N. 1996. Population dynamics of the wood mouse *Apodemus sylvaticus* (Rodentia:Muridae) in a Sitka spruce successional mosaic. *Journal of Zoology, London.* 239: 717-730.

Field J. 1998. Small mammal abundance in organic and intensive farmland in Gloucestershire and Wiltshire, UK. [MSc Thesis.] UK: University of Reading.

Firbank LG, Arnold HR, Eversham BC, Mountford JO, Radford GL, Telfer MG, Treweek, JR, Webb NRC, Wells TCE. 1993. *Managing set-aside land for wildlife*. ITE research publication 7. London: HMSO.

Fitzgibbon CD. 1997. Small mammals in farm woodlands: the effects of habitat, isolation and surrounding land use patterns. *Journal of Applied Ecology*, 34: 530-539.

Flanagan P. 1993. Badgers and bovine tuberculosis: epidemiology and case studies. In: Hayden TJ, ed. *The Badger*. Dublin: Royal Irish Academy. 131-138.

Fleming TH, Heithaus ER. 1986. Seasonal foraging behaviour of the frugivorous bat *Carollia perspicillata*. *Journal of Mammalogy* 67: 660-671.

Fletcher MR, Hunter K. 1992. Pesticide poisoning of animals 1992: investigations of suspected incidents in the United Kingdom. London: Ministry of Agriculture, Fisheries and Food Publications.

Fletcher MR, Hunter K, Barnett EA. 1994. *Pesticide poisoning of animals 1993: investigations of suspected incidents in the United Kingdom*. London: Ministry of Agriculture, Fisheries and Food Publications.

Fletcher MR, Hunter K, Barnett EA. 1995. *Pesticide poisoning of animals 1994: investigations of suspected incidents in the United Kingdom*. London: Ministry of Agriculture, Fisheries and Food Publications.

Fletcher MR, Hunter K, Barnett EA, Sharp EA. 1996. *Pesticide poisoning of animals 1995: investigations of suspected incidents in the United Kingdom*. London: Ministry of Agriculture, Fisheries and Food Publications.

Fletcher MR, Hunter K, Barnett EA, Sharp EA. 1997. *Pesticide poisoning of animals 1996: investigations of suspected incidents in the United Kingdom*. London: Ministry of Agriculture, Fisheries and Food Publications.

Fletcher MR, Hunter K, Barnett EA, Sharp EA. 1998. *Pesticide poisoning of animals 1997: investigations of suspected incidents in the United Kingdom*. London: Ministry of Agriculture, Fisheries and Food Publications.

Fletcher MR, Hunter K, Barnett EA, Sharp EA. **1999**. *Pesticide poisoning of animals 1998: investigations of suspected incidents in the United Kingdom.* London: Ministry of Agriculture, Fisheries and Food Publications.

Fletcher MR, Hunter K, Quick MP, Thompson HM, Greig-Smith PW. **1991**. *Pesticide Poisoning of Animals 1990: Investigations of Suspected Incidents in Great Britain.* London: Ministry of Agriculture, Fisheries and Food Publications.

Flowerdew JR, Tattersall FH. (in press). *Wood mouse.* In: Harris S, ed. *The Handbook of British Mammals* Oxford: Blackwell Scientific Publications.

Flowerdew JR, Gurnell J, Gipps JHW, eds. **1985**. *The Ecology of Woodland Rodents: bank voles and wood mice.* Oxford: Oxford University Press.

Flowerdew JR, Trout RC, Ross JR. **1992**. Myxomatosis: rabbit population dynamics and its ecological effects in the United Kingdom. *OIE Scientific and Technical Review* **11(4)**: 1109-1113.

Forster JA. **1975**. Electric fencing to protect sandwich terns against foxes. *Biological Conservation* **7**: 85

Foster-Turly P, Macdonald SM, Mason CF, eds. **1990**. *Otters: an action plan for their conservation.* Gland, Switzerland: IUCN.

Fowler HW. **1981**. Modern English Language 2nd Edn: Oxford: Oxford University Press.

Fowler J, Cohen L. **1990**. *Practical Statistics for Field Biology.* Milton Keynes: Open University Press.

Frank LG. **1979**. Selective predation and seasonal variation in the diet of the fox (*Vulpes vulpes*) in N.E. Scotland. *Journal of Zoology* **189**: 526-532.

Frylestam B. **1986**. Agricultural land use effects on the winter diet of brown hares in southern Sweden. *Mammal Review* **16**: 157-161.

Fuller RJ, Gregory RD, Gibbons DW, Marchant JH, Wilson JD, Baillie SR, Carter N. **1995**. Population declines and range contractions among lowland farmland birds in Britain. *Conservation Biology* **9**: 1425-1441.

Furlong MJ. **1999**. The impact of a generalist predator, the red fox (*Vulpes vulpes*), on its main prey populations. [PhD thesis.] UK:University of Bristol.

Garnett BT, Delahay RJ, Roper TJ. **2002**. Use of cattle farm resources by badgers (*Meles meles*) and risk of bovine tuberculosis (*Mycobacterium bovis*) transmission to cattle. *Proceedings of the Royal Society (13)* **269**: 1487-1491.

Garthwaite DG, De'Ath A, Thomas MR. **1999**. *Rodenticide usage on farms in Great Britain growing grassland and fodder crops.* Pesticide Usage Survey Report No. 154. London: Ministry of Agriculture, Fisheries and Food Publications

Gill RMA. **1990**. *Monitoring the Status of European and North American Cervids.* GEMS Information Series, 8. Global Environment Monitoring Systems. Nairobi: United Nations Environmental Programme.

Gill RMA. **1992**. A review of damage by mammals in north temperate forests. I Deer. *Forestry*, **65**: 145-169

Gill RMA, Gurnell J, Trout RC. **1995**. Do woodland mammals threaten the development of new woods? In: Ferris-Kahn, ed. *The Ecology of Woodland Creation.* Chichester: John Wiley. 201-224.

Gipps JHW, Alibhai SK. **1991**. Field vole *Microtus agrestsi.* In: Corbet GB, Harris S, eds. *The Handbook of British Mammals.* 3rd Ed. Oxford: Blackwell Publications. 203-208.

Goodger J, Nolan A, Russell WP, Dalley DJ, Thorns CJ, Stuart FA, Croston P, Newell DG. **1994**. Serodiagnosis of *Mycobacterium bovis* infection in badgers: development of an indirect ELISA using a 25 kDa antigen. *Veterinary Record* **135**: 82-85.

Gorman ML, Akbar ZBM. **1993**. A comparative study of the wood mouse Apodemus sylvaticus in two contrasting habitats - deciduous woodland and maritime sand-dunes. *Journal of Zoology,* **229**: 385-396.

Gorman ML, Zubaid AMA. **1993**. A comparative study of the ecology of wood mice *Apodemus sylvaticus* in two contrasting habitats: decidious woodland and maritime sand-dunes. *Journal of Zoology,* London **229**: 385-396.

Gough HC. **1955**. Grazing of winter corn by the rabbit. *Annals of Applied Biology.* **43**: 720-734.

Gough HC, Dunnet FW. **1950**. Rabbit damage to winter corn. *Agriculture* **57**: 374-378.

Gough MC, Rushton SP. **2000**. The application of GIS-modelling to mustelid landscape ecology. *Mammal Review* **30**: 197-216.

Greaves JH, Cullen-Ayres PB. **1988**. Genetics of difenacoum resistance in the rat. In: Suttie JW, ed. *Current Advances in Vitamin K Research.* Amsterdam: Elsevier. 389-397

Greaves JH, Shepard DS, Gill JE. **1982**. Field trials of second-generation anticoagulants against difenacoum resistant Norway rat populations. *Journal of Hygiene,* Cambridge **89**: 295-301.

Green RE. 1979. The ecology of wood mice (*Apodemus sylvaticus*) on arable farmland. *Journal of Zoology, London* **188:** 357-377.

Green BH. 1990. Agricultural intensification and the loss of habitat, species and amenity in British grasslands: a review of historical change and assessment of future prospects. *Grass & Forage Science* **45:** 365-372.

Green J, Green R. 1980. *Otter Survey of Scotland 1977-79.* London: The Vincent Wildlife Trust.

Green J, Green R. 1996. *Otter Survey of Scotland 1991-94.* London: The Vincent Wildlife Trust.

Green RE, Etheridge B. 1999. Breeding success of the hen harrier *Circus cyaneus* in relation to the distribution of grouse moors and the red fox *Vulpes vulpes. Journal of Applied Ecology* **36:** 472-483.

Green MG, Taylor KD. 1975. Preliminary experiments in habitat alteration as a means of controlling field rodents in Kenya. In: Hansson L, Nilsson B, eds. *Biocontrol of Rodents, Ecological Bulletins No. 19.* Stockholm: Swedish Natural Science Research Council. 175-186.

Griffin DR. 1971. The importance of atmospheric attenuation for the echolocation of bats (Chiroptera). *Animal Behaviour* **19:** 55-61.

Griffin JM, Martin SW, Thorburn MA, Eves JA, Hammond RF. 1996. A case-control study on the association of selected risk factors with the occurrence of bovine tuberculosis in the Republic of Ireland. *Preventive Veterinary Medicine* **27:** 217-229.

Grolleau G, Lorgue G, Nahas K. 1989. Toxicité secondaire, en laboratoire, d'un rodenticide anticoagulant (bromadiolone) pour des prédateurs de rongeurs champêtres: buse variable (*Buteo buteo*) et hermine (*Mustela erminea*). *EPPO Bulletin* **19:** 633-648.

Gurnell J. 1981. Woodland rodents and tree seed supplies. In: Chapman JA, Pursley D, eds. *The worldwide furbearer conference proceedings.* Falls Chard, Virginia: Donnelly. 1191-1214.

Gurnell J, Flowerdew JR. 1990. *Live Trapping Small Mammals: A Practical Guide.* London: The Mammal Society.

Gurnell J, Gipps JHW. 1989. Inter-trap movement and estimating rodent densities. *Journal of Zoology, London* **217:**241-254.

Gurney JE, Perrett J, Crocker DR, Pascual JA. 1998. Mammals and farming: information for risk assessment. 1998 unpublished report to Ministry of Agriculture, Fisheries and Food. Central Science Laboratory, York.

Halliwell EC. 1997. The ecology of red squirrels in Scotland in relation to Pine marten predation. [PhD thesis.] UK: University of Aberdeen.

Hamilton T, Hartley G. 1997. Gnawing preferences of rabbits (*Oryctolagus cuniculus*) for wooden pegs of different species of tree. *Gibier Faune Sauvage* **14(3):** 509.

Hamon B. 1985. *Recherche d'organochlorés dans du guano de chauve-souris dans l'Est de la France.* Pub. Commission Permanente de la Protection des Eaux Souterraines et des Cavernes. Région Lorraine. 31

Hansson L. 1985. The food of bank voles, wood mice and yellow-necked mice. *Symposium of the Zoological Society,* London **55:** 141-168.

Harde KW, Hammond PM 1984. *A Field Guide in Colour to Beetles.* London: Octopus.

Harding M. 1994. Restoring Redgrave and Lopham Fen. *Enact* **2:2** 12-15.

Hardy AR, Fletcher MR, Stanley PI. 1986. Pesticides and wildlife: twenty years of vertebrate wildlife investigations by Ministry of Agriculture, Fisheries and Food. *State Veterinary Journal* **40:** 182-192.

Harradine JP. 1976. Anticoagulant rodenticides and non-target wildlife: An ecological evaluation of permanent baiting in rural rat control. [PhD thesis.] UK: University of Edinburgh.

Harris S. 1979. Age-related fertility and productivity in red foxes, *Vulpes vulpes*, in suburban London. *Journal of Zoology* **187:** 195-199.

Harris S, Lloyd HG. 1991. Fox *Vulpes vulpes*. In: Corbet GB, Harris S, eds. *The Handbook of British Mammals*, 3rd Ed. Oxford: Blackwell Publications. 351-367.

Harris S, Cresswell WJ, Forde PG, Trewhella WJ, Woollard T, Wray S. 1990. Home-range analysis using radio-tracking data – a review of problems and techniques particularly as applied to the study of mammals. *Mammal Review* **20:** 97-123.

Harris S, Morris P, Wray S, Yalden D. 1995. *A review of British mammals: population estimates and conservation status of British mammals other than cetaceans.* Peterborough: Joint Nature Conservation Committee.

Harrison MDK. 1997. English Environmentally Sensitive Areas. In: *Environmental Benefits from Agriculture.* Helsinki: Organisation for Economic Co-operation and Development. 141.

Helldin JO. 1998. Pine marten (*Martes martes*) population limitation. [PhD thesis.] Uppsala: Swedish University of Agricultural Sciences.

Hewson R. 1948. Some observations on the Orkney Vole. *The North Western Naturalist.* **23:** 7-10.

Hewson R. 1983. The food of wild cats (*Felis silvestris*) and red foxes (*Vulpes vulpes*) in west and north-east Scotland. *Journal of Zoology, London* **200:** 283-289.

Hewson R. 1984. Scavenging and predation upon sheep and lambs in west Scotland. *Journal of Applied Ecology* **21:** 843-868.

Hewson R. 1985. Lamb carcasses and other food remains at fox dens in Scotland. *Journal of Zoology, London* **206:** 291-296.

Hewson R. 1990. *Victim of myth.* London: League Against Cruel Sports.

Hewson R, Leitch AF. 1983. The food of foxes in forests and the open hill. *Scottish Forestry* **37:** 39-50.

Hewson R, Kolb HH, Knox AG. 1975. The food of foxes in Scottish forests. *Journal of Zoology , London* **176:** 287-292.

Heydon MJ, Reynolds JC. 2000a. Fox (*Vulpes vulpes*) management in three contrasting regions of Britain, in relation to agricultural and sporting interests. *Journal of Zoology, London* **251:** 237-252.

Heydon MJ, Reynolds JC. 2000b. Demography of rural foxes (*Vulpes vulpes*) in relation to cull intensity in three contrasting regions of Britain. *Journal of Zoology, London* **251:** 265-276.

Heydon MJ, Reynolds JC, Short MJ. 2000. Variation in abundance of foxes (*Vulpes vulpes*) between three regions of rural Britain, in relation to landscape and other variables. *Journal of Zoology, London* **251:** 253-264.

Hindmarch C, Pienkowski M. 2000. *Land management: the hidden costs.* Ecological Issues Series, British Ecological Society. Oxford: Blackwell Scientific Publishing.

HMSO. 2000a. *UK Biodiversity: memoranda relating to the enquiry submitted to the Environment Sub-committee.* Environment, Transport and Regional Affairs Committee, Environment Sub-committee. House of Commons, Session 1999-2000. London: HMSO.

HMSO. 2000b. *UK Biodiversity: interim report. Vol I report and proceedings of the Environment Committee.* Environment, Transport and Regional Affairs Committee, Environment Committee. House of Commons, Session 1999-2000. London: HMSO.

Hooper MD. 1974. Hedgerow removal. *Biologist* **21(2):** 81-86.

Howes CA. 1974. Notes on the food of foxes on Spurn Peninsula. *Naturalist, Leeds* **99:** 131-133.

Howes CA. 1978. Notes on the food of foxes at Gibraltar Point, Lincolnshire. *Naturalist, Leeds* **103:** 25-26.

Howes CA. 1980. The seasonal food of foxes on Spurn Peninsula. *Spurn Bird Observatory Report for 1980.* 74-75.

Hudson P. 1992. *Grouse in space and time: the population biology of a managed gamebird.* Fordingbridge: The Game Conservancy Trust

Hutchings MR, Harris S. 1996. *The current status of the brown hare* (Lepus europaeus) *in Britain.* JNCC Report. Peterborough: Joint Natural Conservancy Council.

Hutchings MR, Harris S. 1997. Effects of farm management practices on cattle grazing behaviour and the potential for transmission of bovine tuberculosis from badgers to cattle. *Veterinary Journal* **153:** 149-162.

Hutson AM. 1993. Action plan for the conservation of bats in the United Kingdom. London: The Bat Conservation Trust.

IAMBH. 2000. First round evidence submission to the Committee of Inquiry into Hunting with Dogs. In: Burns T, Edwards V, Marsh J, Soulsby L, Winter M, eds. on the CD accompanying the *Report of the Committee of Inquiry into Hunting with Dogs, CM 4763.* Norwich: The Stationary Office.

Iason GR, Hester AJ. 1999. What are the consequences of reducing sheep density on rabbit-grazed pastures? In: Duncan P, Micol T, eds. *Herbivore feeding strategies, population processes and impact on biodiversity.* Proceedings of ungulate research group, 6-9 July 1999. Chize: France. 79-80.

Inglis IR, Shepherd DS, Smith P, Haynes P, Bull DS, Cowan DP, Whitehead D. 1996. Foraging behaviour of wild rats (*Rattus norvegicus*) towards new foods and bait containers. *Applied Animal Behaviour Science* **47:** 175-190.

Jefferies DJ. 1969. Causes of badger mortality in eastern counties of England. *Journal of Zoology, London* **157:** 429-436.

Jefferies DJ. 1972. Organochlorine insecticide residues in British bats and their significance. *Journal of Zoology, London* **166**: 245-263.

Jefferies DJ. 1989. The changing otter population of Britain 1700-1989. *Biological Journal of the Linnean Society* **38**: 61-69.

Jefferies DJ. 1991. Some observations on Scottish wildcats *Felis silvestris* based on the results of autopsies. *Glasgow Naturalist* **22**: 11-20.

Jefferies DJ. 1996. Decline and recovery of the otter – a personal account. *British Wildlife* **7**: 353-364.

Jefferies DJ, French MC. 1976. Mercury, cadmium, zinc, copper and organochlorine insecticide levels in small mammals trapped in a wheat field. *Environmental Pollution* **10**: 175-182.

Jefferies DJ, Morris PA, Mulleneux JE. 1989. An enquiry into the changing status of the water vole *Arvicola terrestris* in Britain. *Mammal Review* **19**: 111-131.

Jefferies DJ, Stainsby B, French MC. 1973. The ecology of small mammals in arable fields drilled with winter wheat and the increase in their dieldrin and mercury residues. *Journal of Zoology, London* **171**: 513-539.

Jenkins D. 1981. Ecology of otters in Northern Scotland IV. A model system for otter *Lutra lutra* L. conservation in a freshwater system in Aberdeenshire. *Biological Conservation* **20**:123-132.

Jenkins D, Watson A, Miller GR. 1964. Predation and red grouse populations. *Journal of Applied Ecology* **1**: 183-195.

Jenson AG. 1979. *Proofing buildings against rats and mice.* Technical Bulletin **12**. London: Ministry of Agriculture, Fisheries and Food Publications.

Johnson DH. 1980. The comparison of usage and availability measurements for evaluating resource preference. *Ecology* **61**: 65-71.

Johnson IP, Flowerdew JR, Hare R. 1991. Effects of broadcasting and of drilling methiocarb molluscicide pellets on field populations of wood mice, *Apodemus sylvaticus*. *Bulletin of Environmental Contamination and Toxicology* **46**: 84-91.

Jones G. 1990. Prey selection by the greater horseshoe bat (*Rhinolophus ferrumequinum*): optimal foraging by echolocation? *Journal of Animal Ecology* **59**: 587-602.

Jones G, Billington G. 1999. *Radio-tracking study of greater horseshoe bats at Cheddar, North Somerset.* Taunton: English Nature.

Jones G, Morton M. 1992. Radio-tracking studies on the habitat use by greater horseshoe bats (*Rhinolophus ferrumequinum*). In: Priede IG, Swift SW, eds. *Wildlife telemetry, remote monitoring and tracking of animals,* Chichester: Ellis Horwood. 521-537.

Jones G, Rayner JMV. 1989. Foraging behaviour and echolocation of wild horseshoe bats *Rhinolophus ferrumequinum* and *R. hipposideros. Behavioural Ecology and Sociobiology* **25**: 183-191.

Jones G, Duvergé PL, Ransome RD. 1995. Conservation biology of an endangered species: field studies of greater horseshoe bats. *Symposium of the Zoological Society of London* **67**: 309-324.

Joyce KM, McLaren RP, Horrobin CA, Slater JA. 1998. The Water Fringe Areas. In: McLaren RP, ed. *Monitoring of the Water Fringe Area Option of the Habitat Scheme 1994-97.* London: ADAS report to Ministry of Agriculture, Fisheries and Food. 23-47.

Kaluzinski J. 1982a. Dynamics and structure of a field roe population. *Acta theriologica* **27**: 385-408.

Kaluzinski, J. 1982b. Composition of the food of roe deer living in field and the effects of their feeding on plant production. *Acta Theriologica,* **27**: 457-70.

Kayne GG. 1937. *The Control of Tuberculosis in England: Past and Present.* Oxford: Oxford Medical Publications.

Kean JM, Hickling GJ, Barlow ND. 1999. Evaluating potential sources of bovine tuberculosis infection in a New Zealand cattle herd. *New Zealand Journal of Agricultural Research* **42**: 101-106.

Key G, Moore NP, Hart J. 1997. Impact and management of deer in farm woodlands. In: Goldspink CR, King S, Putman RJ, eds. *Population Ecology, Management and Welfare of Deer.* Manchester: Manchester Metropolitan University & Universities' Federation for Animal Welfare. 44-47.

King CM. 1980a. The weasel (*Mustela nivalis*) and its prey in an English woodland. *Journal of Animal Ecology* **49**: 127-159.

King CM. 1980b. Population biology of the weasel *Mustela nivalis* on British game estates. *Holarctic Ecology* **3**: 160-168.

King CM. 1985. Interactions between woodland rodents and their predators. *Symposia of the Zoological Society of London,* **55**: 219-247.

King C. 1989. *The natural history of weasels and stoats.* London: Christopher Helm.

King CM. 1991a. Stoat. In: Corbet GB, Harris S, eds. *The Handbook of British Mammals,* 3rd. Edition. Oxford: Blackwell Scientific Publications. 377-387.]

King, CM. 1991b. Weasel. In: Corbet GB, Harris S eds. *The Handbook of British Mammals,* 3rd. Edition. Oxford: Blackwell Scientific Publications. 387-396.

King CM, Moors PJ. 1979. The life history tactics of mustelids, and their significance for predator control and conservation in New Zealand. *New Zealand Journal of Zoology* **6**: 619-622.

King EJ, Lovell DJ, Harris S. 1999. Effect of climate on the survival of *Mycobacterium bovis* and its transmission to cattle herds in south-west Britain. In: Cowan DP, Feare CJ, eds. *Advances in Vertebrate Pest Management.* Fürth, Germany:Filander Verlag. 147-161.

Kjellander P, Ahlquist I. 1996. *Electric fences to prevent wildlife damage on cereals; experiments on oat fields in Orebro County, 1995.* Internal report; Swedish University of Agricultural Sciences, Department of Conservation Biology.

Kolb HH, Hewson R. 1979. Variation in the diet of foxes in Scotland. *Acta Theriologica* **24**: 69-83.

Kolb HH, Hewson R. 1980a. The diet and growth of fox cubs in two regions of Scotland. *Acta Theriologica* **25**: 325-331.

Kolb HH, Hewson R. 1980b. A study of fox populations in Scotland from 1971 to 1976. *Journal of Applied Ecology* **17**: 7-19.

Kotzageorgis GC, Mason CF. 1997. Small mammal populations in relation to hedgerow structure in an arable landscape. *Journal of Zoology, London* **242**: 425-434.

Krebs, J R, Anderson R, Clutton-Brock T, Morrison I, Young D, Donelly C, Frost S, Woodroffe R. 1997. *Bovine Tuberculosis in Cattle and Badgers.* Report by the Independent Scientific Review Group. London: Ministry of Agriculture, Fisheries and Food Publications.

Krebs CJ. 1989. Ecological Methodology. London: Harper and Row.

Krebs JR, Wilson JD, Bradbury RB, Siriwardena GM. 1999. The Second Silent Spring? *Nature,* 400: 611-612.

Kruuk H. 1978. Spatial organization and te rritorial behaviour of the European badger *Meles meles. Journal of Zoology* **184**: 1-19.

Kruuk H. 1995. *Wild Otters - Predation and populations.* Oxford: Oxford University Press.

Kruuk H, Parish T. 1982. Factors affecting population density, group size and territory size of the European badger, *Meles meles. Journal of Zoology, London* **196**: 31-39.

Kuramoto T. 1972. Studies on bats at the Akiyoshi-dai Plateau, with special reference to the ecological and phylogenic aspects. *Bulletin of the Akiyoshi-dai Science Museum* **8**: 7-119.

Kurki S, Nikula A, Helle P, Linden H. 1998. Abundances of red fox and pine marten in relation to the composition of boreal forest landscapes. *Journal of Animal Ecology* **67**: 874-886.

Lampkin N. 1990. *Organic farming.* Ipswich: Farming Press.

Langley PJW, Yalden DW. 1977. The decline of the rarer carnivores in Great Britain during the nineteenth century. *Mammal Review* **7**: 95-116.

Large ARG, Petts GE. 1997. Rehabilitation of river margins. In: Petts GE, Calow P, eds. *River Restoration.* Oxford: Blackwell Science Ltd. 106-123.

Latham RM. 1952. The fox as a factor in the control of weasel populations. *Journal of Wildlife Management* **16**: 516-517.

Lawton JH, Woodroffe GL. 1991. Habitat and distribution of water voles: why are there gaps in a species' range? *Journal of Animal Ecology* **60** (1): 79-91.

Leckie FM, Thirgood SJ, May R, Redpath SM. 1998. Variation in the diet of red foxes on Scottish moorland in relation to prey abundance. *Ecography* **21**: 599-604.

Lenton EJ, Chanin PRF, Jefferies DJ. 1980. *Otter Survey of England 1977-78.* London: Nature Conservancy Council.

Lever RJAW. 1959. The diet of the fox since myxomatosis. *Journal of Animal Ecology* **28**: 359-375.

Lewis T. 1969a. The distribution of flying insects near a low hedgerow. *Journal of Applied Ecology* **6**: 443-452.

Lewis T. 1969b. The diversity of the insect fauna in a hedgerow and neighbouring fields. *Journal of Applied Ecology* **6**: 453-458.

Lewis T. 1970. Patterns of distribution of insects near a windbreak of tall trees. *Annals of Applied Biology* **65:** 213-220.

Lim JL, Fenn MGP, Buckle AP, Visvalingam M. 1993. Prey selection by barn owls (*Tyto alba*) and its impact on rat control in oil palm plantations. In: Yusof B, ed. *Proceedings 1991 PORIM International Palm Oil Development Conference – Agriculture*. Malaysia: Palm Oil Research Institute of Malaysia. 458-465.

Limpens HJGA, Kapteyn K. 1991. Bats, their behaviour and linear landscape elements. *Myotis* **29:** 39-48.

Lindström ER, Brainerd SM, Helldin JO, Overskaug K. 1995. Pine marten-red fox interactions: a case of intraguild predation? *Annales Zoologici Fennici* **32:** 123-130.

Little TWA, Naylor PF, Wilesmith J W. 1982. Laboratory studies of *Mycobacterium bovis* infection in badgers and calves. *Veterinary Record* **111:** 550-557.

Little TWA, Swan C, Thompson HV, Wilesmith JW. 1982. Bovine tuberculosis in domestic and wild mammals in an area of Dorset. II. The badger population, its ecology and tuberculosis status. *Journal of Hygiene* **89:** 211-224.

Lloyd HG. 1970. Post-myxomatosis rabbit populations in England and Wales. *European and Mediterranean Planr Protection Organisation Publications Series A.* **58:** 197-215.

Lloyd HG. 1980. *The Red Fox*. London: BT Batsford Ltd.

Lockie JD. 1956. After myxomatosis: notes on the food of some predatory animals in Scotland. *Scottish Agriculture* **36:** 65-69.

Lockie JD. 1963. Eagles, foxes and their food supply in Wester Ross, Scotland. *Scottish Agriculture* **42:** 1-4.

Lockie JD. 1964. The breeding density of the Golden eagle and fox in relation to food supply in Wester Ross, Scotland. *Scottish Naturalist* **71:** 67-77.

Longland WS, Price MV. 1991. Direct observations of owls and heteromyid rodents: can predation risk explain microhabitat use? *Ecology,* **72:** 2261-2273.

Lund M, Rasmussen AM. 1986. Secondary poisoning hazards to stone martens (*Martes foina*) fed bromadiolone-poisoned mice. *Nordisk Veterinaer Medicin* **38:** 241-243.

Luttik R, Clook MA, Taylor MR, Hart ADM. 1999. Regulatory aspects of the ecotoxicological risk assessment of rodenticides. In: Cowan DP, Feare CJ, eds. *Advances in Vertebrate Pest Management*. Fürth: Filander Verlag. 369-385.

Macdonald DW. 1984. A questionnaire survey of farmers' opinions and actions towards wildlife. In: Jenkins D, ed. *Agriculture and the Environment*. Cambridge: ITE publications. 171-177.

Macdonald DW. 2001. Bartering biodiversity: what are the options. In: Helm D, ed. *Oxford Review of Economic Policy*. Oxford: Oxford University Press. 142-171.

Macdonald DW, Johnson PJ. 1996. The impact of sport hunting: a case study. In: Dunstone N, Taylor V, eds. *The exploitation of mammal populations*. London: Chapman & Hall. 160-207.

Macdonald DW, Johnson PJ. 2000. Farmers and the custody of the countryside: trends in loss and conservation of non-productive habitats 1981-1998. *Biological Conservation* **94:** 221-234.

MacDonald SM, Mason CF. 1983. Some factors affecting the distribution of otters (*Lutra lutra*). *Mammal Review* **13(1):** 11-23.

Macdonald DW, Smith H. 1991. New perspectives in agro-ecology: between theory and practice in the agricultural ecosystem. In: Firbank LG, Carter N, Derbyshire JF, Potts GR, eds. *The Ecology of Temperate Cereal Fields*. Oxford: Blackwell Scientific Publications. 413-448.

Macdonald DW, Strachan R. 1999. *The Mink and the Water Vole: Analyses for Conservation*. Oxford: Wildlife Conservation Research Unit and the Environment Agency.

Macdonald DW, Mace GM, Barretto GR. 1999. The effects of predators on fragmented prey populations: a case study for the conservation of endangered prey. *Journal of Zoology, London* **247:** 487-506.

Macdonald DW, Mace G, Rushton S. 1998. *Proposals for future monitoring of British mammals*. London: Department of the Environment, Transport and the Regions.

Macdonald DW, Mace GM, Rushton S. 2000b. Conserving British Mammals: Is There A Radical Future. In: Entwistle A, Dunstone N, eds. *Priorities For The Conservation Of Mammalian Diversity: Has The Panda Had Its Day?* London: FFI. 175-205.

Macdonald DW, Tew TE, Todd IA. 1993. The arable wood mouse. *NERC news,* July 1993.16-19.

Macdonald DW, Johnson PJ, Feber RE, Tattersall FH. 2000a. Habitat heterogeneity on farmland. In: Hutchings M, ed. *Ecological Consequences of Habitat Heterogeneity*. Oxford: Blackwell Scientific Publishing. 357-378.

Macdonald DW, Tattersall FH, Johnson PJ, Carbone C, Reynolds JC, Langbein J, Rushton SP, Shirley MDF. 2000c. *Management and control of populations of foxes, deer, hares, and mink in England and Wales, and the impact of hunting with dogs.* Report to Lord Burns' Inquiry into Hunting with Dogs. Wildlife Conservation Research Unit. Oxford: University of Oxford.

Macdonald DW, Tew TE, Todd AI, Garner JP, Johnson PJ. 2000d. Arable habitat use by wood mice *Apodemus sylvaticus.* 3: A farm-scale experiment on the effects of crop rotation. *Journal of Zoology, London* **250**: 313-320.

MacNicoll AD, Gill JE. 1987. The occurence and significance of rodenticide resistance in the UK. In: *British Crop Protection Council Monograph No. 37 Stored Products Pest Control.* Thornton Heath: British Crop Protection Council Publications. 89-95.

MacNicoll AD, Kerins GM, Dennis NJ, Gill JE. 1996. The distribution and significance of anticoagulant-resistant Norway rats (*Rattus norvegicus*) in England and Wales, 1988-95. In: Timm R, Crabb AC, eds. *Proceedings of the 17th Vertebrate Pest Conference,* Davis, California: University of California. 179-185.

MacVicker HJ. 1998. The ecotoxicology of rodenticide use on farms. [Unpublished PhD thesis.] UK: University of Leicester.

MAFF. 1979. *Bovine Tuberculosis in Badgers: Third Report.* London: Ministry of Agriculture, Fisheries & Food.

MAFF. 1996. *The code of practice for prevention of rodent infestation in poultry flocks.* London: Ministry of Agriculture, Fisheries and Food Publications.

MAFF. 1998. *Arable Areas Payments 1998/1999, Explanatory Guide: Parts 1 & 2.* London: Ministry of Agriculture, Fisheries and Food Publications.

MAFF. 1999. *Local Environmental Risk Assessments For Pesticides: A practical guide.* London: Ministry of Agriculture, Fisheries and Food Publications.

MAFF. 2000a. *Conservation grants for farmers, 2000.* London: Ministry of Agriculture, Fisheries and Food Publications.

MAFF. 2000b. *Arable Area Payments Scheme. Explanatory Guide: Part II.* London: Ministry of Agriculture, Fisheries and Food Publications.

MAFF. 2000c. http://www.maff.gov.uk/animalh/tb. 29 November 2000.

MAFF, SERAD, DARD (NI), NAWAD. 1999. *Agriculture in the United Kingdom: 1999.* London: HMSO.

MAFF, SOAEFD, DANI, WO. 1998. *Agriculture in the United Kingdom: 1998.* London: HMSO.

Mallorie H, Flowerdew JR. 1994. Woodland small mammal population ecology in Britain. A preliminary review of the Mammal Society survey of wood mice (*Apodemus sylvaticus*) and bank voles (*Clethrionomys glareolus*), 1982-87. *Mammal Review* **24**: 1-15.

Marchant J, Hudson R, Carter SP, Whittington P. 1990. *Population trends in British Breeding Birds.* Tring: British Trust for Ornithology.

Mason CF. 1989. Water pollution and otter distribution: a review. *Lutra* **32**: 97-131.

Mason CF. 1998. Decline in PCB levels in otters (*Lutra lutra*). *Chemosphere* **36**: 1969-1971.

Mason CF, Macdonald SM. 1986. *Otters: Ecology and Conservation.* Cambridge: Cambridge University Press.

Mayle BA. 1994. Falling foul of fallow. *Timber Grower* **1994**: 34-36.

Mayle BA, Putman RJ, Wyllie I. 2000. The use of trackway counts to establish an index of deer presence. *Mammal Review* **30**: 233-237.

Mayot P, Patillault JP, Stahl P. 1998. Effect of predator removal on survival of pen-rered and wild ring-necked pheasants (*Phasianus colchicus*) released in the Yonne region. *Gibier Faune Sauvage* **15**: 1-19.

McDonald RA, Harris S. 1999. The use of trapping records to monitor populations of stoats *Mustela erminea* and weasels *M. nivalis*: the importance of trapping effort. *Journal of Applied Ecology* **36**: 679-688.

McDonald RA, Harris S. 2000. The use of fumigants and rodenticides by gamekeepers in Great Britain. *Mammal Review* **30**: 57-64.

McDonald RA, Haris S. 2002. Population biology of Stoats *Mustela erminea* and weasels *Mustela nivalis* on game estates in Great Britain. *Journal of Applied Ecology* **39** (in press).

McDonald RA, Murphy EC. 2000. A comparison of the management of stoats and weasels in Great Britain and New Zealand. In: Griffiths HI, ed. *Mustelids in a modern world.* Leiden: Backhuys Publishers. 21-40.

McDonald R, Baker P, Harris S. 1997. *Is the fox a pest? The ecological and economic impact of foxes in Britain.* Bristol: University of Bristol, Electra Press.

McDonald R, Bright PW, Harris S. 1994. *Baseline survey of pine martens in Wales.* Bangor: Report to Countryside Council for Wales.

McDonald RA, Webbon C, Harris S. 2000. The diet of stoats (*Mustela erminea*) and weasels (*Mustela nivalis*) in Great Britain. *Journal of Zoology, London* **252:** 363-371.

McDonald RA, Harris S, Turnbull G, Brown P, Fletcher M. 1998. Anticoagulant rodenticides in stoats (*Mustela erminea*) and weasels (*M. nivalis*) in England. *Environmental Pollution* **103:** 17-23.

McInerney JP. 1987. Assessing the policy of badger control and its effects on the incidence of bovine tuberculosis. *Proceedings of the Society for Veterinary Epidemiology and Preventive Medicine*, April 1987. 133-147.

McKillop IG, Fox SM, Pugh BD, Langton SD. 1996. Developing a model to predict yield loss of winter wheat due to grazing by European wild rabbits. *Brighton Crop Protection Conference Pests and Diseases Council, vol. 1.* 145-156.

McNeilage AJ. 1995. Mountain gorillas in the Virunga volcanoes: ecology and carrying capacity. [PhD thesis.] UK: University of Bristol.

Meek E, Rebecca G, Ribbands B, Fairclough K. 1998. Orkney hen harriers: a major population decline in the absence of persecution. *Scottish Birds* **19:** 290-298.

Meharg AA, Wright J, Osborn D. 1998. The frequency of Environmental Quality Standard (EQS) exceedance for chlorinated organic pollutants in rivers of the Humber catchments. *Science of the Total Environment* **210:** 219-228.

Messenger J, Birks J, Jefferies D. 1997. What is the status of the pine marten in England and Wales? *British Wildlife* **8:** 273-279.

Meyer AN. 1994. Rodent control in practice: food stores. In: Buckle AP, Smith RH, eds. *Rodent Pests and their Control*, Wallingford, Oxford: CAB International. 273-290.

Meyer AN, Shankster A, Langton SD, Jukes J. 1995. National commensal rodent survey 1993. *Environmental Health* **103:** 127-135.

MFHA. 2000. First round evidence submission to the Committee of Inquiry into Hunting with Dogs. In: Burns T, Edwards V, Marsh J, Soulsby L, Winter M, (authors) *The Report of the Committee of Inquiry into Hunting with Dogs, CM 4763.* On the accompanying CD. Norwich: The Stationary Office.

Millais JG. 1904. On a new British vole from the Orkney Islands. *The Zoologist* **8:** 241-246.

Mills S. 1986. Rabbits breed a growing controversy. *New Scientist* **1498:** 50-54.

Ministry of Agriculture and Forestry. 1999. httpi/www.maf.govt.nz/MAFnet/ articles-man/posat/ posat001.htm). 16 February 1998.

Mitchell-Jones AJ, ed. 1987. *The Bat Worker's Manual.* UK: Nature Conservancy Council.

Mitchell-Jones AJ. 1995. The status and conservation of horseshoe bats in Britain. *Myotis* **32/33:** 271-284

Mitchell-Jones AJ, Cooke AS, Boyd IL, Stebbings RE. 1989. Bats and remedial timber treatment chemicals – a review. *Mammal Review* **19:** 93-110.

MLURI. 1988. *The Land Cover of Scotland.* Aberdeen: Macaulay Land Use Research Institute.

Monneret RJ. 1987. *Le faucon pèlerin.* Paris: Editions du Point-Vétérinaire.

Montgomery WI. 1989a. Population regulation in the wood mouse, *Apodemus sylvaticus* I. Density-dependence in the annual cycle of abundance. *Journal of Animal Ecology* **58:** 465-475.

Montgomery WI. 1989b. Population regulation in the wood mouse, *Apodemus sylvaticus* II. Density-dependence in the spatial distribution and reproduction. *Journal of Animal Ecology* **58:** 477-494.

Montgomery WI, Dowie M. 1993. The distribution and population regulation of the wood mouse, *Apodemus sylvaticus* on field boundaries of pastoral farmland. *Journal of Applied Ecology* **30:** 783-791.

Morris, PA. 1991. Common dormouse. In: Corbett GB, Harris S, eds. *The Handbook of British Mammals*, Oxford: Blackwell Publications. 259-264.

Morris PA. 1993. *A red data book for British mammals.* London: The Mammal Society.

Morrison WI, Bourne FJ, Cox DR, Donnelly CA, Gettinby G, McInerney JP, Woodroffe R. 2000. Pathogenesis and diagnosis of infections with *Mycobacterium bovis* in cattle. *Veterinary Record* **146:** 236-242.

Muirhead RH, Gallagher J, Burn KJ. 1974. Tuberculosis in wild badgers in Gloucestershire: Epidemiology. *Veterinary Record* **95:** 552-555.

Mulder JL. 1990. The stoat *Mustela erminea* in the Dutch dune region its local extinction and a possible cause: the arrival of the fox *Vulpes vulpes.* *Lutra* **33:** 1-21.

Murphy EC, Clapperton BK, Bradfield PMF, Speed HJ. 1998a. Effects of rat-poisoning operations on abundance and diet of mustelids in New Zealand podocarp forests. *New Zealand Journal of Zoology* **25:** 315-328.

Murphy EC, Clapperton BK, Bradfield PMF, Speed HJ. 1998b. Brodifacoum residues in target and non-target animals following large-scale poison operations in New Zealand podocarp-hardwood forests. *New Zealand Journal of Zoology* 25: 307-314.

Neuweiler G. 1989. Foraging ecology and audition in echolocating bats. *Trends in Ecology and Evolution* 4: 160-166.

Newell DG, Hewinson RG. 1995. Control of bovine tuberculosis by vaccination. *Veterinary Record* 136: 459-463.

Newell DG, Clifton-Hadley RS, Cheeseman CL. 1997. The kinetics of serum antibody responses to natural infections with *Mycobacterium bovis* in one badger social group. *Epidemiology and Infection* 118: 173-180.

Newsome AE, Parer I, Catling PC. 1989. Prolonged prey suppression by carnivores – predator-removal experiments. *Oecologia* 78: 458-467.

Newton I, Wyllie I. 1992. Recovery of a Sparrowhawk population in relation to declining pesticide contamination. *Journal of Applied Ecology* 29: 476-484.

Newton I, Wyllie I, Freestone P. 1990. Rodenticides in British Barn Owls. *Environmental Pollution* 68: 101-117.

Newton I, Shore RF, Wyllie I, Birks JDS, Dale L. 1999. Empirical evidence of side-effects of rodenticides on some predatory birds and mammals. In: Cowan DP, Feare CJ, eds. *Advances in vertebrate pest management*. Fürth: Filander Verlag. 347-367.

Nix J. 1996. *Farm management pocketbook*. Wye: Wye College Press.

Nix J, Hill P. 2000. *Farm management pocketbook*. Wye: Wye College Press.

Norberg UM, Rayner JMV. 1987. Ecological morphology and flight in bats (Mammalia; Chiroptera): wing adaptations, flight performance, foraging strategy and echolocation. *Philosophical Transactions of the Royal Society, London B* 316: 335-427.

O'Connor RJ, Shrubb M. 1990. *Farming and birds*. Cambridge: Cambridge University Press.

O'Mahoney D, Lambin X, MacKinnon JL, Coles CF. 1999. Fox predation on cyclic field vole populations in Britain. *Ecography* 22: 575-581.

O'Mairtin DO, Williams DH, Griffin JM, Dolan LA, Eves JA. 1998. The effect of a badger removal programme on the incidence of tuberculosis in an Irish cattle population. *Preventative Veterinary Medicine* 34: 47-56.

O'Neill BD, Pharo HJ. 1995. The control of bovine tuberculosis in New Zealand. *New Zealand Veterinary Journal* 43: 249-255.

O'Sullivan WM. 1996. Otter conservation: factors affecting survival, with particular reference to drainage and pollution within an Irish river system. In: Reynolds JD, ed. *The Conservation of Aquatic Systems*. Dublin: Royal Irish Academy. 117-133.

Obrtel R, Holisova V. 1983. Assessment of the damage done to a crop of maize (*Zea mays*) by roe deer (*Capreolus capreolus*). *Folia Zoologica, Brno,* 32: 109-118.

Obrtel R, Holisova V, Kozena I. 1984. Deer damage to sugar beet leaves. *Folia Zoologica, Brno.* 33: 99-108.

Ogilvy SE. 1996. LINK-IFS - An integrated approach to crop husbandry. *Aspects of Applied Biology: Rotations and Cropping* 47: 335-352.

Oksanen T, Oksanen L, Fretwell SD. 1985. Surplus killing in the hunting strategy of small predators. *American Naturalist* 126: 328-346.

Olney NJ, Garthwaite DG. 1992. *Rodenticide usage on farms in England growing arable crops 1990*. Pesticide Usage Survey Report No. 95. London: Ministry of Agriculture, Fisheries and Food Publications.

Olney NJ, Garthwaite DG. 1994. *Rodenticide usage in Great Britain on farms growing arable crops 1992*. Pesticide Usage Survey Report No. 113. London: Ministry of Agriculture, Fisheries and Food Publications.

Olney NJ, Davis RP, Thomas MR, Garthwaite DG. 1991a. *Rodenticide usage on farms in England growing arable crops 1988*. Pesticide Usage Survey Report No. 89. London: Ministry of Agriculture, Fisheries and Food Publications.

Olney NJ, Davis RP, Thomas MR, Garthwaite DG. 1991b. *Rodenticide usage on farms in England and Wales growing grassland and fodder crops 1989*. Pesticide Usage Survey Report No. 90. London: Ministry of Agriculture, Fisheries and Food Publications.

Olney NJ, Thomas MR, Garthwaite DG. 1994. *Rodenticide usage on farms in Great Britain growing grassland and fodder crops 1993*. Pesticide Usage Survey Report No. 122. London: Ministry of Agriculture, Fisheries and Food Publications.

Ovenden GN, Swash ARH, Smallshire D. 1998. Agri-environment schemes and their contribution to the conservation of biodiversity in England. *Journal of Applied Ecology* **35**: 955-960.

Packer JJ, Birks JDS. 1999. An assessment of British farmers' and gamekeepers' experiences, attitudes and practices in relation to the European Polecat *Mustela putorius*. *Mammal Review* **29**: 75-92.

Pain DJ, Pienkowski MW. 1997. Conclusions: a future for farming and birds? In: Pain DJ, Pienkowski MW, eds. *Farming and Birds in Europe*. NY: Academic Press.

Palomares F, Caro TM. 1999. Interspecific killing among mammalian carnivores. *American Naturalist* **153**: 492-508.

Pech R, Hood GM, McIlroy J, Saunders G. 1997. Can foxes be controlled by reducing their fertility? *Reproduction, Fertility & Development.* **9**: 41-50.

Pech RE, Sinclair ARE, Newsome AE, Catling PC. 1992. Limits to predator regulation of rabbits in Australia: evidence from predator-removal experiments. *Oecologia* **89**: 102-112.

Pelfrene AF. 1991. Synthetic Organic Rodenticides. In: Hayes WJ, Laws ER, eds. *Handbook of Pesticide Toxicology*. San Diego: Academic Press. 1271-1316.

Pelz H-J. 1989. Ecological aspects of damage to sugar-beet seeds by *Apodemus sylvaticus*. In: Putman RJ, ed. *Mammals as Pests*. London: Chapman & Hall. 34-48.

Peng R. 1991. The influence of microclimate on the spatial distribution of flying insects. [PhD thesis.] UK: University of Leeds.

Peng R, Sutton SL, Fletcher CR. 1994. Distribution patterns of some species of Scatopsidae (Insecta: Diptera) and the effect of microclimate on their flight activity. *Journal of Zoology, London* **232**: 585-594.

Pepper HW. 1992. *Forest Fencing*. Forestry Commission Bulletin 102. London: HMSO.

Pepper HW. 1998. *The prevention of rabbit damage to trees in woodland*. Forestry Commission Practice Note 2. London: HMSO.

Pepper HW. 1999. *The Development and Testing of Deterrents for Deer in both Rural and Urban Fringe Areas*. London: Technical review for the Ministry of Agriculture Fisheries and Foods.

Pepper HW, Chadwick AH, Butt RW. 1992. *Electric Fencing against Deer*. Forestry Commission Research Information Note 235. London: HMSO.

Petersen RC, Petersen BM, Lacoursiére J. 1992. A building block model for Stream restoration. In: Boon PJ, Calow P, Petts GE, eds. *River Conservation and Management*. Chichester: J.Wiley & Sons Ltd. 293-309.

Pether JVS, Lloyd G. 1993. The clinical spectrum of human hantavirus infection in Somerset, UK. *Epidemiology and Infection* **111**; 171-175.

Philcox CK, Grogan AL, Macdonald DW. 1999. Patterns of otter *Lutra lutra* road mortality in Britain. *Journal of Applied Ecology* **36**: 748-762.

Phillips WM. 1953. The effect of rabbit grazing on a re-seeded pasture. *Journal of the British Grassland Society*, **8**: 169-181.

Picozzi N. 1984. Breeding biology of polygynous hen *harriers Circus cyaneus* in Orkney. *Ornis Scandanavica*. **15**: 1-10.

Plesner Jensen S. 1993. Temporal changes in food preferences of wood mice, *Apodemus sylvaticus*, L. *Oecologia* **94**: 76-82.

Plesner Jensen S, Honess P. 1995. The influence of moonlight on vegetation height preference and trappability of small mammals. *Mammalia* **59**: 35-42.

Poché RM. 1988. Rodent tissue residue and secondary hazard studies with bromadiolone. *EPPO Bulletin* **18**: 323-330.

Pollard E, Relton J. 1970. Hedges V: A study of small mammals in hedges and cultivated fields. *Journal of Applied Ecology*, **7**: 549-557.

Potts GR. 1980. The effects of modern agriculture, nest predation and game management on the population ecology of partridges (*Perdix perdix* and *Alectoris rufa*). *Advances in Ecological Research* **11**: 1-82.

Povey FD, Smith H, Watt T. 1993. Predation of annual grass weed seeds in arable field margins. *Annals of Applied Biology*, **122**: 323-328.

Power AP, Watts BGA. 1987. *The badger control policy: an economic assessment*. Government Economic Service Working Paper No. 96. London: Ministry of Agriculture Fisheries and Food.

Prend J, Granada-Lorencio C. 1996. The relative importance of riparian habitat structure on Otter *Lutra lutra* (L.) sprainting activity in a small Mediterranean catchment. *Biological Conservation* **76**: 9-15

Prescott CV, El-Amin M, Smith RH. 1992. Calciferols and bait shyness in the laboratory rat. In: Borrecco JE, Marsh RE, eds. *Proceedings of the 15th Vertebrate Pest Conference.* Davis, California: University of California. 218-223.

Prickett AJ. 1992. Recent surveys of post-harvest pest problems in farm and commercial grain stores in the UK. In: *Brighton Crop Protection Conference.* Brighton, England: Proceedings of British Crop Protection Council Conference, November 1992. 271-280.

Produce Studies. 1995. *Farmers attitudes to Fox Control.* Newbury, Berkshire: Produce Studies Ltd.

Produce Studies. 2000. A national survey of hunts in England and Wales, First round evidence submission to the Committee of Inquiry into Hunting with Dogs. In: Burns T, Edwards V, Marsh J, Soulsby L, Winter M, eds. *The Report of the Committee of Inquiry into Hunting with Dogs, CM 4763.* Norwich: The Stationary Office. On the accompanying CD.

Putman RJ. 1986. Foraging by roe deer in agricultural areas and impact on arable crops. *Journal of Applied Ecology,* **23:** 91-99.

Putman, RJ. 1988. *The Natural History of Deer.* London: Chapman and Hall.

Putman RJ. 1989. *Mammals as Pests.* London: Chapman and Hall.

Putman RJ. 1994. Damage by deer in coppice woodlands: an analysis of factors affecting the severity of damage and options for management. *Quarterly Journal of Forestry,* **88:** 45-54.

Putman RJ. 1995. *Status and Impact of Deer in the Lowlands and Options for Management.* London: Technical review for Ministry of Agriculture, Fisheries and Foods.

Putman RJ. 1996. *Deer on National Nature Reserves: Problems and Practices.* English Nature Research Reports 173. Peterborough: English Nature.

Putman, RJ. 1998. The potential role of habitat manipulation in reducing deer impact. In: Goldspink CR, King S, Putman RJ, eds. *Population Ecology, Management and Welfare of Deer,* Manchester: Manchester Metropolitan University/ Universities' Federation for Animal Welfare. 95-101.

Putman RJ, Moore NP. 1998. Impact of deer in lowland Britain on agriculture, forestry and conservation habitats. *Mammal Review,* **28:** 141-164.

Quy RJ, Cowan DP, Swinney T. 1993. Tracking as an activity index to measure gross changes in Norway rat populations. *Wildlife Society Bulletin* **21:** 122-127.

Quy RJ, Shepherd DS, Inglis IR. 1992a. Bait avoidance and effectiveness of anticoagulant rodenticides against warfarin – and difenacoum-resistant populations of Norway rats (*Rattus norvegicus*). *Crop Protection* **11:** 14-20.

Quy RJ, Cowan DP, Morgan C, Swinney T. 1996. Palatability of rodenticide baits in relation to their effectiveness against farm populations of the Norway rat. In: Timm RM, Crabb AC, eds. *Proceedings of the 17th Vertebrate Pest Conference,* Davis, California: University of California. 133-138.

Quy RJ, Cowan DP, Haynes P, Inglis IR, Swinney T. 1992b. The influence of stored food on the effectiveness of farm rat control. In: *Proceedings of British Crop Protection Council Conference.* 291-300.

Quy RJ, Cowan DP, Haynes P, Inglis IR, Swinney T. 1994. Predicting the outcome of rodenticide trials against Norway rats living on farms. In: Halverson WS, Crabb AC, eds. *Proceedings of the 16th Vertebrate Pest Conference,* Davis, California: University of California. 133-137.

Quy RJ, Cowan DP, Haynes PJ, Sturdee AP, Chalmers RM, Bodley-Tickell AT, Bull SA. 1999. The Norway rat as a reservoir host of *Cryptosporidium parvum. Journal of Wildlife Diseases* **35:** 660-670.

Quy RJ, Cowan DP, Prescott CV, Gill JE, Kerins GM, Dunsford G, Jones A, MacNicoll AD. 1995. Control of a population of Norway rats resistant to anticoagulant rodenticides. *Pesticide Science* **45:** 247-256.

Rands MRW. 1985. Pesticide use on cereals and the survival of grey partridge chicks: a field experiment. *Journal of Applied Ecology* **22:** 49-54.

Ransome, RD. 1991. Greater horseshoe bat. In: Corbett GB, Harris S, eds. *The Handbook of British Mammals,* Oxford: Blackwell Publications. 88-94.

Ransome RD. 1996. *The management of feeding areas for greater horseshoe bats.* English Nature Research Reports 174. Peterborough: English Nature.

Ransome RD. 1997. *The management of greater horseshoe bat feeding areas to enhance population levels.* English Nature Research Reports 241. Peterborough: English Nature.

Ransome RD. 1998. *The impact of maternity roost conditions on populations of greater horseshoe bats.* English Nature Research Reports 292. Peterborough: English Nature.

Ratcliffe DA. 1980. *The Peregrine Falcon.* Calton: T&AD Poyser.

Raven PJ, Holmes NTH, Davison EH, Fox PJA, Everard M, Fozzard IR, Rouen KJ. 1998. *River Habitat Quality: the physical character of rivers and streams in the UK and the Isle of Man.* Bristol: Environment Agency.

Redfern R, Gill JE. 1978. The development and use of a test to identify resistance to the anticoagulant difenacoum in the Norway rat (*Rattus norvegicus*). *Journal of Hygiene, Cambridge* **81**: 427-431.

Redfern R, Gill JE. 1980. Laboratory evaluation of bromadiolone as a rodenticide for use against warfarin-resistant and non-resistant rats and mice. *Journal of Hygiene, Cambridge* **84**: 263-268.

Reimoser F, Gossow H. 1996. Impact of ungulates on forest vegetation and its dependence on the silvicultural system. *Forest Ecology and Management* **88**: 107-119.

Rennison BD, Dubock AC. 1978. Field trials of WBA 8119 (PP581, brodifacoum) against warfarin resistant infestations of *Rattus norvegicus*. *Journal of Hygiene, Cambridge* **80**: 77-82.

Reynolds JC. 2000. *Fox Control in the Countryside.* The Game Conservancy Trust, Fordingbridge, Hants, UK.

Reynolds P. 1992. The impact of changes in land-use in Orkney on the vole *Microtus arvalis orcadensis* and its avian predators. [Unpublished PhD thesis.] UK: University of Aberdeen.

Reynolds JC, Aebischer NJ. 1991. Comparison and quantification of carnivore diet by faecal analysis: a critique, with recommendations, based on a study of the fox *Vulpes vulpes*. *Mammal Review* **21**: 97-122.

Reynolds N, Batley J. 1999. Questing for conies. *Shooting & Conservation* Spring 1999. 40-43.

Reynolds JC, Tapper SC. 1995a. Predation by foxes *Vulpes vulpes* on brown hares *Lepus europaeus* in central southern England, and its potential impact on annual population growth. *Wildlife Biology* **1**: 145-58.

Reynolds JC, Tapper SC. 1995b. The ecology of the red fox *Vulpes vulpes* in relation to small game in rural southern England. *Wildlife Biology* **1**: 105-119.

Reynolds JC, Tapper SC. 1996. Control of mammalian predators in game management and conservation. *Mammal Review* **26**: 127-156.

Richards DF. 1977. Observations on the diet of the red fox (*Vulpes vulpes*) in south Devon. *Journal of Zoology, London* **183**: 495-504.

Richards CGJ. 1981. Field trials of bromadiolone against infestations of warfarin resistant *Rattus norvegicus*. *Journal of Hygiene, Cambridge* **86**: 363-36

Richards CGJ. 1989. The pest status of rodents in the United Kingdom. In: Putman RJ, ed. *Mammals as pests.* London: Chapman and Hall. 21-33.

Robertson PA. 1988. Survival of released pheasants, *Phasianus colchicus*, in Ireland. *Journal of Zoology, London* **214**: 683-695.

Robertson PA. 1991. Estimating the nesting success and productivity of British pheasants *Phasianus colchicus* from nest-record schemes. *Bird Study* **38**: 73-79.

Rogers LM, Gorman ML. 1995. The population dynamics of small mammals living in set-aside and surrounding semi-natural and crop land. *Journal of Zoology, London* **236**: 451-464.

Rogers LM, Cheeseman CL, Mallinson PJ, Clifton-Hadley R. 1997. The demography of a high density badger (*Meles meles*) population in the west of England. *Journal of Zoology, London* **242**: 705-728.

Rogers LM, Delahay RJ, Cheeseman CL, Langton S, Smith GC, Clifton-Hadley RS. 1998. Movement of badgers (*Meles meles*) in a high density population: individual, population and disease effects. *Proceedings of the Royal Society London. B.* **265**: 1-8.

Rogers LM, Delahay RJ, Cheeseman CL, Smith GC, Clifton-Hadley RS. 1999. The increase in badger (*Meles meles*) density at Woodchester Park, south west England: a review of the implications for disease (*Mycobacterium bovis*) prevalence. *Mammalia* **63**: 183-192.

Rogers LM, Delahay RJ, Hounsome TD, Cheeseman CL. 2000. Changes in badger (*Meles meles*) social organisation in response to increasing population density at Woodchester Park, south west England. In: Griffiths HI, ed. *Mustelids in the Modern World.* Hull: Hull University Press. 267-279.

Ross JR, Page RJC, Nadian AK, Langton SD. 1998. The develpoment of a carbon monoxide producing cartridge for rabbit control. *Wildlife Research* **25**: 305-314.

Ross J, Tittensor AM, Fox AP, Sanders MF. 1989. Myxomatosis in farmland rabbit populations in England and Wales. *Epidemiological Infection* **103**: 333-357.

Rothera S. 1998. *Breckland Natural Area Profile.* Norwich: English Nature.

Rowe FP, Swinney T. 1977. Population dynamics of small rodents in farm buildings and on arable land. *EPPO Bulletin* **7**: 431-437.

RSPB. 1991. A future for environmentally sensitive farming. Royal Society for the Protection of Birds submission to the UK Review of Environmentally Sensitive Areas.

Sargeant AB. 1978. Red fox prey demands and implications to prairie duck production. *Journal of Wildlife Management* 42: 520-527.

Saunders G, Greentree C, McLeod L. 1997. *Fox predation: impact and management on agricultural land and associated remnant habitats,* NSW, Australia: Final Report to the Bureau of Resource Sciences, Dept of Primary Industries and Energy.

Scholey G. 1995. Return of the "drowners"? *Enact* 3:(1) 10-11.

Scottish Natural Heritage 1994. *Red deer and the natural heritage.* SNH policy paper. Battleby: Scottish Natural Heritage.

Scott, Lord Justice. 1942. *Report of the committee on land utilisation in rural areas, Cmnd 6378.* Ministry of Works and Planning. London: HMSO

Shaw G, Livingstone J. 1992. The pine marten - its reintroduction and subsequent history in the Galloway Forest Park. *Transactions of the Dumfriesshire and Galloway Natural History and Antiquarian Society* 67: 1-7.

Sheail J. 1971. *Rabbits and their history.* Newton Abbott, Devon: David & Charles.

Sheail, J. 1976. *Nature in Trust, the History of Nature Conservation in Britain.* London: Blackie.

Shore RF, Birks JDS, Freestone P. 1999. Exposure of non-target vertebrates to second-generation rodenticides in Britain, with particular reference to the polecat *Mustela putorius. New Zealand Journal of Ecology* 23: 199-206.

Shore RF, Birks JDS, Freestone P, Kitchener AC. 1996. Second generation rodenticides and polecats (*Mustela putorius*) in Britain. *Environmental Pollution* 91: 279-282.

Shore RF, Myhill DG, French MC, Leach DV, Stebbings RE. 1991. Toxicity and tissue distribution of pentachlorophenol and permethrin in pipistrelle bats experimentally exposed to treated timber. *Environmental Pollution* 73: 101-118.

Shore RF, Feber RE, Firbank LG, Fishwick SK, Macdonald DW, Norum U. 1997. The impacts of molluscicide pellets on spring and autumn populations of wood mice *Apodemus sylvaticus. Agriculture, Ecosystems and Environment* 64: 211-217.

Short MJ, Reynolds JC. 2001. Physical exclusion of non-target species in tunnel trapping of mammalian pests. *Biological Conservation* 98: 139-147.

Simonetti JA. 1989. Microhabitat use by small mammal in central Chile. *Oikos* 56: 309-318.

Singleton GR, Leirs H, Hinds LA, Zhang Z. 1999. Ecologically-based management of rodent pests – re-evaluating our approach to an old problem. In: Singleton GR, Hinds LA, Liers H, Zhang Z, eds. *Ecologically-based rodent management.* Canberra: Australian Centre for International Agricultural Research.17-29.

Sleeman DP. 1990. Dens of Irish Stoats. *Irish Naturalists Journal* 23: 202-203.

Smith GC. 1997. An analysis of the form of density dependence in a simulation model of a seasonal breeder undergoing control. *Ecological Modelling* 95: 181-189.

Smith RH. 1999. Population biology and non-target effects of rodenticides: trying to put the eco into ecotoxicology. In: Cowan DP, Feare CJ, eds. *Advances in Vertebrate Pest Management.* Fürth: Filander Verlag. 331-346.

Smith H, Macdonald DW. 1992. The impacts of mowing and sowing on weed populations and species richness in field margin set-aside. In: Clarke J, ed. *Set-aside. British Crop Protection Council Monograph 50.* Farnham: British Crop Protection Council Publications. 117-122.

Smith GC, Trout RC. 1994. Using matrices to determine rabbit population growth and the potential for control. *Journal of Applied Ecology.* 31: 223-230.

Smith P, Berdoy M, Smith RH. 1994b. Body weight and social dominance in anticoagulant-resistant rats. *Crop Protection* 13: 311-315.

Smith GC, Cheeseman CL, Clifton-Hadley RS. 1997. Modelling the control of bovine tuberculosis in badgers in England: culling and the release of lactating females. *Journal of Applied Ecology* 34: 1375-1386.

Smith GC, Pugh B, Trout RC. 1994a. Age and sex bias in wild rabbit (*Oryctolagus cuniculus*) samples. *New Zealand Journal of Zoology.* 22: 115-121.

Smith P, Townsend MG, Smith RH. 1991. A cost of resistance in the brown rat? Reduced growth rate in warfarin-resistant lines. *Functional Ecology* 5: 441-447.

Smith P, Berdoy M, Smith RH, Macdonald. 1993. A new aspect of warfarin resistance in wild rats: benefits in the absence of poison. *Functional Ecology* **7**: 190-194.

Smith GC, Richards MS, Clifton-Hadley RS, Cheeseman CL. 1995. Modelling bovine tuberculosis in badgers in England: Preliminary results. *Mammalia* **59**: 639-650.

Sotherton NW. 1991. Conservation Headlands: a practical combination of intensive cereal farming and conservation. In: Firbank LG, Carter N, Darbyshire JF, Potts GR, eds. *The Ecology of Temperate Cereal Field.* Oxford: Blackwell Scientific Publications. 373-397.

Sotherton NW. 1998. Land use changes and the decline of farmland wildlife: an appraisal of the set-aside approach. *Biological Conservation* **83**: 259-268.

Sotherton NW, Moreby SJ, Langley MG. 1987. The effects of the foliar fungicide pyrazophos on beneficial arthropods in barley fields. *Annals of Applied Biology* **111**: 75-87

Southern HN. 1940. The ecology and population dynamics of the wild rabbit. *Annals of Applied Biology.* **27**: 509-526.

Southern, HN, Lowe VPW. 1982. Predation by tawny owls *(Strix aluco)* on bank voles *(Clethrionomys glareolus)* and wood mice *(Apodemus sylvaticus). Journal of Zoology, London* **198**: 83-102.

Southern HN, Watson JS. 1941. Summer food of the red fox *(Vulpes vulpes)* in Great Britain: a preliminary report. *Journal of Animal Ecology* **10**: 1-11.

Stebbings RE. 1982. Radio-tracking greater horseshoe bats with preliminary observations on flight patterns. *Symposium of the Zoological Society of London* **49**: 161-173.

Stebbings RE. 1988. *Conservation of European Bats.* London: Christopher Helm.

Stebbings RE, Griffith F. 1986. *Distribution and status of bats in Europe.* Huntingdon: Natural Environment Research Council, Institute of Terrestrial Ecology.

Stephens MN. 1957. *The Natural History of the Otter.* London: Universities Federation for Animal Welfare.

Stephens DW, Krebs JR. 1986. *Foraging theory.* Princeton: Princeton University Press.

Stoate C. 1996. The changing face of lowland farming and wildlife, part 2, 1945-1995. *British Wildlife* **7** (3): 162-172.

Stoate C. 1999. The influence of field boundary structure on breeding territory establishment of whitethroat *Sylvia communis* and yellowhammer *Emberiza citrinella*. Field margins and buffer zones: ecology, management and policy. *Aspects of Applied Biology,* **54**: 125-130.

Stoate C, Tapper SC. 1993. The impact of three hunting methods on the brown hare *(Lepus europaeus)* population in Britain. *Giber Faune Sauvage* **10**: 229-240.

Stockdale EA, Lampkin NH, Hovi M, Keatinge R, Lenartsson EKM, Macdonald DW, Padel S, Tattersall FH Wolfe MS, Watson C. (2000). Agronomic and environmental implications of organic farming systems. *Advances in Agronomy.* **70**: 261-327.

Storch I. 1993. Habitat selection by capercaillie in summer and autumn: is bilberry important? *Oecologia* **95**: 257-265.

Storch I, Lindström ER, De Jounge J. 1990. Diet and habitat selection of the pine marten in relation to competition with the red fox. *Acta Theriologica* **35**: 311-320.

Strachan R. 1997. *Water Voles.* London: Whittet.

Strachan R. 1998. *Water Vole Conservation Handbook.* Oxford: English Nature, the Environment Agency and the Wildlife Conservation Research Unit.

Strachan R, Jefferies DJ. 1993. The water vole *Arvicola terrestris* in Britain 1989-1990: its distribution and changing status. London: The Vincent Wildlife Trust.

Strachan R, Jefferies DJ. 1996. *Otter Survey of England 1991-1994: a report on the decline and recovery of the otter in England and its distribution, status and conservation in 1991-1994.* London: The Vincent Wildlife Trust.

Strachan R, Jefferies DJ, Chanin PRF. 1996. *Pine marten survey of England and Wales 1987-1988.* Peterborough: Joint Nature Conservation Committee.

Strachan C, Strachan R, Jefferies DJ. 2000. *Preliminary report on the status and distribution of the water vole* (Arvicola terrestris) *in Britain 1996-1998.* London: The Vincent Wildlife Trust.

Strong L. 1993. Overview. The impact of avermectins on pastureland ecology. In: Herd R, Strong L, Wardaugh K, eds. *Environmental impact of avermectin usage in livestock. Veterinary Parasitology* **48** (Special issue): 3-18.

Strong L, James S. 1993. Some effects of ivermectin on the yellow dung fly, *Scatophaga stercoraria*. In: Herd R, Strong L, Wardaugh K, eds. *Environmental impact of avermectin usage in livestock. Veterinary Parasitology* **48** (Special issue): 181-192.

Summers D. 1994. Live stock and stream banks. *Enact* **2(4)**: 21-23.

Sumption KJ, Flowerdew JR. 1985. The ecological effects of the decline in rabbits (*Oryctolagus cuniculus*) due to myxomatosis. *Mammal Review* **15**: 151-186.

Swash A, Grice PV, Smallshire D. 2000. The contribution of the UK Biodiversity Action Plan and agri-environmental schemes to the conservation of farmland birds in England. In: Aibischer NJ, Evans AD, Grice PV, Vickery JA, eds. *Ecology and Conservation of Lowland Farmland Birds*. Tring: British Ornithologists Union. 36-42.

Swinton J, Tuyttens F, Macdonald D, Nokes DJ, Cheeseman CL, Clifton-Hadley R. 1997. A comparison of fertility control and lethal control of bovine tuberculosis in badgers: the impact of perturbation induced transmission. *Philosophical Transactions of the Royal Society of London (B)* **352**: 619-631.

Tapper SC. 1979. The effect of fluctuating vole numbers (*Microtus agrestis*) on a population of weasels (*Mustela nivalis*) on farmland. *Journal of Animal Ecology* **48**: 603-617.

Tapper SC. 1991. Brown Hare. In: Corbett GB, Harris S, eds. *The Handbook of British Mammals*. Oxford: Blackwell Publications.

Tapper SC. 1992. *Game Heritage: an ecological review from shooting and gamekeeping records*. Fordingbridge: Game Conservancy Ltd.

Tapper SC, ed. 1999. *A Question of Balance. Game animals and the role in the British Countryside*. Fordingbridge: The Game Conservancy Trust.

Tapper SC, Barnes RFW. 1986. Influence of farming practice on the ecology of the brown hare (*Lepus europaeus*). *Journal of Applied Ecology* **23**: 39-52.

Tapper SC, Parsons N. 1984. The changing status of the brown hare (*Lepus capensis* L.) in Britain. *Mammal Review* **14**: 57-70.

Tapper SC, Green RE, Rands MRW. 1982. Effects of mammalian predators on partridge populations. *Mammal Review* **12**: 159-167.

Tapper S, Brockless M, Potts D. 1991. The Salisbury Plain predation experiment: the conclusion. *Game Conservancy Annual Review* **22**: 87-91.

Tapper SC, Potts D, Brockless M. 1996. The effect of an experimental reduction in predation pressure on the breeding success and population density of grey partridges *Perdix perdix*. *Journal of Applied Ecology* **33**: 965-978.

Tarrant KA, Westlake GE. 1988. Laboratory evaluation of the hazard to wood mice, (*Apodemus sylvaticus*), from the agricultural use of methiocarb molluscicide pellets. *Bulletin of Environmental Contamination and Toxicology* **40**: 147-152.

Tarrant KA, Johnson IP, Flowerdew JR, Greig-Smith PW. 1990. Effects of pesticide applications on small mammals in arable fields, and the recovery of their populations. In: *Brighton Crop Protection Conference-Pests and Diseases*. Farnham: The British Crop Protection Council. 173-182.

Tattersall FH, Hart BJ, Manley WJ, Macdonald DW, Feber RE. 1999a. Small mammals on set-aside blocks and margins. *Aspects of Applied Biology* **54**: 131-138.

Tattersall FH, Macdonald DW, Feber RE, Hart BJ, Manley WJ. 2001. Habitat use by wood mice in a changeable arable landscape. *Journal of Zoology, London* **255**: 487-494.

Tattersall F, Fagiano AL, Bembridge JD, Edwards P, Macdonald DW, Hart BJ. 1999b. Does the method of set-aside establishment affect its use by wood mice? *Journal of Zoology, London* **249**: 472-476.

Tattersall FH, Macdonald DW, Hart BJ, Johnson P, Manley W, Feber R 2002. Is habitat 'linearity' important for small mammals on farmland? *Journal of Applied Ecology* **39**: 643-652.

Tattersall F, Macdonald D, Manley W, Gates S, Feber R, Hart B. 1997. Small mammals on one year set-aside. *Acta Theriologica* **42**: 329-334.

Taylor J. 1969. Salmonella in wild animals. *Symposium Zoological Society, London* **24**: 51-73.

Taylor KD. 1975. Competitive displacement as a possible means of controlling commensal rodents on islands. In: Hansson L, Nilsson B, eds. *Biocontrol of Rodents*, Ecological Bulletins No. 19. Stockholm: Swedish Natural Science Research Council. 187-194.

Taylor KD, Quy RJ. 1978. Long distance movements of a common rat (*Rattus norvegicus*) revealed by radio-tracking. *Mammalia* **42**: 63-71.

Tew TE. 1992. Radio-tracking arable-dwelling wood mice (*Apodemus sylvaticus*). In: Priede IG, Swift SM, eds. *Wildlife Telemetry*. Chichester: Ellis Horwood. 561-569.

Tew TE. 1994. Farmland hedgerows: habitat, corridors or irrelevant? A small mammal's perspective. In: Watt TA, ed. *Hedgerow Management and Nature Conservation.* Wye: Wye College Press. 80-94.

Tew TE, Macdonald DW. 1993. The effects of harvest on arable wood mice, *(Apodemus sylvaticus).* *Biological Conservation,* **65:** 279-283.

Tew TE, Macdonald DW. 1994. Dynamics of space use and male vigour amongst wood mice, Apodemus sylvaticus, in the cereal ecosystems. *Behavioural Ecology and Sociobiology,* **34:** 337-345.

Tew TE, Macdonald DW, Rands MRW. 1992. Herbicide application affects microhabitat use by arable wood mice *(Apodemus sylvaticus).* *Journal of Applied Ecology* **29:** 532-540.

Tew TE, Todd IA, Macdonald DW. 2000. Habitat use of the arable ecosystem by wood mice *(Apodemus sylvaticus).* 2. Microhabitat. *Journal of Zoology, London* **250:** 305-311.

Thomas MR, Wild S. 1996. *Rodenticide usage in Great Britain on farms growing arable crops 1994.* Pesticide Usage Survey Report No. 130. London: Ministry of Agriculture, Fisheries and Food Publications.

Thomas MR, Garthwaite DG, Banham AR. 1997. *Arable Farm Crops In Great Britain 1996.* Pesticide Usage Survey Report 141. London: Ministry of Agriculture, Fisheries and Food Publications.

Thomas MB, Wratten SD, Sotherton NW. 1991. Creation of 'island' habitats in farmland to manipulate populations of beneficial arthropods: predator densities and emigration. *Journal of Applied Ecology* **28:** 906-917.

Thompson HV, King CM. 1994. *The European Rabbit: the history and biology of a successful coloniser.* Oxford: Oxford University Press.

Thompson HV, Worden AN. 1956. *The rabbit.* London: Collins New Naturalist Series.

Todd IA, Tew TE, Macdonald DW. 2000. Habitat use of the arable ecosystem by wood mice, *(Apodemus sylvaticus).* 1. Macrohabitat. *Journal of Zoology, London* **250:** 299-303.

Tompkins DM, Draycott RAH, Hudson PJ. 2000. Field evidence for apparent competition mediated via the shared parasites of two gamebird species. *Ecology Letters* **3:** 10-14.

Townsend MG, Bunyan PJ, Odam EM, Stanley PI, Wardall HP. 1984. Assessment of secondary poisoning hazard of warfarin to least weasels. *Journal of Wildlife Management* **48:** 628-632.

Treweek J, José P, Benstead P, eds. 1997. *The Wet Grassland Guide: managing floodplain and coastal wet grasslands for wildlife.* Bedfordshire: Royal Society for the Protection of Birds, Institute of Terrestrial Ecology, English Nature.

Trout RC. 1994. Don't let rabbits beet your profits down to the ground. *British Sugar Beet Review* **62:** 30-33.

Trout RC. 1999. The results of studies on rabbit haemorrhagic disease in Britain. In: Jarvis BDW, ed. *Rabbit control, RCD; dilemmas and implications.* Wellington: Royal Society of New Zealand. 67-71.

Trout RC, MacVicker HJ. 1994. Making rabbits and small herbivores dislike young trees. In: *XXIst Congress of the International Union of Game Biologists.* Halifax: Canada. 72.

Trout RC, Smith GC. 1995. The reproductive productivity of the wild rabbit *(Oryctolagus cuniculus)* in southern England, *Journal of Zoology, London.* **237:** 411-422.

Trout RC, Tittensor AM. 1989. Can predators regulate wild rabbit *Oryctolagus cuniculus* population density in England and Wales. *Mammal Review* **19:** 153-173.

Trout RC, Tapper SC, Harradine J. 1986. Recent trends in the rabbit population in Britain. *Mammal Review* **16:** (3/4) 117-123.

Trout RC, Langton S, Smith GC, Haines-Young RH. 2000. Factors affecting the abundance of rabbits *(Oryctolagus cuniculus)* in England and Wales. *Journal of Zoology* **252:** 227-238.

Trout RC, Putman RJ, Moore N, Hart J. 1994. *A Review of Lowland Deer.* Report to Countryside Division Land Use Conservation & Countryside, Ministry of Agriculture, Fisheries and Foods. London.

Tuyttens FA, Macdonald DW, Rogers LM, Cheeseman CL, Roddam A. 2000. Comparative study of the consequences of culling badgers *(Meles meles)* on biometrics, movement and population dynamics. *Journal of Animal Ecology* **69:** 567-580.

Tyler GA, Green RE, Casey C. 1998. Survival and behaviour of corncrake *Crex crex* chicks during the mowing of agricultural grassland. *Bird Study* **45:** 35-50.

UK Biodiversity Group. 1998a. Tranche 2 Action Plans: *Volume I – vertebrates and vascular plants.* Peterborough: English Nature.

UK Biodiversity Group. 1998b. Tranche 2 Action Plans: *Volume II – terrestrial and freshwater habitats.* Peterborough: English Nature.

Vaughan N, Jones G, Harris S. 1997. Habitat use by bats (Chiroptera) assessed by means of a broad-band acoustic method. *Journal of Applied Ecology* **34:** 716-730.

Velander KA. 1983. *Pine marten survey of Scotland, England and Wales 1980-82.* London: The Vincent Wildlife Trust.

Vordermeier HM, Cockle PC, Whelan A, Rhodes S, Palmer N, Bakker D, Hewinson RG. 1999. Development of diagnostic reagents to differentiate between *Mycobacterium bovis* BCG vaccination and *M. bovis* infection in cattle. *Clinical and Diagnostic Laboratory Immunology* **6:** 675-682.

Wade PM. 1997. Management of macrophytic vegetation. In: Petts GE, Calow P, eds. *River Restoration.* Oxford: Blackwell Science Ltd. 144-166.

Wai-Ping V, Fenton MB. 1989. Ecology of spotted bat (*Euderma maculatum*) roosting and foraging behavior. *Journal of Mammalogy* **70:** 617-622.

Wall R, Strong L. 1987. Environmental consequences of treating cattle with the antiparasitic drug Ivermectin. *Nature* **327:** 418-421.

Walton KC. 1964. The distribution of the polecat (*Putorius putorius*) in England, Wales and Scotland, 1959–62. *Proceedings of the Zoological Society of London* **143:** 333-336.

Walton KC. 1968. The distribution of the polecat, *Putorius putorius*, in Great Britain, 1963–67. *Journal of Zoology, London* **155:** 237-240.

Walton KC. 1970. The polecat in Wales. in: Lacey WS, ed. *Welsh Wildlife in Trust.* Bangor: North Wales Wildlife Trust. 98-108.

Ward D, Holmes N, José P. 1994. *The New Rivers and Wildlife Handbook.* Bedfordshire: Royal Society for the Protection of Birds, National Rivers Authority & The Wildlife Trusts.

Waters DA, Jones G. 1995. Echolocation call structure and intensity in five species of insectivorous bats. *Journal of Experimental Biology* **198:** 475-489.

Watkins RW, Whiterow A, Bull DS, Cowan DP. 1999. The use of familiar odors to reduce the impact of container neophobia on the control of Norway rats (*Rattus norvegicus*) In: Johnston, ed. *Advances in Chemical Signals In Vertebrates* 13, New York: Plenum Press. 655-661.

Watson A. 1955. The winter food supply of six highland foxes. *Scottish Naturalist* **67:** 123-124.

Watson A. 1976. Food remains in the droppings of foxes (*Vulpes vulpes*) in the Cairngorms. *Journal of Zoology, London* **180:** 495-496.

Webster JA. 2001. A review of the historical evidence of the habitat of the pine marten in Cumbria. *Mammal Review* **31:** 17-32.

Webster JP, Macdonald DW. 1995. Parasites of wild brown rats (*Rattus norvegicus*) on UK farms. *Parasitology* **111:** 247-255.

White GC, Garrott RA. 1985. *Analysis of wildlife radio-tracking data.* London: Academic Press.

White PCL, Harris S. 1995a. Bovine tuberculosis in badger (*Meles meles*) populations in south-west England: an assessment of past, present and possible future control strategies using simulation modelling. *Philosophical Transactions of the Royal Society of London (B)* **349:** 415-432.

White PCL, Harris S. 1995b. Bovine tuberculosis in badger (*Meles meles*) populations in southwest England: the use of a spatial stochastic simulation model to understand the dynamics of the disease. *Philosophical Transactions of the Royal Society of London (B)* **349:** 391-413.

White PCL, Whiting SJ. 2000. Public attitudes towards badger culling to control bovine tuberculosis in cattle. *Veterinary Record* **147:** 179-184.

White PCL, Lewis AJG, Harris S. 1997. Fertility control as a means of controlling bovine tuberculosis in badger (*Meles meles*) populations in south-west England: predictions from a spatial stochastic simulation model. *Proceedings of the Royal Society of London (B)* **264:** 1737-1747.

White P, Baker P, Newton Cross G, Smart J, Moberly R, McLaren G, Ansell R, Harris S. 2000. *Management of the populations of foxes deer, hares and mink, and the impact of hunting with dogs.* A report to the Burns Inquiry into Hunting with hounds in England and Wales. Bristol: University of Bristol.

Whitten AJ. 1990. Recovery: a proposed programme for Britain's protected species. *Nature Conservancy Council, CSD Report no. 1089.*

Wilesmith JW, Little TWA, Thompson HV, Swan C. 1982. Bovine tuberculosis in domestic and wild mammals in an area of Dorset. 1. Tuberculosis in cattle. *Journal of Hygiene, Cambridge* **89:** 195-210.

Wilkins KT. 1982. Highways as barriers to rodent dispersal. *Southwestern Naturalist* **27:** 459-460.

Wilkinson L, Hill M, Welna JP, Birkenbeuel GK. 1992. *SYSTAT for Windows, Version 5 Edition.* SYSTAT, Inc. Evanston, IL.

Wilson WL. 1990. Changes in the diet of foxes (*Vulpes vulpes*) on the sands of Forvie National Nature Reserve, during the eider nesting period, 1974-1988. *Journal of Zoology, London* **221**: 305-308.

Wilson PJ. 1992a. Britain's Arable Weeds. *British Wildlife* **3(3)**: 149-161

Wilson, PJ. 1992b. The natural regeneration of vegetation under set-aside in southern England. In: *Set-aside. British Crop Protection Council Monograph No. 50*: Farnham: British Crop Protection Council British. 73-78.

Wilson PJ. 1993. Conserving Britain's cornfield flowers. In: *Brighton Crop Protection Conference – Weeds.* Farnham, The British Crop Protection Council. 411-417.

Wilson GJ, 1998. Patterns of population change in the Eurasian badger *Meles meles* in Britain, 1988-1997. [PhD thesis.] UK: University of Bristol.

Wilson G, Harris S, McLaren G. 1997. *Changes in the British badger population, 1988 to 1997.* London: People's Trust for Endangered Species.

Wilson WL, Montgomery WI, Elwood RW. 1993. Population regulation in the wood mouse *Apodemus sylvaticus* (L.). *Mammal Review,* **23**: 65-92.

Wilson JD, Evans J, Browne SJ, King JR. 1997. Territory distribution and breeding success of skylarks *Alauda arvensis* on organic and intensive farmland in southern England. *Journal of Applied Ecology* **34**: 1462-1478.

Wilson J, Evans A, Poulsen JG, Evans J. 1995. Wasteland or Oasis? The use of set-aside by breeding and wintering birds. *British Wildlife* **6(4)**: 214-223.

Wolton RJ. 1985. The ranging and nesting behaviour of wood mice, *Apodemus sylvaticus* (Rodentia, Muridae), as revealed by radiotracking. *Journal of Zoology, London* (A) **206**: 203-24.

Wolton RJ, Flowerdew JR. 1985. Spatial distribution and movements of wood mice, yellow-necked mice and bank voles. *Symposia of the Zoological Society of London* **55**: 249-75.

Wood-Gee V. 1994. Water worries on livestock farms. *Farming & Conservation,* October 1994. 12-14.

Woodroffe R, Frost D, Clifton-Hadley RS. 1999. Attempts to control tuberculosis in cattle by removing infected badgers: constraints imposed by live test sensitivity. *Journal of Applied Ecology* **36**: 494-501.

Yalden DW. 1999. *The History of British Mammals.* London: Poyser Natural History.

Young C, Morris C, Andrews C. 1995. Agriculture and the Environment in the UK: towards an understanding of the role of 'farming culture'. *Greener Management International* **12**: 63-80.

Zar JH. 1984. *Biostatistical Analysis.* 2nd. Edition. New Jersey: Prentice-Hall International, Inc.

Zejda J, Zapletal M 1969. Habitat requirements of the water vole (*Arvicola terrestris* Linn.) along water streams. *Zoologiké Listy* **18(3)**: 225-238.

Zuckerman, Lord. 1980. *Badgers, Cattle and Tuberculosis.* London: HMSO.